AF842994

Annette Summers Engel (Ed.)
Microbial Life of Cave Systems
Life in Extreme Environments

Life in Extreme Environments

Edited by
Dirk Wagner

Volume 3

Microbial Life of Cave Systems

DE GRUYTER

Editor
Annette Summers Engel
Department of Earth and Planetary Sciences
University of Tennessee
1412 Circle Drive
Knoxville, TN 37996-1410, USA
aengel1@utk.edu

ISBN 978-3-11-033499-9
e-ISBN (PDF) 978-3-11-033988-8
e-ISBN (EPUB) 978-3-11-038952-4
ISSN 2197-9227

Library of Congress Cataloging-in-Publication Data
A CIP catalog record for this book has been applied for at the Library of Congress.

Bibliographic information published by the Deutsche Nationalbibliothek
The Deutsche Nationalbibliothek lists this publication in the Deutsche Nationalbibliografie; detailed bibliographic data are available on the Internet at http://dnb.dnb.de.

© 2015 Walter de Gruyter GmbH, Berlin/Boston
Cover image: Alex_11/iStock/Thinkstock
Typesetting: le-tex publishing services GmbH, Leipzig
Printing and binding: Hubert & Co. GmbH & Co. KG, Göttingen
♾ Printed on acid-free paper
Printed in Germany

www.degruyter.com

Preface

In 2013, Dirk Wagner invited me to edit a new topical volume in the De Gruyter book series *Life in Extreme Environments*, after Diana Northup recommended me for the project. The volume title, "Microbial Life of Cave Systems," excited me. No books on this topic had been published, despite the noticeable increase in peer-reviewed literature describing cave microbiology over the past 30 years. A book like this would be the first authoritative, comprehensive account of the microbiology of caves, and potentially draw in new researchers, from students to academic and technology professionals. I was humbled by the invitation and grateful for Diana's recommendation.

As is true for all subsurface habitats, cave habitats are naturally bound on all sides by rock, and darkness is the most ubiquitous feature. However, nearly all of the chapters in this book reveal how pervasive extreme environmental conditions are to cave habitats. I do not mean extreme in the context of the customary microbiology definition, being conditions that humans consider abnormal. I mean extreme in the context of drastic variability. Differential movement of water and air through rock, sometimes quickly in conduits or fissures or slowly through the rock matrix, and the nature of hydrological connection to the surface or proximity to groundwater, affect the availability of organic substrates and nutrients needed for energy and metabolism. For some types of caves, organic matter and nutrients can be abundant and accumulate over time because decomposition rates are slower than deposition. In other types of caves, organic carbon and inorganic nutrient limitation is evident, but microbial activities associated with nutrient acquisition and cycling, degradation of organic matter, the production of new organic matter in the absence of photosynthetically-derived material, and induction of mineral precipitation can overcome those limitations and serve to sustain biologically diverse ecosystems. Water and air movement through rock also affects environmental conditions like salinity, temperature, and pH, which all can, and often do, span a full range of conditions. Many cave systems take thousands to millions of years to form, and as their environmental conditions stabilize over time, the endemic organisms adapt to those conditions. As such, it is clear that cave habitats are particularly susceptible to disturbance. Human-induced disturbances, like contamination events or even converting a cave into a tourist attraction, significantly alter environmental conditions and change the metabolic foundation and evolutionary trajectory of a cave ecosystem. Microbes have evolved strategies to survive through and compete in extreme environmental conditions, and may also evolve strategies to survive disturbances. These strategies can be explored for biotechnology potential, a theme that resonates throughout the book. Collectively, the book chapters build a complex picture of cave habitat extremes and of the multifaceted roles of microbial life in caves.

I am thankful to all the enthusiastic, contributing authors. As I discuss in the introductory chapter, many of us are kindred spirits who came into the science as part of a few original teams. We are passionate about the subject matter and about caves, which might seem 'extreme' to those outside of this research area! What should become clear from each chapter in the book is that none of the research would be possible if someone had not ventured into the darkness to study the microbiology of caves or if we had not convinced others to explore with us. In many ways, our ability to enter caves makes our research unique from the types of work done in most other subterranean habitats, excluding artificial tunnels and mines, that must be accessed from wells and boreholes. Chapter by chapter, the research not only brings to focus the caves, and the microbiology of caves, but also emphasizes the collegiality of cavers.

> "If it form the one landscape that we, the inconstant ones,
> Are consistently homesick for, this is chiefly
> Because it dissolves in water. Mark these rounded slopes
> With their surface fragrance of thyme and, beneath,
> A secret system of caves and conduits; hear the springs
> That spurt out everywhere with a chuckle,
> Each filling a private pool...; when I try to imagine a faultless love
> Or the life to come, what I hear is the murmur
> Of underground streams, what I see is a limestone landscape."
>
> – W. H. Auden, excerpt from *In Praise of Limestone*

Annette Summers Engel

Contents

Contributing authors

Xavier Ariño
Unidad de Botanica
Universidad Autonoma de Barcelona
Barcelona, Spain

Hazel A. Barton
Department of Biology and Department of Geosciences
University of Akron
Akron, OH, USA
e-mail: bartonh@uakron.edu

David Benavente
Laboratorio de Petrologia Aplicada
Universidad de Alicante
Alicante, Spain

Suzanna L. Bräuer
Department of Biology
Appalachian State University
Boone, NC, USA

Juan Carlos Cañaveras
Laboratorio de Petrologia Aplicada
Universidad de Alicante
Alicante, Spain

Sarah K. Carmichael
Department of Geology
Appalachian State University
Boone, NC, USA
e-mail: carmichaelsk@appstate.edu

Yin Chen
School of Life Sciences
University of Warwick
Coventry, United Kingdom

Andrew Cronan
Global Underwater Explorers
Australia

Soledad Cuezva
Laboratorio de Petrología Aplicada
Departamento de Ciencias de la Tierra y del Medio Ambiente
Universidad de Alicante
Alicante, Spain

Liam D. H. Elbourne
Department of Chemistry and Biomolecular Sciences
Macquarie University
Sydney, NSW, Australia

Angel Fernandez-Cortes
Geomnia Natural Resources SLNE
Madrid, Spain

Elena Garcia-Anton
Museo Nacional de Ciencias Naturales
Madrid, Spain

Michael R. Gillings
Department of Biological Sciences
Macquarie University
Sydney, NSW, Australia

Alexandra Hillebrand
Emil Racovita Institute of Speleology
Bucharest, Romania
e-mail: alexandra.hillebrand@iser.ro

Andrew J. Holmes
School of Molecular Biosciences
Sydney University
Sydney, NSW, Australia

Daniel S. Jones
BioTechnology Institute & Dept. of Earth Sciences
University of Minnesota
Minneapolis, MN, USA
e-mail: dsjones@umn.edu

Valme Jurado
Microbiologia Ambiental y Patrimonio Cultural
Instituto de Recursos Naturales y Agrobiologia
Sevilla, Spain

Deepak Kumaresan
School of Earth and Environment
The University of Western Australia
Perth, Australia
e-mail: deepak.kumaresan@uwa.edu.au

Kathleen Lavoie
Faculty of Arts and Science
Plattsburgh State University of New York
Plattsburgh, NY, USA
e-mail: lavoiekh@plattsburgh.edu

Antje Legatzki
Dr. von Hauner Children's Hospital
Ludwig Maximilians University
Munich, Germany

Raina M. Maier
Department of Soil, Water and Environmental Science
University of Arizona
Tucson, AZ, USA

Pedro Martin-Sanchez
Microbiologia Ambiental y Patrimonio Cultural
Instituto de Recursos Naturales y Agrobiologia
Sevilla, Spain

Ana Z. Miller
Microbiologia Ambiental y Patrimonio Cultural
Instituto de Recursos Naturales y Agrobiologia
Sevilla, Spain

Janze Mulec
Karst Research Institute
Postojna, Slovenia
e-mail: janez.mulec@guest.arnes.si

Colin Murrell
School of Environmental Sciences
University of East Anglia
Norwich, UK
e-mail: j.c.murrell@uea.ac.uk

Julia W. Neilson
Department of Soil, Water and Environmental Science
University of Arizona
Tucson, AZ USA

Diana E. Northup
Department of Biology
University of New Mexico
Albuquerque, NM, USA
e-mail: dnorthup@unm.edu

Marianyoly Ortiz
Universidad de Puerto Rico en Ponce
Ponce, Puerto Rico
e-mail: moo@email.arizona.edu

Audrey Paterson
Department of Earth and Planetary Sciences
University of Tennessee
Knoxville, TN, USA
e-mail: apaters1@utk.edu

Ian T. Paulsen
Department of Chemistry and Biomolecular Sciences
Macquarie University
Sydney, NSW, Australia
e-mail: ian.paulsen@mq.edu.au

Miguel Angel Rogerio-Candelera
Microbiologia Ambiental y Patrimonio Cultural
Instituto de Recursos Naturales y Agrobiologia
Sevilla, Spain

Cesareo Saiz-Jimenez
Microbiologia Ambiental y Patrimonio Cultural
Instituto de Recursos Naturales y Agrobiologia
Sevilla, Spain
e-mail: saiz@irnase.csic.es

Sergio Sanchez-Moral
Museo Nacional de Ciencias Naturales
Madrid, Spain

Jason Stephenson
School of Life Sciences
University of Warwick
Coventry, United Kingdom

Annette Summers Engel
Department of Earth and Planetary Sciences
University of Tennessee
Knoxville, TN, USA
e-mail: aengel1@utk.edu

Sasha G. Tetu
Department of Chemistry and Biomolecular Sciences
Macquarie University,
Sydney, NSW, Australia
e-mail: sasha.tetu@mq.edu.au

Daniela Wischer
School of Environmental Sciences
University of East Anglia
Norwich, United Kingdom

Annette Summers Engel

1 Bringing Microbes into Focus for Speleology: An Introduction

Abstract: Early cave microbiology research focused on growing microbes from caves. Knowledge about microorganisms in caves, and of microbial diversity in subsurface habitats, increased after the widespread use of molecular genetics methods. This chapter introduces the history of cave microbiology research, as well as some of the researchers who essentially started the field. Publication rates for culture-dependent and culture-independent (i.e. molecular genetics) research are presented, as well as collaboration networks based on co-authorship of culture-independent studies published since 1997. Major themes from this book, the first census of microbial life in caves, are introduced. The chapter ends with ideas about future research directions.

1.1 Introduction

Most life on Earth that comprises the familiar tree of life, now and throughout geologic time, is microscopic. Microorganisms evolved more than three billion years ago. For much of Earth's history, they dominated the planet in biomass and metabolic function. Some of these microbes, including Bacteria, Archaea, Fungi, and single-celled eukaryotes, are familiar to us because they can cause disease or inflection to plants and animals. Others are less familiar but we know of their presence because they cause slime coatings on pebbles in a stream, create gases such as oxygen, methane, and hydrogen sulfide as a function of their metabolic activity, and even stimulate the formation or destruction of minerals. Microbes occur on and inside larger animals as beneficial symbionts, especially in the gastrointestinal tract. Microorganisms are responsible for the biogeochemical cycling of elements like carbon, nitrogen, sulfur, phosphorus, and iron. Some microbes use solar energy to convert carbon dioxide into organic carbon through photosynthesis, while others break down organic matter produced by photosynthesis as heterotrophs. Some microbes in extreme habitats, such as deep-sea hydrothermal vents and caves, produce organic carbon from chemical reactions, referred to as chemosynthesis. In the past, what could not be seen with the naked eye was ignored or considered inconsequential to the other scientific disciplines, such as chemistry and geology. However, mounting evidence now suggests that microorganisms serve as the foundation of all ecosystems and have a profound impact on the chemical reactions taking place at the Earth's surface, in the shallow subsurface, and even deep within the oceanic and continental crusts. In fact, the subsurface teems with microbial life.

Caves represent a type of natural, rocky subsurface habitat that can be solutionally or collapse enlarged. Caves around the world host some of the most exotic landscapes, minerals, and mineral formations [1–3], as well as cave-adapted organisms [4]. Most caves form by dissolution of soluble sedimentary rocks (► Fig. 1.1 (a)), including carbonates (e.g. limestone, dolomite) and evaporates (e.g. gypsum, halite) [3, 5]. Solution caves, in combination with underground passages and self-evolving flow systems, form a karst terrain. Karst landscapes comprise 15–20% of the ice-free Earth's surface [2]. Caves formed in basaltic rock are referred to as lava tubes or lava tube caves (► Fig. 1.1 (b)), and their formation involves the flow and differential cooling of magma [3]. Sea caves (► Fig. 1.1 (c)), and collapse or solution caves, can also develop in any type of rock, including metamorphic (i.e. marble) or igneous (i.e. granite, ► Fig. 1.1 (d)) rocks. Such terrain is referred to as pseudokarst [6]. Caves can also form in ice, as glacial caves (► Fig. 1.1 (e)), and some caves can contain ice year-round (► Fig. 1.1 (f)). Depending on the cave formation process, or speleogenesis, the surface and subsurface can be hydrologically connected from sinking surface water at entrances or through fissures and fractures in the rock that extend from the surface soil zone into the rock. Entrances establish distinct environmental zonation based on light, from the twilight zone at an entrance to the dark zone within a cave's interior.

The physical and chemical properties of hosting rocks, with the geologic and hydrologic setting of a cave, affect the taxonomic and functional diversity of organisms found within a cave system. When most people think about cave life, usually the pigmentless, eyeless, leggy oddities from the animal world come to mind. These charismatic animals, from invertebrates to bats, have been formally studied since the middle of the nineteenth century. Modern biospeleological investigations focus on the taxonomic diversity and richness of animals in caves across biogeographic distances and on the evolutionary mechanisms that lead to obligate adaptions from living in the subsurface, or troglomorphy [7]. But from ongoing discoveries of new, obligate cave and subterranean fauna from aquifers, vadose zone habitats like seepage springs, and epikarst, attention has turned to understanding cave ecology, and specifically how organisms interact with each other and the cave environment. Although microbes are known to be at the energetic and nutritional base of cave ecosystems, the extent of microbial diversity in caves, and of their roles in cave and karst ecosystems, has not been fully realized. Surprisingly, of all of the types of caves found worldwide, our knowledge about microbial colonization, diversity, and metabolic function in caves has been limited to only a few examples [8]. Knowing more about the microbiology of caves will help us to understand other areas of speleology, including microbial impacts on water and contaminant flow into and through the subsurface, mineralogy, geochemistry, ecosystem nutrient flux, and species adaptation and evolution. New discoveries in the areas of biogeochemistry, microbial ecophysiology, and symbiosis are possible from studying cave microbiology.

Fig. 1.1. Examples of different types of caves and entrances. (a) Large entrance of a carbonate rock cave with flowing river, Planinska Jama near Postojna, Slovenia. (b) Small lava tube cave entrance in basalt, El Malpais National Monument, New Mexico (USA). (c) Sea cave entrance formed in igneous rock, Anemone Cave at Schooner Overlook in Acadia National Park, Maine (USA). (d) Granite cave entrance, Maine (USA). (e) Entrance to glacier cave, Kennicott Glacier, Alaska (USA). (f) Permeant ice and ice formations at the entrance to Peştera Scărişoara, Apuseni Mountains, Romania. All of the images were taken by Annette Summers Engel.

This book provides the first census of microbial life in caves. Despite all of the recent investigations and collaborative efforts, our knowledge of microbial life in caves is still far behind what we know of the ocean, subseafloor habitats, or other terrestrial settings like soils. The hope is that this book will lead to more research. Each chapter is organized around common themes related to methodological developments and the identification of habitat constraints that affect, and perhaps control, microbial diversity and metabolism. Numerous case studies are presented and reveal how microbes colonize reactive air, water, and rock interfaces, with specific examples from carbonate caves and basaltic lava tube caves. A range of geochemical conditions in diverse habitats are also presented, including acidic cave walls associated with hydrogen sulfide-laden waters, alkaline surfaces of speleothems, secondary deposits rich in manganese and iron, methane-rich waters, waters rich with dissolved nitrogen compounds, and nutrient-poor oligotrophic conditions on carbonate and lava tube cave walls and sediments. Research being conducted from other types of rocky habitats, like quartzite and sandstone [9, 10], granite [11], gypsum [12], and even from ice caves, caves with permanent ice, or glacier caves [13], is underrepresented in the book. Chapter content not only reviews prior research but also includes original work done to evaluate cave microbial diversity comprehensively using culture-based methods and molecular genetics methods, so-called culture-independent approaches. Some of these earlier studies were completed at the advent of, and since, technological and experimental advances in molecular biology involving isotopic labeling and "omics" methods and their meta-counterparts, including genomics and transcriptomics [14–19].

In general, our knowledge of cave microbiology has kept pace with methodological and technological advances. New methods now permit studies of microbial diversity and functional activities in ways previously impossible. Consequently, the rate of publication for studies using solely culture-dependent methods, versus using a combination of culture-dependent and -independent methods, or solely culture-independent methods describing 16S rRNA gene sequence surveys has increased since 1997 (▸ Fig. 1.2). Booms in publication numbers in recent years are linked to major conferences, including the 2009 and 2013 International Congresses of Speleology [20, 21] and International Symposium for Environmental Biogeochemistry in 2011. The papers presented at these conferences made their way into the peer-reviewed literature, including a special issue on the biogeochemistry and microbial ecology of cave systems for the *Geomicrobiology Journal* in 2014 [22]. Before 1997, there were about a dozen publications describing the microbial life of caves, predominately based on microscopic and culture-based descriptions associated with mineral deposits or groundwater [23–27].

In this chapter, I summarize the current state of knowledge of cave microbial diversity, point to knowledge gaps, and introduce the overall book content. My goal in the chapter is to bring the people responsible for cave research, and their methods used to study cave microbes, into focus for other scientific disciplines within speleology. The number of individuals contributing to our knowledge of cave microbial diversity is

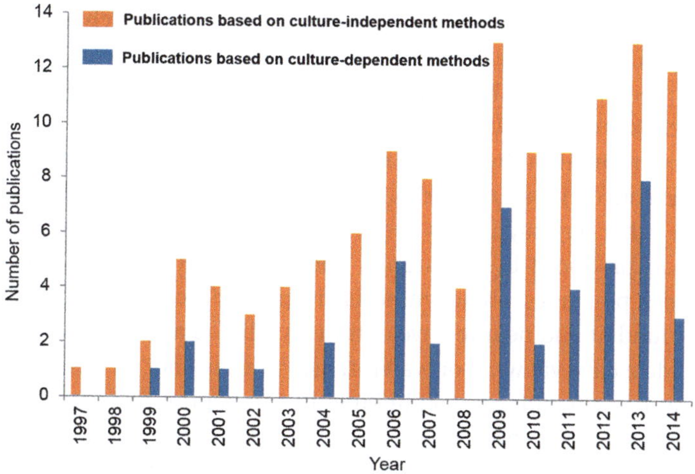

Fig. 1.2. A nonexhaustive summary of peer-reviewed publications in reputable journals that describe microbial life of caves, differentiated as culture-dependent and culture-independent or molecular genetics work involving 16S rRNA gene surveys, since 1997. Most of this research has focused on bacterial diversity. There has been generally an increase in the number of culture-independent publications since the first one [17]. Prior to the Vlasceanu et al. study in 1997 from the Movile Cave, Romania, there were several dozen peer-reviewed publications describing microbial life of caves, predominately based on microscopic and culture-based descriptions.

relatively small, and most of the research to date has centered on bacterial diversity, although recent attention is being placed on archaeal and microeukaryotic, including fungal, diversity. Undoubtedly, exciting discoveries will continue to increase our knowledge of these different major taxonomic groups, as well as of cave geological and ecological processes related to cave formation and modification, speleothem formation and deterioration, food web structure, and ecosystem change. One area where new discoveries await is related to using cave microbes to help identify novel biological compounds, such as antibiotic, antimicrobial, and anticancer compounds [28–31]; this research is mentioned, but underrepresented, throughout the book. Several chapters in the book are dedicated to applied microbiology associated with cave tourism and anthropogenic impact. This research overlaps work intended to protect and conserve archeological places, so information gained from cave microbiology can be applied to other sensitive habitats worldwide.

1.2 Energy to Sustain Subsurface Ecosystems

From a purely ecological standpoint, the abiotic conditions of the subsurface and caves are somewhat predictable: total darkness and relatively stable climate influenced by the ambient temperature of a region. But conditions can change more rapidly near an entrance or in the shallow subsurface, such as from daily fluctuations of light, or within cave passages if there is periodic flooding. Seasonal temperature changes, which bring warm or cold air into a cave, and more or less water into the subsurface based on the amount of precipitation, affect the stability of the habitat and the types of biological communities that can develop and be sustained in a cave.

There are two primary mechanisms by which a microbe can obtain carbon, energy, and nutrients: autotrophs convert inorganic carbon (CO_2, HCO_3^-) to organic carbon for cellular growth, through either photosynthesis or chemosynthesis, and heterotrophs assimilate organic carbon produced by autotrophs and that already exists in an ecosystem. Total darkness precludes photosynthesis deep inside caves, but phototrophic microbes are found at and near to cave entrances. Microbial cells can also be transported deep inside a cave via wind or water; these cells may not be metabolically active but still retain genetic information linking them to a prior phototrophic lifestyle. Also, for some shallow subsurface cave systems, the sun can still provide allochthonous energy and nutrient sources because plant roots might penetrate the unsaturated zone looking for water, and photosynthetically produced dissolved and particulate organic matter can be washed into the subsurface. Changes in the supply of surface material into the subsurface can result in energy and nutrient limitations that lead to nutrient-poor, oligotrophic conditions in some caves [32]. Investigating energy and nutrient availability in the subsurface has been a difficult endeavor because much of the subsurface is inaccessible for study [33–35]. Nevertheless, low food stores have likely influenced the evolution of subsurface life, and could be a reason for the generally lower diversity and abundance of animals in most caves. This area of research, specifically focused on the diversity of microbes adapted to low nutrient levels or oligotrophy, and of genetic adaptations to low nutrient conditions, is described in Chapter 4 in this book.

Prior to the discovery of the chemosynthesis at the deep-sea hydrothermal vents in the late 1970s, all life on Earth was thought to depend on photosynthetically produced energy. Chemosynthesis is the conversion of inorganic carbon into organic carbon using inorganic chemical energy and does not require sunlight. Redox-sensitive compounds in rock and water serve as energy sources for microorganisms, called chemolithoautotrophs ("self-feeding rock-eaters"), that belong exclusively to the domains Bacteria and Archaea [36]. Reactive compounds serving as electron donors can exist as reduced iron and manganese in minerals, or as gases like methane, hydrogen sulfide, molecular hydrogen, and carbon monoxide. Important electron acceptors include oxygen, which can be rapidly consumed in the shallow subsurface by aerobic organisms, as well as nitrate, manganese, iron, sulfate, and carbon dioxide. Utilization of non-oxygen, alterative electron acceptors is referred to as anaerobic respiration. There are also microorganisms termed chemoorganotrophs that gain cellular energy from chemicals but use organic carbon compounds. Chemosynthetically based systems, including caves and deep karst aquifer systems, are known to have high biological diversity [34, 37–44]. The subject of chemolithoautotrophy permeates this book, and research findings can be found in more than half of the chapters.

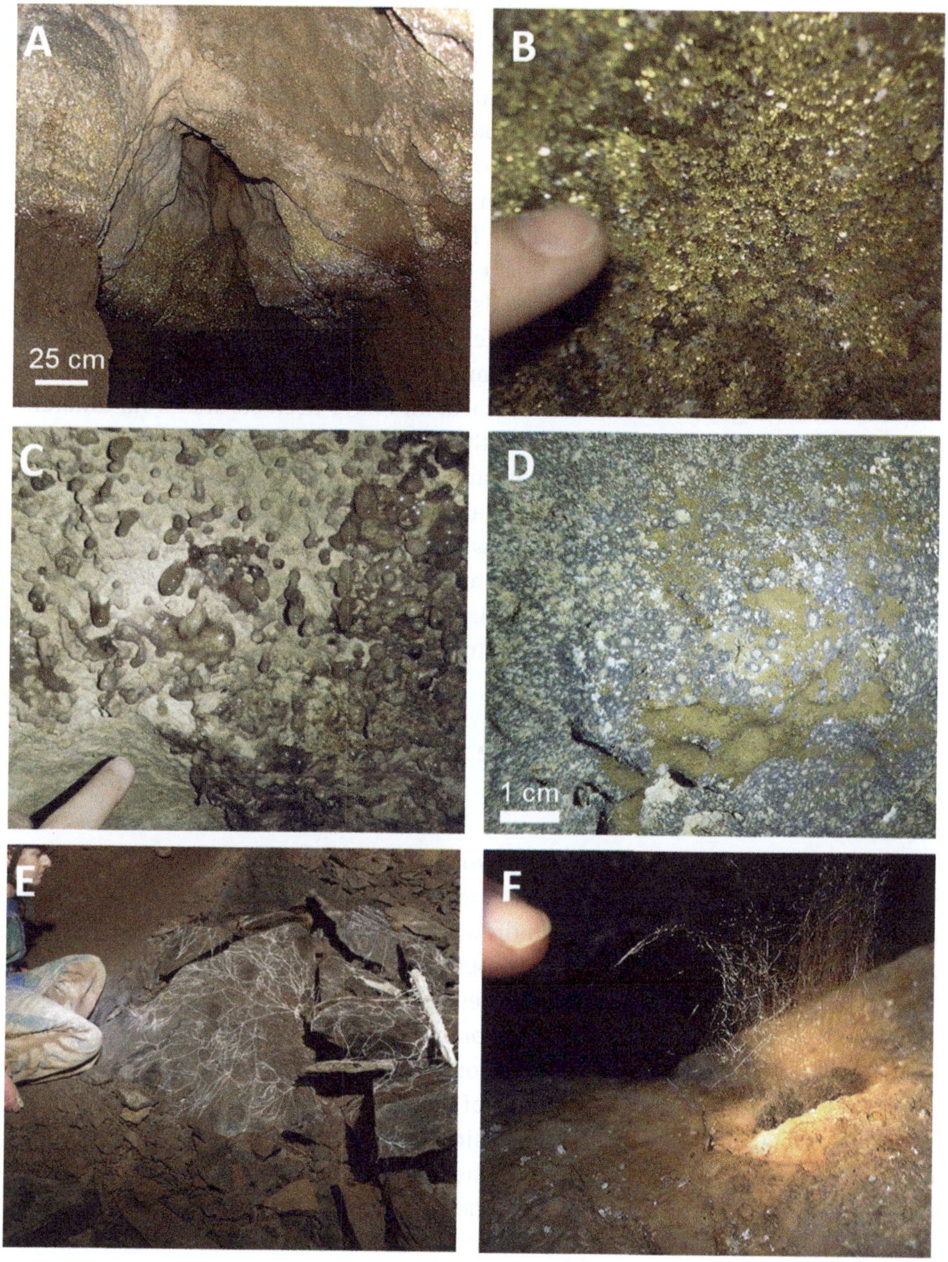

Fig. 1.3. Images of microbial colonies from caves. (a) White and yellow colonies that cover the walls and ceiling of a passage where water ponds frequently (i.e. the lower half of the picture is without colonies because the wall surfaces flood). (b) Close-up of yellow colonies commonly observed in caves. (c) Thick microbial biofilm oozing and hanging from a cave ceiling. (d) Blue colonies with white centers from a cave wall. (e) Fungal colony, originating from a piece of wood that washed into the cave. (f) Fungal colonization of raccoon feces in a cave. All of the images were taken by Annette Summers Engel.

Once established in a cave environment, microorganisms form macroscopic colonies. One of the most recognizable microbial features on moist surfaces in caves, including bare rock walls, speleothems, and sediments, are shiny gold, white, pink, yellow, or white colonies that look like blotches or patches (▸ Fig. 1.3 (a)). Sometimes these colonies have droplets of water on them because water condenses on the hydrophobic surface of the colonies, which comprise millions of individual cells of microbes (▸ Fig. 1.3 (b)). Colonies can also form mucous coatings or biofilms on cave walls and ceilings, either under alkaline (▸ Fig. 1.3 (c)) or acidic conditions; mucous-like biofilms hanging from acidic cave walls and ceilings are common in caves formed by sulfuric acid speleogenesis and are referred to as snottites [45–48]. Colonies can also be other colors, such as black, gray, and blue (▸ Fig. 1.3 (d)). Microbial colonies can form from fungi or Bacteria and Archaea, and have been studied from different types of caves, including pristine caves with low human visitation to highly impacted touristic caves, as well as from carbonate and basaltic caves [49–52]. Fungal colonies can be quite large, and spread more than a meter from the initial inoculation source (▸ Fig. 1.3 (e)), and are usually associated with organic matter (▸ Fig. 1.3 (f)). From the contents of the book, fungi and other microeukaryotes, like yeasts and slime molds, may appear to be understudied from caves, but these microbial groups have been the focus of numerous diversity and ecology studies [53–61].

1.3 Historical Framework of Cave Microbiology Research and Collaboration

The development and application of microbiological methods to study microbes in various environments has a long history, not just for cave research. The earliest studies from caves required microscopes and drew upon culturing methods, predominately based on human microbiology and pathogen research. From the 1900s through the 1940s, the focus was on determining whether there was a microbiological component to cave formation or microbiological origin for secondary nitrate deposits, called saltpetre or saltpeter, and carbonate speleothems [23, 62–64]. Through the 1960s and early 1990s, research on the microbial life of caves turned to culturing and trying to associate microbial processes with specific mineralogical phenomena [24, 65–69]. Other research focused on describing microbial metabolism as it is related to food web structure and cave ecosystem development. Conclusions derived from these earlier studies were that microbes could not be of great importance to most caves, and that microbes would be generally inactive and have low biomass due to nutrient limitation because nearly all microbes identified were similar to, if not identical in function to, soil communities [24, 70, 71]. For most speleologists, the ecological role for microbes was relegated to being exclusively food sources for cave animals and heterotrophs, more or less because heterotrophic media were used to isolate and grow microbes and geochemical evidence was insufficient to conclude that microbial activity was essential to the systems studied at that time. Stable isotope systematics started to inform about the importance of microbial metabolic processes to ecosystems associated with sulfur cycling or from geochemically stratified systems, such as anchialine caves [72–74].

1.3.1 Research following the advent of molecular genetics techniques

Our understanding of microbial diversity and community structure in caves did not truly start until the development and application of molecular genetics techniques in the 1990s. Chapters 2 and 4 describe the success (or lack thereof) of this approach to informing us about the diversity of microbes in caves. With molecular methods, our concept of life in caves started to change rapidly from the early 1990s through the early 2000s. The first culture-independent studies of cave microbial communities were carried out from microbial mats in sulfidic cave systems, after successful extraction of DNA, and PCR amplification and cloning of rRNA gene sequences from environmental samples [75–77], including extreme environments like hot springs [78] and low-biomass water from deep basaltic aquifers [37], were published. The two cave studies were carried out nearly at the same time. One study from Movile Cave in Romania [17] was started in the mid-1990s and was completed by Luminiţa Vlasceanu and Radu Popa, both graduate students from Romania who worked with Professor Brian Kinkle at the University of Cincinnati (USA), and who were collaborators with Serban Sarbu, also a Romanian graduate student at the University of Cincinnati. Sarbu was one of the first researchers to explore Movile Cave after its discovery by Cristian Lascu in 1986 [79]. The other study from Sulphur River in Parker Cave, Kentucky [14], started in 1995 and was completed in 1998 by collaborators Diana Northup from the University of New Mexico (USA) and Norman Pace from the University of Colorado, Boulder (USA), with graduate students at the time, Esther Angert and Andrew Peek, and postdoctoral researchers Anna-Louise Reysenbach and Brett Goebel.

These projects spawned two distinguishable clusters of cave microbiology researchers, which are clearly identified from a modeled collaboration network based on co-authorship for 135 microbial diversity publications utilizing molecular genetics from 1997 until the time of this writing (▸ Fig. 1.4 (a)). One cluster resulted from Kinkle (shown in bold and as a purple node) and the other clusters from Pace and Northup (shown in bold and as red-orange nodes). These groups separately began collaborating with different people, training new students and postdoctoral researchers, and started to develop mature collaboration networks over time. By 1999 through 2001, the groups expanded to include Sarbu (lavender node) and Engel (dark blue node), both graduate students at the University of Cincinnati, and Barton (orange node) who was a postdoctoral researcher with Professor Pace. Interestingly, the collaboration expansion mostly focused on sulfidic cave ecosystems initially, but Engel and Barton

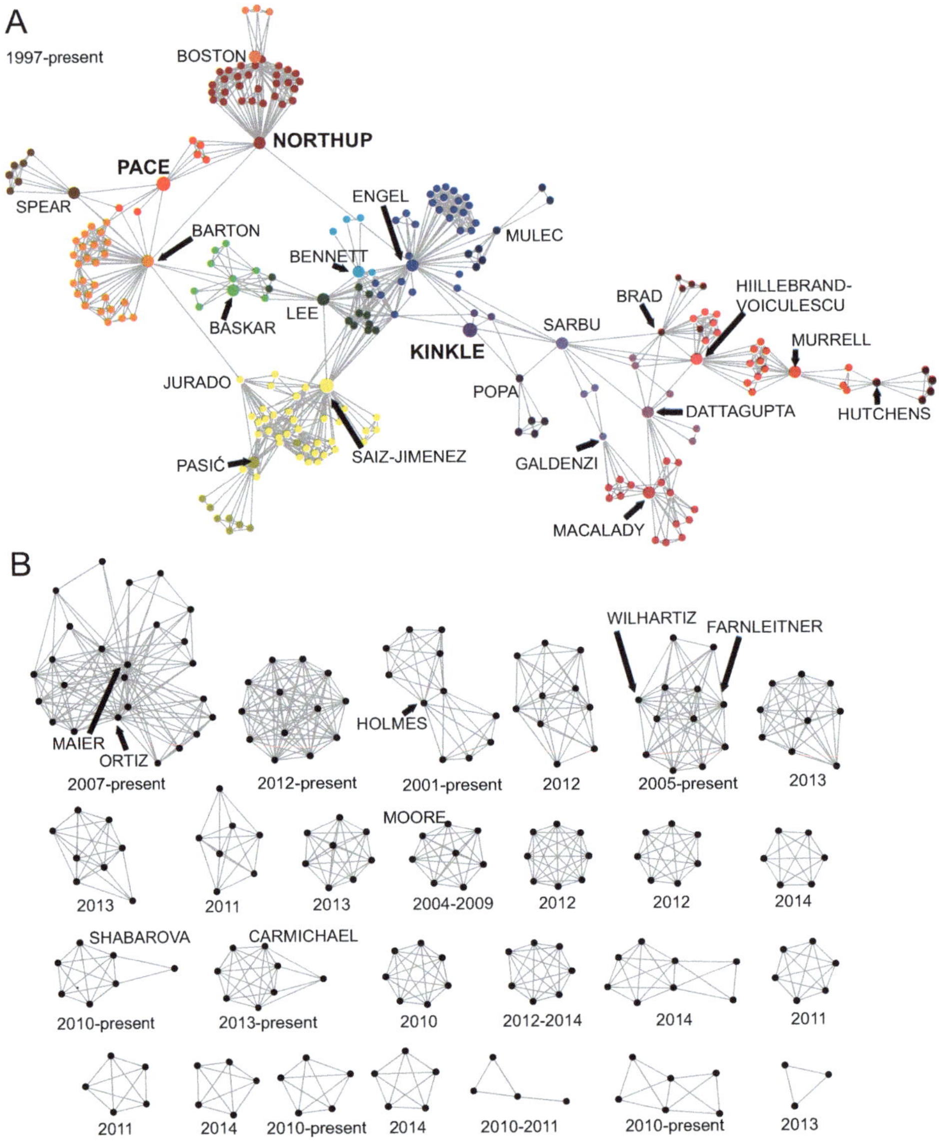

Fig. 1.4. (a) A co-author network or map generated using the Sci[2] Tool (version 1.1) [80] from 135 publications of cave microbiology research utilizing molecular genetics methods since 1997. Nodes are people, and node size is based on the relative contribution of that individual to forming a cluster. Nodes are colored and shaded to show specific clusters and subsequent expansion. Edges (lines) highlight co-authorship linkages and are unweighted based on the number of papers (i.e. line length does not equal more co-authored publications). Blank spaces represent areas where collaboration can be made in the future. (b) Co-author networks created during the same analysis of the 135 publications used for part A. These co-authors have mature collaborations within their clusters, but have not yet collaborated with any other research groups. Nodes for lead investigators of some clusters with more than one publication are noted. Edges (lines) only highlight co-authorship linkages and are unweighted to the number of publications. The timing of publications for each of the clusters is noted.

started collaborating with new people and formed their own clusters by expanding into different types of cave systems.

Other co-author clusters arose independently in the late 1990s and early 2000s, including one from Saiz-Jimenez (yellow node) focusing on show caves and applied microbiology, and one from Murrell (bright pink node) based on research from Movile Cave with Elena Hutchens, a postdoctoral researcher from Romania. These two groups started to collaborate with others. Either independently or after collaboration, the network of co-authors grew in the mid-2000s through the early 2010s to include Lee, Macalady, Dattagupta, and Hillebrand-Voiculescu, again mostly through sulfidic karst research. Clusters from Pašić and Baskar represent expansion of independent, country-based groups.

In addition, 26 other co-author clusters formed resulting from publications of culture-independent research (▸ Fig. 1.4 (b)). All of these clusters remain separate from the large network, although some have mature research collaborations that developed over nearly a decade or more. The fact that these clusters have not co-authored publications with anyone in the large network may be because most of the groups have been working in one location or cave system almost exclusively. An interesting point is that these isolated groups have not been working in sulfidic karst. Another reason for isolated clusters may be that the groups are mostly represented by only one main investigator or graduate student focused on cave research, and who utilizes molecular genetics methods regularly. The smaller clusters with less than six or seven investigators represent recent growth in cave microbiology research, mostly in Asian countries.

1.3.2 Sulfidic cave research

More than 30% of the publications resulting from culture-independent studies since 1997 have been from sulfidic cave systems (▸ Fig. 1.2), predominately from studies in the Movile Cave [16], the Frasassi caves in Italy [81–83], Cueva de Villa Luz in Mexico [45, 46], Lower Kane Cave in Wyoming [15, 84–87], as well as other caves with active hydrogen sulfide waters [40, 88]. A summary of research from sulfidic karst systems can be found in Chapter 9, and detailed research from the Movile Cave is in Chapter 10. Caves in the Nullarbor Plain of Australia, with high concentrations of dissolved nitrogen compounds, were also studied [89] and described in Chapter 6 of this book. Except for the Nullarbor Plain caves, the rest of the systems are hypogenic, formed from rising, usually mineralized (and containing sulfide) fluids that made their way to the shallow subsurface (in contrast to epigenic caves formed from nonmineralized fluids that flow from the surface into the subsurface). Geologists and hydrologists were interested in sulfidic hypogene caves from the standpoint of sulfuric acid speleogenesis [90–94], which was considered to be a purely abiotic process until experimental evidence from Lower Kane Cave demonstrated the role of microbes [95].

Microbial communities from the sulfidic systems were found to be dominated by sulfur-oxidizing bacterial groups that were taxonomically and functionally similar to groups found at deep-sea hydrothermal vents. This led to research focused on specific taxonomic groups like the Epsilonproteobacteria in karst caves and springs [42, 96–99]. Other prevalent metabolic groups, predominately chemolithoautotrophs, included methylotrophs, methanotrophs, and ammonia-oxidizing Archaea [16, 98, 100, 101]. The Movile Cave ecosystem was the first terrestrial chemosynthetic ecosystem identified [38, 102], and several other chemolithoautotrophically based ecosystems were documented or suspected [15, 40, 42, 85]. Sulfidic cave ecosystems were also found to be rich with troglobitic and stygobitic fauna, with diversity and richness as high as or higher than most nonsulfidic caves [41, 103–105]. Animals consume bacteria for food [106] in an otherwise resource-limited environment, and therefore rely on chemolithoautotrophic primary productivity at the base of the ecosystem [38, 81]. Probably one of the most biologically significant discoveries made in recent time was of the sulfidic Ayyalon Cave in Israel and its endemic, cave-adapted fauna [41, 105]. Although some research findings are detailed in Chapter 10 of this book, our knowledge of the microbial diversity and ecosystem function is limited currently to only microscopic descriptions. Thus far, the Movile Cave sulfidic groundwater ecosystem remains the most diverse sulfidic ecosystem, with 33 of 48 different invertebrate species being endemic to the region [107].

Fundamental cave ecosystem research also led to looking for chemosymbiotic associations between microbes and invertebrates in sulfidic cave systems. The Frasassi cave system was the first to have a documented chemosymbiotic association between sulfur-oxidizing *Thiothrix* spp. ectosymbionts and *Niphargus* amphipods [108–110]. Microscopic observations and molecular evidence now suggest that the *Thiothrix–Niphargus* association from the Frasassi caves may be more widespread than previously thought and expand to include other sulfidic karst groundwater systems, including the Movile Cave [111]. Interestingly, novel chemosymbiotic associations are now being identified in other systems, including anchialine caves that have stratified water columns with distinct dissolved oxygen and sulfide gradients and unique fauna [112, 113]. Considerable future research is likely to follow from these studies.

Sulfidic, hypogenic caves, particularly the Frasassi caves [47, 82], have also been interesting places to study acidic cave-wall microbial communities and evolutionary adaptation to, and genetic regulation of, acidic conditions. Acidic cave-wall biofilms have also been studied from other caves [15, 46]. *Acidithiobacillus thiooxidans*, with lesser abundances of uncultivated Thermoplasmatales archaea and Acidimicrobiaceae bacteria, have been investigated from Frasassi cave metagenomes [114, 115]. A detailed summary of this work, as well as of microbiological and molecular genetics methods, is featured in Chapter 2 of this book.

1.3.3 Other cave research – nonsulfidic cave systems

Despite the apparent emphasis on sulfidic cave systems for nearly two decades, approximately 60% of publications using culture-independent methods since 1997 were about the microbiology of nonsulfidic carbonate caves (► Fig. 1.2). Moreover, only 5% of all microbiological studies using molecular methods since 2011 have been from lava tube caves [51, 52], despite recognition of cave microbial communities in lava tubes for decades [116–119]. Detailed information about lava tube cave microbiology is provided in Chapter 8.

Carbonate caves represent most of the caves known worldwide, as nearly 15–20% of the Earth's ice-free surface is karstic [2]. Research published prior to 1997 resulted from culture-dependent and microscopic investigations of microbial groups associated with groundwater contamination [25, 26], or living on cave walls and sediments [120] and secondary mineral precipitates, such as moonmilk [121, 122]. Microbial communities were also studied to determine whether they were the cause of cave-wall and Paleolithic painting biodeterioration [123, 124]. A complete summary of research prior to the mid-1990s was completed by Barton and Northup [71].

One of the first culture-independent study from nonsulfidic carbonate caves was published in 2000 by Northup and co-authors [125] after analyzing ferromanganese deposits and corrosion residues from Lechuguilla Cave in Carlsbad Caverns National Park, New Mexico (USA). The cave is >6 million years old and formed from sulfuric acid speleogenesis but is not actively affected by sulfidic waters [92]. The Northup et al. study presented the first genetic evidence to indicate a microbial role in iron and manganese mineral precipitation after carbonate rock corrosion in caves. A more detailed follow-up study in 2003 described additional taxonomic diversity among manganese- and iron-oxidizing bacteria from different proteobacterial classes, as well as actinomycetes. Other nitrogen-cycling bacteria and archaeal groups were linked to rock and mineral geochemistry in Lechuguilla and Spider caves, New Mexico [126]. In the past decade, several new research teams have focused on iron and manganese geomicrobiology of caves. Details about this research direction are provided in Chapter 7 of this book.

The research group of Professor Cesareo Saiz-Jimenez in Spain published some of their first molecular studies describing bacterial phylogenetic diversity from Spanish Paleolithic paintings in Altamira Cave [127] and Tito Bustillo Cave [128] in 2002. This research followed after investigations of dripwaters and stalactites in 1999 and 2000 [124], and from collaboration with people who studied the microbial diversity of biodeteriorated wall paintings in the Castle of Herberstein, Austria, and St Martin's church in Germany [129]. Prior to the molecular research, it was becoming increasingly clear that tourism in the show caves was causing cave-wall degradation and negatively impacting the cave paintings [130]. Microbes played a part [122]. A detailed history of microbiological research performed from show caves, mines, tunnels, and buildings is described in Chapter 11, and details about microbiological research from

Altamira Cave can be found in Chapter 14. Lascaux Cave in France is another historically important cave receiving intense microbiological study, which is described in Chapter 13.

From all of these investigations, the undeniable effects of anthropogenic activities on the cave environment, of cave-wall and speleothem surfaces, and of Paleolithic paintings and other archeological remains, were clearly demonstrated. First, artificial lighting causes the formation of "maladie verte" or lampenflora, green stains created by algae, diatoms, cyanobacteria, and other phototrophs growing around lights in the cave. Although green, phototrophic microbial growth is expected at cave entrances, this phenomenon deep inside a cave was first described from show caves, such as in Lascaux Cave, as far back as at least the 1960s. Details about lampenflora research are in Chapter 12. Second, opening (i.e. through artificial entrances) and closing a cave (i.e. from covering the overlying rock with artificial soil and concrete) to surface water infiltration, as well as installing air conditioners and other devices to circulate outside air into a cave, also affect microbial growth. Fungal colonization, forming white, black, gray, and gray-bluish biofilms on cave walls, sediments, and paintings, can develop [131–133]. Yellow colonies and biofilms, which can also be common in caves not heavily impacted by tourism [50, 132, 134], also form. Lastly, research showed that fungi and bacteria are brought into the cave from outside, and that microbial diversity and physiology appear to be distinct between heavily impacted and pristine areas. Also, foreign microbes can out-compete the indigenous populations, likely due to increased organic matter introduced from anthropogenic activity (e.g. paint, lint, etc.) [135, 136].

Identifying the potential differences in microbial community compositions between show caves and pristine caves has also been studied from other systems, such as Mammoth Cave in Mammoth Cave National Park, Kentucky (USA), and Kartchner Caverns in Arizona (USA). This research is described in Chapters 3 and 5, respectively. It is clear that our knowledge of this topic is still growing, even in the area of whether basic cave exploration by small groups of individuals can also negatively impact microbial communities in caves (i.e. as opposed to large groups of people visiting the same areas of a show cave several times a day). For instance, early research on this topic came from the expedition camps in Lechuguilla Cave where cave explorers, surveyors, and scientists spent days underground. *Escherichia coli* was found in drinking pools near the camps, which was taken to mean that the pools were contaminated by the people [137, 138]. The findings were criticized [139], but there has been almost no follow-up work on the topic.

1.4 The Future of Cave Microbiology Research

This book summarizes research performed to understand the microbiology of caves, drawing from historical studies and presenting current state-of-the-art work. However, despite the decades of research performed to understand "who's there?" in different types of cave systems worldwide, we still have much to learn [140]. Compared to other habitats being investigated, it is safe to say that advances in environmental microbiology and ecology of caves still lag behind advances being made in other habitats, including the submarine ocean crust, one of the remotest places on the planet and far more difficult to access than most caves. Although we continue to explore and discover new caves and habitats, research progress is slow. The pace is set by the one large, and many small, collaboration networks of researchers (▸ Fig. 1.4). Of the 27 teams, less than a third has been actively publishing at a rate of at least one publication per year for the past decade. This has resulted in cumulative publication curves that show an overall increase in the number of publications per year, but also spurts and gaps likely due to the starting or stopping research collaboration caused by politics and/or funding limitations (▸ Fig. 1.2). One advantage of having larger collaboration networks is that resources can be pooled to accomplish research tasks.

Also, from the co-author network, only two generations of research collaborations, and no more than three collaboration linkages through time, appear to have developed in the almost two decades of time that have elapsed. Specifically, graduate students and postdoctoral researchers trained by some of the early-cluster-forming individuals have not yet started to form new collaborations. This could also be due to limitations associated with funding and/or the availability of professional positions for these younger individuals, especially considering that the average period of time for a graduate student or postdoctoral researcher at a university is five or less years. Clearly, growth and expansion for cave microbiology research will come as more collaborations are made and as new researchers obtain positions where they can initiate new work.

New research will also come from changing the types of questions we ask. Specifically, the research community needs to transition from addressing the question, "who's there?", to the question, "what are they doing?" This research will rely on technological advances that link geochemical and genetic data to understand genes, enzymes, and metabolites, such as from "omics" (i.e. transcriptomics, proteomics) of RNA, oligonucleotide rRNA probing, microautoradiography, and stable-isotope probing. Currently, results from only a few cave metagenomics studies have been published, or are currently being analyzed, which means much more work remains [18, 19, 115, 141]. Some of these findings are reported on in this book, in Chapters 2, 5, 6, and 10.

Acknowledgments

The collaboration network analysis done for this chapter was inspired by conversations with Audrey T. Paterson and Scott A. Engel. Funding from the Jones Endowment for Aqueous Geochemistry at the University of Tennessee–Knoxville is acknowledged.

References

[1] White W. Geomorphology and Hydrology of Karst Terrains. New York, Oxford University Press, 1988.
[2] Ford DC, Williams P. Karst Hydrology and Geomorphology: Wiley, 2007.
[3] Palmer AN. Cave Geology. Dayton, Ohio, Cave Books, 2007.
[4] Culver DC, Pipan T. The Biology of Caves and Other Subterranean Habitats. Oxford, Oxford University Press, 2009.
[5] White WB, Culver DC. Cave, definition of. In: Culver DC, White WB, eds. Encyclopedia of Caves. Burlington, MA, Elsevier Academic Press, 2000, 81–5.
[6] Halliday WR. Pseudokarst in the 21st century. J Cave Karst Stud 2007, 69, 103–13.
[7] Pipan T, Culver DC. Convergence and divergence in the subterranean realm: a reassessment. Biol J Linn Soc 2012, 107, 1–14.
[8] Lee NM, Meisinger DB, Aubrecht R, et al. Caves and Karst Environments. In: Bell EM, ed. Life at Extremes; Environments, Organisms and Strategies for Survival Oxfordshire, UK, CAB International, 2012, 320–44.
[9] Aubrecht R, Brewer-Carías C, Šmída B, Audy M, Kováčik Ľ. Anatomy of biologically mediated opal speleothems in the world's largest sandstone cave: Cueva Charles Brewer, Chimantá Plateau, Venezuela. Sed Geol 2008, 203, 181–95.
[10] Barton HA, Giarrizzo JG, Suarez P, et al. Microbial diversity in a Venezuelan orthoquartzite cave is dominated by the Chloroflexi (Class Ktedonobacterales) and Thaumarchaeota Group I.1c. Front Microbiol 2014, 5.
[11] Sallstedt T, Ivarsson M, Lundberg J, R. S, Vidal Romaní JR. Speleothem and biofilm formation in a granite/dolerite cave, Northern Sweden. Int J Speleol 2014, 43, 305–13.
[12] Cacchio P, Ercole C, Contento R, et al. Involvement of bacteria in the origin f a newly described spelethem in the gypsum cave of Grave Grubbo (Crotone, Italy). J Cave Karst Stud 2012, 74, 7–18.
[13] Tebo BM, Davis RE, Anitori RP, Connell LB, Schiffman P, Staudigel H. Microbial communities in dark oligotrophic volcanic ice cave ecosystems of Mt. Erebus, Antarctica. Front Microbiol 2015, 6.
[14] Angert ER, Northup DE, Reysenbach AL, Peek AS, Goebel BM, Pace NR. Molecular phylogenetic analysis of a bacterial community in Sulphur River, Parker Cave, Kentucky. Am Mineral 1998, 83, 1583–92.
[15] Engel AS, Porter ML, Kinkle BK, Kane TC. Ecological assessment and geological significance of microbial communities from Cesspool Cave, Virginia. Geomicrobiol J 2001, 18, 259–74.
[16] Hutchens E, Radajewski S, Dumont MG, McDonald IR, Murrell JC. Analysis of methanotrophic bacteria in Movile Cave by stable isotope probing. Environ Microbiol 2004, 6, 111–20.
[17] Vlasceanu L, Popa R, Kinkle B. Characterization of *Thiobacillus thioparus* LV43 and its distribution in a chemoautotrophically based groundwater ecosystem. Appl Environ Microbiol 1997, 63, 3123–7.
[18] Tetu SG, Breakwell K, Elbourne LDH, Holmes AJ, Gillings MR, Paulsen IT. Life in the dark: metagenomic evidence that a microbial slime community is driven by inorganic nitrogen metabolism. ISME J 2013, 7, 1227–36.
[19] Ortiz M, Legatzki A, Neilson JW, et al. Making a living while starving in the dark: metagenomic insights into the energy dynamics of a carbonate cave. ISME J 2014, 8, 478–91.
[20] White WB. Proceedings of the 15th International Congress of Speleology, July 19–26, Kerrville, Texas. International Union of Speleology, 2009.
[21] Filippi M, Bosák P. Proceedings of the 16th International Congress of Speleology, July 21–28, Brno. Volume 1. Praha: Czech Speleological Society, 2013.

[22] Dattagupta S. Introduction to the special issue "Biogeochemistry and Microbial Ecology of Cave Systems." Geomicrobiol J 2014, 31, 173–4.
[23] Faust B. The formation of saltpeter in caves. Nat Speleol Soc Bull 1949, 11, 17–23.
[24] Caumartin V. Review of the microbiology of underground environments. Nat Speleol Soc Bull 1963, 25, 1–14.
[25] Brigmon RL, Martin HW, Morris TL, Bitton G, Zam SG. Biogeochemical ecology of *Thiothrix* spp. in underwater limestone caves. Geomicrobiol J 1994, 12, 141–59.
[26] Rusterholtz K, Mallory LM. Density, activity and diversity of bacteria indigenous to a karstic aquifer. Microb Ecol 1994, 28, 79–99.
[27] Mikell AT, Smith CL, Richardson JC. Evaluation of media and techniques to enumerate heterotrophic microbes from karst and sand aquifer springs. Microb Ecol 1996, 31, 115–24.
[28] Nakaew N, Sungthong R, Yokota A, Lumyong S. *Nonomuraea monospora* sp. nov., an actinomyceteisolated from cave soil in Thailand, and emended description of the genus *Nonomuraea*. Int J Syst Evol Microbiol 2012, 62, 3007–12.
[29] Bhullar K, Waglechner N, Pawlowski A, et al. Antibiotic resistance is prevalent in an isolated cave microbiome. PLoS ONE 2012, 11, e34953.
[30] Cheeptham N. Cave Microbiomes: A Novel Resource for Drug Discovery. SpringerBriefs in Microbiology 1. New York, Springer, 2013.
[31] Cheeptham N, Sadoway T, Rule D, et al. Cure from the cave: volcanic cave actinomycetes and their potential in drug discovery. Int J Speleol 2013, 42, 35–47.
[32] Poulson TL, Lavoie KH. The trophic basis of subsurface ecosystems. In: Wilkens H, Culver DC, Humphreys WF, eds. Subterranean Ecosystems Ecosystems of the World 30. Amsterdam, Elsevier Academic Press, 2000, 231–50.
[33] Pedersen K. Exploration of deep intraterrestrial microbial life: current perspectives. FEMS Microbiol Lett 2001, 185, 9–16.
[34] Birdwell JE, Engel AS. Characterization of dissolved organic matter in cave and spring waters using UV-Vis absorbance and fluorescence spectroscopies. Org Geochem 2010, 41, 270–80.
[35] Simon KS, Pipan T, Ohno T, Culver DC. Spatial and temporal patterns in abundance and character of dissolved organic matter in two karst aquifers. Fund Appl Limnol 2010, 177, 81–92.
[36] Engel AS. Chemoautotrophy. In: White WB, Culver DC, eds. Encyclopedia of Caves, 2nd edn. Oxord, UK, Academic Press, imprint of Elsevier, 2012, 125–34.
[37] Stevens T, McKinely J. Lithoautotrophic microbial ecosystems in deep basalt aquifers. Science 1995, 270, 450–4.
[38] Sarbu SM, Kane TC, Kinkle BK. A chemoautotrophically based cave ecosystem. Science 1996, 272, 1953–5.
[39] Goldscheider N, Hunkeler D, Rossi P. Review: Microbial biocenoses in pristine aquifers and an assessment of investigative methods. Hydrogeol J 2006, 14, 926–41.
[40] Opsahl SP, Chanton JP. Isotopic evidence for methane-based chemosynthesis in the Upper Floridan aquifer food web. Oecologia 2006, 150, 89–96.
[41] Por FD. Ophel: a groundwater biome based on chemoautotrophic resources. The global significance of the Ayyalon Cave finds, Israel. Hydrobiologia 2007, 592, 1–10.
[42] Porter ML, Engel AS, Kane TC, Kinkle BK. Productivity-diversity relationships from chemolithoautotrophically based sulfidic karst systems. I J Speleol 2009, 38, 27–40.
[43] Hutchins BT, Schwartz BF, Engel AS. Environmental controls on organic matter production and transport across surface-subsurface geochemical boundaries in the Edwards Aquifer, Texas, USA. Acta Carsologica 2013, 42/2–3, 245–59.
[44] Jin J, Zimmerman AR, Moore PJ, Martin JB. Organic and inorganic carbon dynamics in a karst aquifer: Santa Fe River Sink-Rise system, north Florida, USA. J Geophys Res Biogeosci 2014, 119, 340–57.

[45] Hose L, Pisarowicz J. Cueva de Villa Luz, Tobasco, Mexico: Reconnaissance study of an active sulfur spring cave and ecosystem. J Cave Karst Stud 1999, 6, 13–21.
[46] Hose LD, Palmer AN, Palmer MV, Northup DE, Boston PJ, DuChene HR. Microbiology and geochemistry in a hydrogen-sulphide rich karst environment. Chem Geol 2000, 169, 399–423.
[47] Macalady JL, Jones DS, Lyon EH. Extremely acidic, pendulous cave wall biofilms from the Frasassi cave system, Italy. Environ Microbiology 2007, 9, 1402–14.
[48] Jones DS, Schaperdoth I, Macalady JL. Metagenomic evidence for sulfide oxidation in extremely acidic cave biofilms. Geomicrobiol J 2014, 31, 194–204.
[49] Barton HA, Taylor NM, Kreate MP, Springer AC, Oehrle SA, Bertog JL. The impact of host rock geochemistry on bacterial community structure in oligotrophic cave environments. Int J Speleol 2007, 36, 93–104.
[50] Portillo MC, Gonzalez JM, Saiz-Jimenez C. Metabolically active microbial communities of yellow and grey colonizations on the walls of Altamira Cave, Spain. J Appl Microbiol 2008, 104, 681–91.
[51] Northup DE, Melim LA, Spilde MN, et al. Lava cave microbial communities within mats and secondary mineral deposits: Implications for life detection on other planets. Astrobiology 2011, 11, 601–18.
[52] Hathaway JJM, Garcia MG, Balasch MM, et al. Comparison of bacterial diversity in Azorean and Hawai'ian lava cave microbial mats. Geomicrobiol J 2014, 31, 205–20.
[53] Bastian F, Alabouvette C, Saiz-Jimenez C. The impact of arthropods on fungal community structure in Lascaux Cave. J Appl Microbiol 2009, 106, 1456–62.
[54] Jurado V, Fernandez-Cortes A, Cuezva S, et al. The fungal colonisation of rock-art caves: experimental evidence. Naturwissenschaften 2009, 96, 1027–34.
[55] Bindschedler S, Millière L, Cailleau G, Job D, Verrecchia EP. Calcitic nanofibres in soils and caves: a putative fungal contribution to carbonatogenesis. Geol Soc London Spec Publ 2010, 336, 225–388.
[56] Wang W, Ma X, Ma Y, et al. Molecular characterization of airborne fungi in caves of the Mogao Grottoes, Dunhuang, China. Int Biodeter Biodegr 2011, 65, 726–31.
[57] Docampo S, Trigo MM, Recio M, Melgar M, García-Sánchez J, Cabezudo B. Fungal spore content of the atmosphere of the Cave of Nerja (southern Spain): diversity and origin. Sci Total Environ 2011, 409, 835–43.
[58] Vaughan MJ, Maier RM, Pryor BM. Fungal communities on speleothem surfaces in Kartchner Caverns, Arizona, USA. Int J Speleol 2011, 40, 65–77.
[59] Connell L, Staudigel H. Fungal diversity in a dark oligotrophic volcanic ecosystem (DOVE) on Mount Erebus, Antarct Biol 2013, 2, 798–809.
[60] Vanderwolf K, David Malloch JD, McAlpine DF, Forbes GJ. A world review of fungi, yeasts, and slime molds in caves. Int J Speleol 2013, 42, 77–96.
[61] Vaughan MJ, Nelson WM, Soderlund CA, Maier RM, Pryor BM. Assessing fungal community structure from mineral surfaces in Kartchner Caverns using multiplexed 454 pyrosequencing. Microb Ecol 2015, DOI 10.1007/s00248-014-0560-9.
[62] Hess WH. The origin of nitrates in cavern earths. J Geol 1900, 8, 129–34.
[63] Dudich E. Biologie der aggteleker tropfsteinhohle "Baradla" in ungaarn. Speläol Mon (Vienna) 1932, 13, 246
[64] Høeg OA. Cyanophyceae and bacteria in calcareous sediments in the interior of limestone caves in Nord-Rana, Norway. Nytt Magazin for Naturvidenskapene 1946, 85, 99–104.
[65] Thrailkill JV. Origin of cave popcorn [abs.]. Nat Speleol Soc Bull 1964, 27, 59.
[66] Smyk B, Drzal M. Research on the influence of microorganisms on the development of karst phenomena. Geogr Polonica 1964, 2, 57–60.
[67] Went FW. Fungi associated with stalactite growth. Science 1969, 166, 385–6.

[68] Caumartin V. La corrosion microbienne dans un reseau karstique. Proceedings of the 4th International Congress of Speleology in Yugoslavia, 1965. Ljubljana: Federation Yougoslave de Speleologie 1968, 31–8.

[69] Hubbard DA, Herman JS, Bell PE. The role of sulfide oxidation in the genesis of Cesspool Cave, Virginia, USA. In: 9th International Congress of Speleology, Barcelona, Spain, 1986, 255–7.

[70] Northup DE, Lavoie K. Geomicrobiology of caves: a review. Geomicrobiol J 2001, 18, 199–222.

[71] Barton HA, Northup DE. Geomicrobiology in cave environments: Past, current and future perspectives. J Cave Karst Stud 2007, 69, 163–178.

[72] Bottrell SH, Smart PL, Whitaker F, Raiswell R. Geochemistry and isotope systematics of sulphur in the mixing zone of Bahamian blue holes. Appl Geochem 1991, 6, 97–103.

[73] Stoessell RK, Moore YH, Coke JG. The occurrence and effect of sulfate reduction and sulfide oxidation on coastal limestone dissolution in Yucatan cenotes. Ground Water 1993, 31, 566–75.

[74] Humphreys WF. Physico-chemical profile and energy fixation in Bundera Sinkhole, an anchialine remiped habitat in north-western Australia. J Royal Soc W Aust 1999, 82, 89–98.

[75] Ward DM, Bateson MM, Weller R, Ruff-Roberts AL. Ribosomal RNA analysis in microorganisms as they occur in nature. Adv Microb Ecol 1992, 12, 219–86.

[76] Amann RI, Ludwig W, Schleifer K-H. Phylogenetic identification and in situ detection of individual microbial cells without cultivation. Microb Mol Biol Rev 1995, 59 143–69.

[77] Head IM, Saunders JR, Pickup RW. Microbial evolution, diversity, and ecology: A decade of ribosomal RNA analysis of uncultivated microorganisms. Microb Ecol 1998, 35, 1–21.

[78] Reysenbach AL, Wickham GS, Pace NR. Phylogenetic analysis of the hyperthermophilic pink filament community in Octopus Spring, Yellowstone National Park. Appl Environ Microbiol 1994 60, 2113–9.

[79] Lascu C, Sarbu SM. Pesteri Scufundate. Bucurest, Romania: Academiei Republicii Socialiste, 1987.

[80] Sci^2 Team. Science of science (Sci^2) Tool. Indiana University and SciTech Strategies, http://sci2.cns.iu.edu, 2009.

[81] Sarbu SM, Galdenzi, S., Menichetti, M., and Gentile, G. Geology and biology of Grotte di Frasassi (Frasassi Caves) in Central Italy, an ecological multi-disciplinary study of a hypogenic underground karst system. In: Wilkens H, Culver DC, Humphreys S, eds. Ecosystems of the World: Subterranean Ecosystems. Oxford, Elsevier Science, 2000, 361–81.

[82] Vlasceanu L, Sarbu SM, Engel AS, Kinkle BK. Acidic, cave-wall biofilms located in the Frasassi Gorge, Italy. Geomicrobiol J 2000, 17, 125–39.

[83] Macalady JL, Lyon EH, Koffman B, et al. Dominant microbial populations in limestone-corroding stream biofilms, Frasassi Cave system, Italy. Appl Environ Microbiol 2006, 72, 5596–609.

[84] Engel AS. Geomicrobiology of Sulfidic Karst Systems [Masters]. Cincinnati: University of Cincinnati; 1999.

[85] Porter M. Ecosystem Energetics of Sulfidic Karst [Masters]. Cincinnati: University of Cincinnati; 1999.

[86] Engel AS, Lee N, Porter ML, Stern LA, Bennett PC, Wagner M. Filamentous "*Epsilonproteobacteria*" dominate microbial mats from sulfidic cave springs. Appl Environ Microbiol 2003, 69, 5503–11.

[87] Engel AS, Porter ML, Stern LA, Quinlan S, Bennett PC. Bacterial diversity and ecosystem function of filamentous microbial mats from aphotic (cave) sulfidic springs dominated by chemolithoautotrophic '*Epsilonproteobacteria*'. FEMS Microbiol Ecol 2004, 31–53.

[88] Barton H, Luiszer F. Microbial metabolic structure in a sulfidic cave hot spring: Potential mechanisms of biospeleogenesis. J Cave Karst Stud 2005, 67, 28–38.
[89] Holmes AJ, Tujula NA, Holley M, et al. Phylogenetic structure of unusual aquatic microbial formations in Nullarbor caves, Australia. Environ Microbiol 2001, 3, 256–64.
[90] Hill CA. Sulfuric-acid speleogenesis of Carlsbad Cavern and its relationship to hydrocarbons, Delaware Basin, New-Mexico and Texas. AAPG Bull 1990, 74, 1685–94.
[91] Hubbard D, Herman J, Bell P. Speleogenesis in a travertine scarp: observations of sulfide oxidation in the subsurface. In: Herman J, Hubbard D, eds. Travertine-marl: Stream deposits in Virginia. Charlottesville, VA: Dept. of Mines, Minerals, and Energy, Division of Mineral Resources, Virginia Division of Mineral Resources, 1990, 177–84.
[92] Polyak VJ, McIntosh WC, Guven N, Provencio P. Age and origin of Carlsbad Cavern and related caves from $^{40}Ar/^{39}Ar$ of alunite. Science 1998, 279, 1919–22.
[93] Klimchouk AB, Ford DC, Palmer AN, Dreybrodt W, eds. Speleogenesis: Evolution of Karst Aquifers. Huntsville, Alabama, National Speleological Society, 2000.
[94] Polyak VJ, Provincio P. By-product materials related to H_2S-H_2SO_4-influenced speleogenesis of Carlsbad, Lechuguilla, and other caves of the Guadalupe Mountains, New Mexico. J Cave Karst Stud 2001, 63, 23–32.
[95] Engel AS, Stern LA, Bennett PC. Microbial contributions to cave formation: new insight into sulfuric acid speleogenesis. Geology 2004, 32, 369–72.
[96] Campbell B, Engel AS, Porter ML, Takai K. The versatile "*Epsilon-proteobacteria*": Key players in sulfidic habitats. Nature Rev Microbiol 2006, 4, 456–68.
[97] Porter ML, Engel AS. Diversity of uncultured *Epsilonproteobacteria* from terrestrial sulfidic caves and springs. Appl Environ Microbiol 2008, 74, 4973–7.
[98] Macalady JL, Dattagupta S, Schaperdoth I, Jones DS, Druschel GK, Eastman D. Niche differentiation among sulfur-oxidizing bacterial populations in cave waters. ISME J 2008, 2, 590–601.
[99] Engel AS, Meisinger DB, Porter ML, et al. Linking phylogenetic and functional diversity to nutrient spiraling in microbial mats from Lower Kane Cave (USA). ISME J 2010, 4, 98–110.
[100] Rohwerder T, Sand W, Lascu C. Preliminary evidence for a sulfur cycle in Movile Cave, Romania. Acta Biotechnol 2003, 23, 101–7.
[101] Chen Y, Wu L, Boden R, et al. Life without light: microbial diversity and evidence of sulfur- and ammonium-based chemolithotrophy in Movile Cave. ISME J 2009, 3, 1093–104.
[102] Sarbu SM, Popa R. A unique chemoautotrophically based ecosystem. In: Camacho A, ed. The Natural History of Biospeleology. Madrid: Monografias Museo Nacional de Ciencias Naturales; 1992, 637–66.
[103] Culver DC, Sket B. Hotspots of subterranean biodiversity in caves and wells. J Cave Karst Stud 2000, 62, 11–7.
[104] Engel AS. Observations on the biodiversity of sulfidic karst habitats. J Cave Karst Stud 2007, 69, 187–206.
[105] Por FD, Dimentman C, Frumkin A, Naaman I. Animal life in the chemoautotrophic ecosystem of the hypogenic groundwater cave of Ayyalon (Israel): A summing up. Natural Sci 2013, 5 Paper ID 30808.
[106] Roach KR, Tobler MT, Winemiller KO. Hydrogen sulfide, bacteria, and fish:a unique, subterranean food chain. Ecol 2011, 92, 2056–62.
[107] Sarbu SM. Movile Cave: a chemoautotrophically based groundwater ecosystem. In: Wilkens H, Culver DC, Humphreys WF, eds. Subterranean Ecosystems. Amsterdam, Elsevier, 2000, 319–43.
[108] Dattagupta S, Schaperdoth I, Montanari A, et al. A novel symbiosis between chemoautotrophic bacteria and a freshwater cave amphipod. ISME J 2009, 3, 935–43.

[109] Flot J-F, Wörheide G, Dattagupta S. Unsuspected diversity of *Niphargus* amphipods in the chemoautotrophic cave ecosystem of Frasassi, central Italy. BMC Evol Biol 2010, 10, 171.

[110] Bauermeister J, Ramette A, Dattagupta S. Repeatedly evolved host-specific ectosymbioses between sulfur-oxidizing bacteria and amphipods living in a cave ecosystem. PLoS ONE 2012, 7, e50254.

[111] Flot J-F, Bauermeister J, Brad T, Hillebrand-Voiculescu A, Sarbu SM, Dattagupta S. *Niphargus-Thiothrix* associations may be widespread in sulphidic groundwater ecosystems: evidence from southeastern Romania. Mol Ecol 2014, 23, 1405–17.

[112] Pakes MJ, Mejía-Ortíz L. Chemosynthetic ectosymbiosis reported in the predatory anchialine cave endemic, *Xibalbanus tulumensis* (Yager, 1987) (Remipedia). Crustaceana 2014, 87, 1657–67.

[113] Pakes MJ, Weis AK, Mejía-Ortiz L. Arthropods host intracellular chemosynthetic symbionts, too: cave study reveals an unusual form of symbiosis. J Crustacean Biol 2014, 34, 334–41.

[114] Jones DS, Albrecht HL, Dawson KS, et al. Community genomic analysis of an extremely acidophilic sulfur-oxidizing biofilm. ISME J 2012, 6, 158–70.

[115] Jones DS, Schaperdoth I, Macalady JL. Metagenomic evidence for sulfide oxidation in extremely acidic cave biofilms. Geomicrobiol J 2014, 31, 194–204.

[116] Howarth FG. The cavernicolous fauna of Hawaiian lava tubes, 1. Introduction. Pacific Insects 1973, 15, 139–51.

[117] Howarth FG. Community structure and niche differentiation in Hawaiian lava tubes. In: Mueller-Dombois D, Bridges KW, Carson HL, eds. Island Ecosystems: Biological Organization in Selected Hawaiian Communities US/IBP Synthesis Series 15. Stroudsburg, PA, Hutchinson Ross Publishing Company, 1981, 318–36.

[118] Ashmole NP, Orom P, Ashmole MJ, Martin JL. Primary faunal succession in volcanic terrain: lava and cave studies on the Canary Islands. Biol J Linn Soc 1992, 46, 207–34.

[119] Northup DE, Welbourn WC. Life in the twilight zone: lava tube ecology. New Mexico Bureau of Mines & Mineral Resources Bulletin 1997, 156, 69–82.

[120] Northup DE, Carr DL, Crocker MT, et al. Biological investigations in Lechuguilla Cave. Nat Speleol Soc Bull 1994, 56, 54–63.

[121] Cañaveras JC, Hoyos M, Sanchez-Moral S, et al. Microbial communities associated with hydromagnesite and needle-fiber aragonite deposits in a karstic cave (Altamira, Northern Spain). Geomicrobiol J 1999, 16, 9–25.

[122] Groth I, Vettermann R, Schuetze B, Schumann P, Saiz-Jimenez C. Actinomycetes in karstic caves of northern Spain (Altamira and Tito Bustillo). J Microbiol Meth 1999, 36, 115–22.

[123] Arroyo G, Arroyo I, Arroyo E. Microbiological analysis of Maltravieso Cave (Caceres), Spain. Int Biodeter Biodegr 1997, 40, 131–9.

[124] Laiz L, Groth I, Gonzalez I, Saiz-Jimenez C. Microbiological study of the dripping waters in Altamira Cave (Santillana del Mar, Spain). J Microbiol Meth 1999, 36, 129–38.

[125] Northup DE, Dahm CN, Melim LA, et al. Evidence for geomicrobiological interactions in Guadalupe caves. J Cave Karst Stud 2000, 62, 80–90.

[126] Northup DE, Barns SM, Yu LE, et al. Diverse microbial communities inhabiting ferromanganese deposits in Lechuguilla and Spider Caves. Environ Microbiol 2003, 5, 1071–86.

[127] Schabereiter-Gurtner C, Saiz-Jimenez C, Piñar G, Lubitz W, Rölleke S. Phylogenetic 16S rRNA analysis reveals the presence of complex and partly unknown bacterial communities in Tito Bustillo Cave, Spain, and on its Palaeolithic paintings. Environ Microbiol 2002, 4, 392–400.

[128] Schabereiter-Gurtner C, Saiz-Jimenez C, Piñar G, Lubitz W, Rölleke S. Altamira Cave Paleolithic paintings harbour partly unknown bacterial communities. FEMS Microbiol Lett 2002, 211, 7–11.

[129] Gurtner C, Heyrman J, Piñar J, Lubitz G, Swings J, Rölleke S. Comparative analyses of the bacterial diversity on two different biodeteriorated wall paintings by DGGE and 16S rDNA sequence analysis. Int J Biodeterior Biodegrad 2000, 46, 229–39.
[130] Sanchez-Moral S, Soler V, Cañaveras JC, Sanz-Rubio E, Van Grieken R, Gysels K. Inorganic deterioration affecting the Altamira Cave, N Spain: quantitative approach to wall-corrosion (solutional etching) processes induced by visitors. Sci Total Environ 1999, 243–244, 67–84.
[131] Schabereiter-Gurtner C, Saiz-Jimenez C, Piñar G, Lubitz W, Rölleke S. Phylogenetic diversity of bacteria associated with Paleolithic paintings and surrounding rock walls in two Spanish caves (Llonín and La Garma). FEMS Microbiol Ecol 2004, 47, 235–47.
[132] Pašić L, Kovce B, Sket B, Herzog-Velikonja B. Diversity of microbial communities colonizing the walls of a karstic cave in Slovenia. FEMS Microbiol Ecol 2010, 71, 50–60.
[133] Borsodi AK, Knáb M, Krett G, Makk J. Biofilm bacterial communities inhabiting the cave walls of the Buda thermal karst system, Hungary. Geomicrobiol J 2012, 29, 611–27.
[134] Porca E, Jurado V, Žgur-Bertok D, Saiz-Jimenez C, Pašić L. Comparative analysis of yellow microbial communities growing on the walls of geographically distinct caves indicates a common core of microorganisms involved in their formation. FEMS Microbiol Ecol 2012, 81, 255–66.
[135] Ikner LA, Toomey RS, Nolan G, Neilson JW, Pryor BM, Maier R. Culturable microbial diveristy and the impact of tourism in Kartchner Caverns, Arizona. Microb Ecol 2007, 53, 30–42.
[136] Chelius MK, Beresford G, Horton H, et al. Impacts of alterations of organic inputs on the bacterial community within the sediments of Wind Cave, South Dakota, USA. Int J Speleol 2009, 38, 1–10.
[137] Northup DE, Beck KM, Mallory LM. Human impact on the microbial communities of Lechuguilla cave: Is protection possible during active exploration? J Cave Karst Stud 1997, 59, 166.
[138] Hunter AJ, Northup DE, Dahm CN, Boston PJ. Persistent coliform contamination in Lechuguilla Cave pools. J Cave Karst Stud 2004, 66, 102–10.
[139] Barton HA, Pace NR. Discussion: Persistent coliform contamination in Lechuguilla Cave pools. J Cave Karst Stud 2005, 68, 55–7.
[140] Engel AS, Northup DE. Caves and karst as model systems for advancing the microbial sciences. In: Martin J, White WB, eds. Frontiers in Karst Research, Special Publication 13. Leesburg, Virginia, Karst Waters Institute, 2008, 37–48.
[141] Hamilton TL, Jones DS, Schaperdoth I, Macalady JL. Metagenomic insights into S(0) precipitation in a terrestrial subsurface lithoautotrophic ecosystem. Front Microbiol 2015 DOI: 10.3389/fmicb.2014.00756.

Daniel S. Jones

2 Methods for Characterizing Microbial Communities in Caves and Karst: A Review

Abstract: Microorganisms play a central role in the biogeochemistry of caves and karst. However, characterizing the biodiversity of caves is challenging because environmental microbial communities often have extremely high richness and contain numerous organisms that have neither been isolated nor described using traditional culturing techniques. Fortunately, culture-independent methods can be applied to study cave populations and communities, and are especially powerful if combined with culture-based information. The purpose of this chapter is to provide a general overview of methods to analyze cave microorganisms, with emphasis on commonly used approaches, including cultivation, rRNA-based methods, and microscopy techniques, as well as on emerging technologies such as metagenomics and metatranscriptomics. We discuss specific examples and applications from the cave microbiology literature, and end with a case study on the microbial communities of acidic cave biofilms.

2.1 Introduction

Over the past several decades, we have begun to realize the immense diversity of microbial life on Earth [1]. Together, Bacteria and Archaea are the numerically dominant organisms on our planet, and they are ubiquitous at the Earth's surface, as well as throughout the habitable regions of the subsurface. Caves are no exception. Microbial life is a pervasive feature in caves, and can be found as sparse microbial populations in oligotrophic caves [2, 3], as densely packed cave-wall biofilms in energy-rich sulfidic systems [4], and everything in between [5]. Microorganisms are intimately involved in many fundamental processes in cave ecosystems, including nutrient and element cycling [6, 7], primary production [8], and processes related to the dissolution or precipitation of carbonates [9, 10] and other cave minerals [11, 12]. Furthermore, the subsurface contains immense microbial biomass and novel microbial diversity [13–15], yet it remains largely underexplored. Karst terrains cover approximately 15% of the ice-free Earth's surface [16], and constitute an important reservoir for microbial diversity that could contain new branches in the tree of life, novel microbial metabolisms, and unique sources of genetic information for pharmaceutical or biotechnological applications [17].

To access the microbial diversity of caves and karst, we require methods to characterize environmental microbial communities and describe cave microorganisms. The identification, description, and quantification of microbial populations and communities are requisite first steps to understand and define microbial roles in cave ecology and biogeochemistry. However, environmental microbial communities have extraordinarily high diversity and contain many organisms with unknown physiology [1, 18, 19]. Here, I present an overview of methods to characterize environmental microbial communities, both for assessing community structure and composition, and for probing microbial metabolic and functional processes.

2.2 Culture-based Analyses

For most of the history of environmental microbiology research, our primary information on microbial processes has come from studies of microorganisms and microbial processes in culture. Growing, maintaining, and manipulating microorganisms *in vitro* remain the most direct and effective ways to describe the metabolic processes and physiological requirements of different taxa. Enrichment culturing is the more widely applied technique in which microorganisms are grown in a specific medium under a defined set of conditions that select for organisms of interest. Enrichment begins once a medium is inoculated with environmental material (the inoculum) and the inoculate grows under controlled conditions.

Enrichments might contain different populations, so it is often desirable to obtain pure cultures. A strain is a genetically identical microbial population that originated from a single cell. Once in isolation, a microbial strain is much easier to characterize because any biological processes that alter conditions in the medium can be directly attributed to that organism. To acquire a pure culture, strains are separated from the medium and other organisms to be grown in isolation. Commonly, this separation is achieved by streaking cultures onto a solid medium, usually agar based. Individual colonies are then picked from the solid surface and re-streaked and re-picked multiple times to ensure isolation. However, because not all organisms grow on solid surfaces, isolation by serial dilution in liquid media or via agar dilution tubes is also possible. Readers are referred to Madigan et al. [20] for basic information on cultivation techniques, and to reviews by Leadbetter [21], Keller and Zengler [22], and Epstein [23] for novel strategies to cultivate recalcitrant organisms.

Cultivation-dependent analyses have a long history of use in cave research [24, 25], and have proven enormously valuable. For example, Vlasceanu et al. [26] showed that caves contain indigenous microbiota after isolating and describing *Thiobacillusthioparus* strain LV43 from Movile Cave in Romania. Culture-based studies have also revealed cave microbial diversity more broadly [4, 27, 28] and have been used to identify geomicrobiological interactions, such as microbially induced calcite precipitation [29–31], microbial limestone dissolution [32], and relationships among iron- and manganese-oxidizing bacteria and cave ferromanganese deposits [11, 33, 34].

Culture-based analyses have an important drawback in the context of environmental microbial studies. Organisms that readily grow in enrichments might not be the most abundant microorganisms in the environment, but instead may represent “weeds” that are selected for *in vitro*. In fact, it is commonly estimated that

99% of microorganisms in the environment are not known in culture. This is commonly known as the "great plate count anomaly," a phrase invoked by Staley and Konopka [36] to describe the dramatic discrepancies between direct cell counts from the environment versus enumeration of viable microbes in culture. Culture-based analyses, therefore, present a biased view of microbial diversity in the natural environment [37]. Fortunately, culture-independent analyses developed over the past three decades offer an alternative means to assess microbial diversity, abundance, and environmentally relevant processes directly.

2.3 Culture-independent Analyses Based on rRNA Genes

The most widely applied techniques for culture-independent analysis of environmental microbial communities are based on the analysis of the genes that encode ribosomal RNA (rRNA). rRNA is a central component of the ribosome, which is the cellular molecule that catalyzes the translation of messenger RNA (mRNA) into peptide sequences. Bacterial and archaeal ribosomes contain three rRNA sequences: a 5S rRNA and a 23S rRNA molecule each occur in the large ribosomal subunit and a 16S rRNA molecule in the small ribosomal subunit. The "S" in 5S, 23S, and 16S denotes a sedimentation coefficient that varies with molecular size and shape in Svedberg units. For our purposes, rRNA gene sequence length is a more useful designation than Svedberg units. Gene sequence lengths vary, but a typical bacterium has a 16S rRNA gene sequence that is approximately 1500 base pairs (bp) long, while the 5S and 23S rRNA-gene sequences are roughly 120 and 2900 bp long, respectively.

rRNA gene sequences are among the most conserved nucleic acid sequences known, and thus permit phylogenetic comparisons among distantly related groups of organisms. Analysis of rRNA sequences first led to the discovery of the domain Archaea and revealed the three-domain tree of life [38]. Small subunit rRNA gene sequences, specifically the genes that encode for the 16S rRNA sequence in Bacteria and Archaea and the 18S rRNA sequence in Eukarya, remain the most widely used gene sequences for microbial taxonomic identification to date.

2.3.1 rRNA gene (rDNA) cloning

Environmental rRNA gene sequencing (also referred to as rDNA sequencing) involves the extraction of DNA directly from an environment sample, such as cave sediment, followed by separation and sequencing of rRNA genes. Generally, DNA is extracted from environmental samples by cellular lysis (e.g. via beat beating, repeated freeze-thaw cycles, or other chemical and enzymatic means) followed by separation of the nucleic acids from other cellular components and environmental materials (e.g. via phenol/chloroform/isoamyl alcohol extraction or with other DNA-binding substances) and further DNA purification [39]. Because the DNA is extracted directly from environmental materials, additional steps are often required for samples that contain certain minerals or organic compounds that inhibit DNA extraction or subsequent steps. For example, the presence of iron [40], humic acids [41], or excessive polysaccharides [42] often necessitates modified extraction procedures to achieve successful DNA recovery, quality, and purity.

Cloning, or clone library construction, has been widely applied to analyze environmental rRNA genes over the past few decades. Simply stated, cloning is a technique by which environmental genes or genomic regions are inserted into *Escherichia coli* cells (via transformation) and the *E. coli* are then grown in such a way as to separate the individual environmental genes for sequencing. Clone library construction proceeds as follows (▸ Fig. 2.1): (i) DNA is extracted from an environmental sample; (ii) rRNA genes are amplified via polymerase chain reaction (PCR) from the DNA extract, prior to being (iii) ligated into a plasmid vector that is (iv) transformed into competent *E. coli* cells; commercially available plasmid vectors contain genes for antibiotic resistance, so (v) *E. coli* are grown on an antibiotic-laced agar plate that selects only for *E. coli* that contain a plasmid insert; (vi) individual *E. coli* colonies are then picked (often using blue/white screening to indicate cells that contain an insert); and (vii) environmental gene inserts are replicated to provide adequate copies for sequencing, generally by capillary Sanger technology [43, 44]. Inserts are replicated in step (vii) via either colony PCR, which is amplification of the insert directly by PCR of *E. coli* colonies, or by growing the colonies in liquid medium and extracting plasmids from a larger volume of *E. coli* biomass. Note that separation of the *E. coli* on the agar plate in step (v) effectively isolates individual gene sequences from the mixed environmental sample. The PCR product in step (ii) contains a mixture of 16S rRNA sequences from microbes in the environmental community, but in steps (iii) and (iv), each successfully ligated vector and successfully transformed *E. coli* only receive a single rRNA sequence (▸ Fig. 2.1). Therefore, picking individual *E. coli* colonies in step (vi) is akin to selecting random 16S rRNA genes from the environment. For detailed information, readers are referred to the molecular biology handbook by Sambrook and Russell [39].

Environmental 16S rRNA gene clones are analyzed by comparing newly acquired rRNA genes from a sample to gene sequences previously retrieved from known isolates or other environmental sequences. The number of 16S rRNA genes in public databases has been increasing rapidly in recent years [1]. At the time of writing, one popular and well-curated database (www.arb-silva.de [45]) contains over 500,000 nonredun-

▸ **Fig. 2.1.** Three methods for culture-independent analysis of environmental microbial communities. Both cloning and amplicon sequencing target a specific gene or genetic region of interest, such as the 16S rRNA gene, while metagenomics generates a dataset of genomic DNA sequences from the entire community. The resulting metagenome can contain the complete genetic complement of multiple environmental organisms.

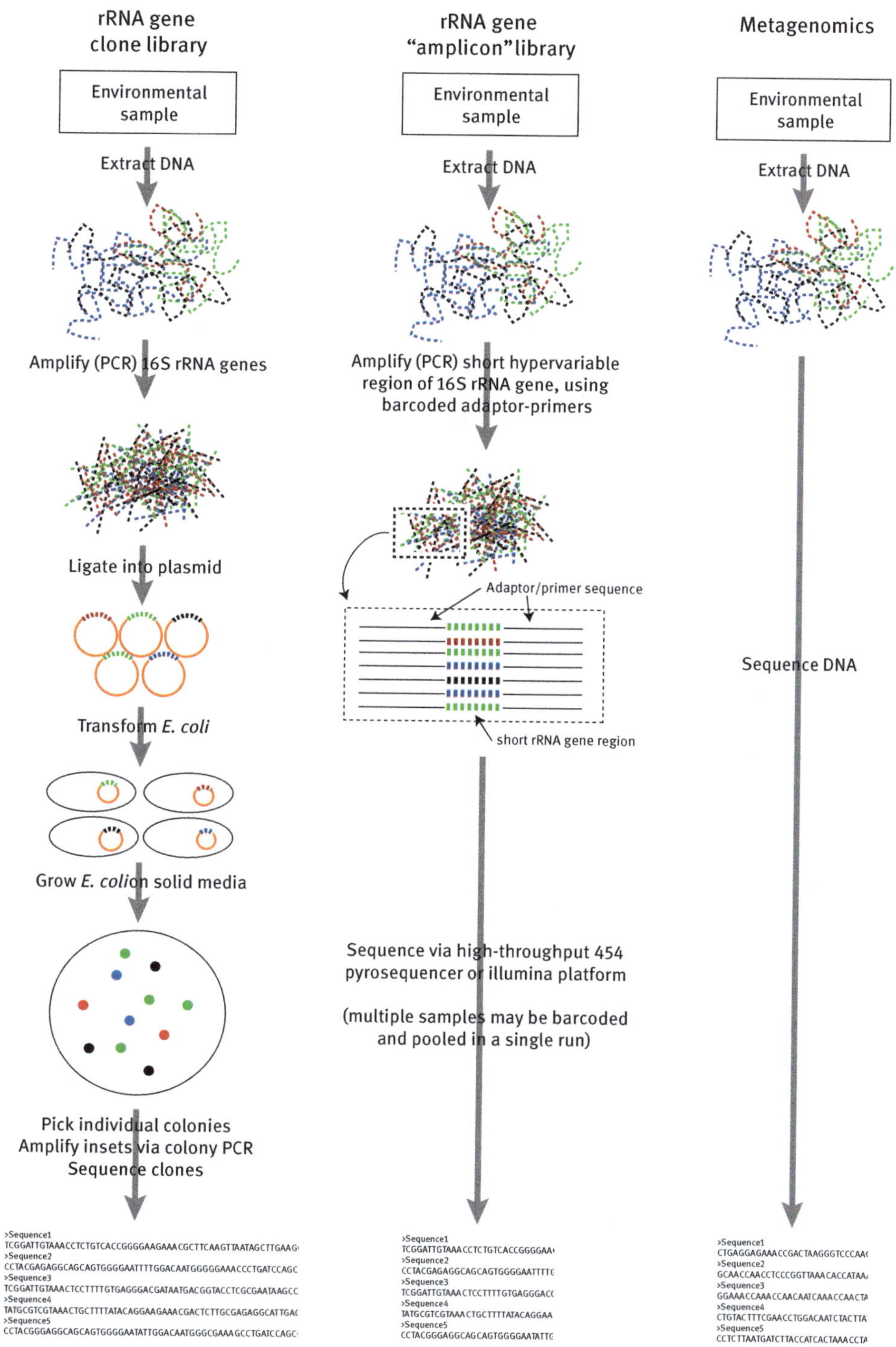
rRNA gene clone library
Environmental sample
Extract DNA
Amplify (PCR) 16S rRNA genes
Ligate into plasmid
Transform E. coli
Grow E. coli on solid media
Pick individual colonies
Amplify insets via colony PCR
Sequence clones
rRNA gene "amplicon" library
Environmental sample
Extract DNA
Amplify (PCR) short hypervariable region of 16S rRNA gene, using barcoded adaptor-primers
Adaptor/primer sequence
short rRNA gene region
Sequence via high-throughput 454 pyrosequencer or illumina platform
(multiple samples may be barcoded and pooled in a single run)
Metagenomics
Environmental sample
Extract DNA
Sequence DNA

dant, nearly full-length 16S rRNA genes, and over four million partial rRNA sequences. The taxonomic affiliation of environmental rRNA genes can be quickly determined by a simple sequence comparison against these databases, or by more rigorous phylogenetic methods. The latter is required if no close relatives are present in public databases. rRNA genes do not directly indicate microbial species [37, 46], but they can serve as proxies. For example, 16S rRNA genes that share 97% sequence similarity are often considered the same "species," although a more appropriate designation is operational taxonomic units (OTUs). OTUs can be considered analogous to species, genera, or higher taxonomic divisions, with the important distinction that they are "operationally" defined and should be interpreted within the constraints of the technique used.

A primary challenge when applying environmental rRNA sequencing is to avoid introducing bias during library creation. Any process that systematically increases or decreases a particular rRNA sequence with respect to its initial proportion in the sample may result in a dataset that inaccurately represents the true composition of the environmental community. For example, microorganisms can have multiple copies of the *rrn* operon [47]. An organism with two *rrn* operons will appear twice as abundant in a 16S rRNA gene clone library as an organism with a single *rrn* operon. This "*rrn* copy bias" is inherent in any 16S rRNA-based study. Different biases can also be introduced during clone library creation, such as from the PCR step if rRNA gene primers are not truly universal (i.e. if a PCR primer contains a mismatch with particular sequences or a group of sequences, see Section 2.8, Case Study), or if the activity of the polymerase enzyme is impeded or slowed by certain sequences [48, 49]. Additionally, bias might occur at the transformation step, as some sequences inhibit *E. coli* growth, or during the DNA extraction step if nucleic acid is more readily extracted from certain organisms than others [50].

This list of potential biases may seem daunting, but it is a reality that environmental microbiologists must acknowledge. It is generally impossible to avoid introducing any sort of bias with environmental techniques. As long as the potential sources for bias are recognized, and the results are cautiously interpreted, then appropriate conclusions can be drawn. It is advantageous to compare multiple techniques for nucleic acid extraction, PCR amplification, and cloning to compare different biases and to confirm results and identify potential artifacts (see Section 2.8). Specifically, steps may be taken to limit bias, such as selecting appropriate PCR primers, minimizing PCR cycles, or combining DNA extractions generated via different lysis procedures.

In recent years, 16S rRNA gene cloning has been widely applied in cave microbiology studies. One of the earliest applications of cloning to cave systems was performed by Angert et al. [51], in which cloning was used to describe a microbial community from a sulfidic stream in Parker Cave, Kentucky, USA. In other early applications, Vlasceanu et al. [52] and Hose et al. [4] used clone libraries to characterize the communities of highly acidic cave "snottites" (see also Section 2.8). Other studies applied cloning to explore the microbiology of ferromanganese corrosion residues in

Lechuguilla Cave, New Mexico [34], unusual "microbial mantles" in Nullarbor Caves, Australia [53], and microbial mats from Cesspool Cave, Virginia, USA [32]. In each of these studies, 16S rRNA gene cloning revealed novel microbial diversity.

2.3.2 High-throughput rRNA amplicon sequencing

Recent developments in high-throughput DNA sequencing technology now allow for large libraries of short DNA sequences, known as "amplicons," to be generated at lower cost and with less human labor compared with cloning. rRNA gene amplicon sequencing, also referred to as "tag" or "pyrotag" sequencing, has become a popular way to characterize environmental microbial communities. Amplicon libraries can contain anywhere from thousands to tens of millions of sequences. Amplicons are short sequences of 100–500 bp length that include one or more "hypervariable" regions of the 16S rRNA gene that have faster mutation rates compared to the rest of the gene sequence [55]. By capturing this variability, and despite their short length, rRNA gene amplicons have enough resolution to distinguish among microbial taxa.

Like cloning, environmental amplicon sequencing involves separation and sequencing of rRNA genes from an environmental DNA extract (▸ Fig. 2.1): (i) DNA extraction; (ii) PCR amplification of short regions of the 16S rRNA gene using primers that include a priming sequence for the rRNA region of interest and adaptors and primers for an Illumina sequencing platform (www.illumina.com) or a 454 Life Sciences pyrosequencer (www.454.com) (▸ Fig. 2.1); (iii) purification of the PCR products; and (iv) direct sequencing on an Illumina or 454 platform. Individual gene sequences are isolated from the mixed PCR product in step (ii) during the generation of 454 or Illumina sequencing libraries. With 454 technology, sequences are isolated as each is ligated to a bead and individually amplified in aqueous "microreactors" formed by a water-in-oil emulsion [54]. Bead-bound sequences are then packed into individual wells in a plate and sequenced in parallel by pyrosequencing. With Illumina technology, individual DNA sequences are separated and bound across the surface of a proprietary flow cell and amplified and sequenced in place. The adaptor/primers added in step (ii) are required for binding to 454 beads or the Illumina flow cell, and commonly include a unique barcode (a "tag") that effectively identifies all the reads from a single sample. Barcoding allows multiple samples to be pooled in a single sequencing run, and thus constitutes a significant cost savings on a per-sample basis.

High-throughput amplicon sequencing was first applied by Sogin et al. [56], who described marine sediment community diversity by using amplicons of the V6 hypervariable region of the 16S rRNA gene. The first applications of amplicon sequencing for cave and karst microbial diversity studies were completed for Mexican cenotes by Sahl et al. [57, 58] from the V1 and V2 regions of 16S rRNA genes. In other applications, Ortiz et al. [59] used amplicon libraries to characterize microbial communities from speleothems in Kartchner Caverns, Arizona, USA, and Gray and Engel [60] used amplicon sequencing to describe microbial communities associated with karst processes in the Edwards Aquifer, Texas.

Amplicon sequencing has advantages and disadvantages compared to cloning. Typical amplicon libraries are much larger than clone libraries, and thus represent a more effective method for describing the uncommon and rare community members. Furthermore, because amplicon library generation requires no *E. coli* transformation step (► Fig. 2.1), there is a significant labor savings. Amplicon sequencing is also free from biases associated with transformation, although the approach is still subject to DNA extraction and PCR amplification biases. Because amplicon sequencing makes use of high-throughput DNA sequencing technologies, it can be cheaper. However, despite these advantages, rRNA amplicon sequencing is not yet a suitable replacement for cloning. With the current technology, rRNA amplicons have short read lengths and only represent a fraction of the full 16S rRNA sequence. Amplicons are generally only considered adequate to differentiate microbial "genera" and not "species" because they have lower taxonomic resolution than the full-length rRNA gene sequence. As such, amplicons are currently not as suitable for robust phylogenetic analysis because short sequences contain fewer variable positions and produce less reliable alignments. Amplicon libraries also have high error rates from sequencing that can generate artificial diversity [61]. Fortunately, increasing read lengths from high-throughput sequencers, as well as recent developments that offer dramatically improved error rates [62–64], offer a very positive outlook for future applications of amplicon sequencing.

2.3.3 Terminal restriction fragment length polymorphism (T-RFLP)

T-RFLP is a technique by which amplified DNA sequences are separated and identified via restriction enzyme digestion. When applied to 16S rRNA genes, T-RFLP is performed by (i) amplifying rRNA genes from an environmental DNA extraction using one or more fluorescently labeled PCR primers; (ii) digesting the amplified 16S rRNA genes with one or more restriction enzymes, which cleave the genes at specific recognition sites; (iii) separating fragments of specific sizes after the restriction enzyme digestion by capillary electrophoresis, after which only sequences that include the fluorescent primer (the 5′- or 3′-end fragments, or "terminal" fragments) are analyzed [65]. In different organisms, restriction enzyme recognition sites occur at different positions along the gene, and these differences in recognition sites form the basis for microbial identification. In step (iii), the length distribution of fluorescently labeled fragments is recorded as an electropherogram, which represents the structure and diversity of the microbial community. A fragment of a particular length represents an OTU and the intensity of that fragment is proportional to the abundance of that OTU. Because taxa are distinguished by differences in restriction fragment binding sites, T-RFLP has less taxonomic resolution than direct sequence analysis. Additionally, full-length rRNA gene sequences are required to link T-RFLP fragments to microbial taxa. However, because no sequencing is directly required, a much larger number of samples can be analyzed by T-RFLP than is typically reasonable by cloning. Readers are referred to Liu et al. [66] and Osborn et al. [67] for additional information on T-RFLP.

T-RFLP has been applied to studies of cave microbial communities, including assessing microbial community response to different inputs of human- and animal-derived carbon in Wind Cave, South Dakota [68], relating microbial community shifts in a Floridan karst aquifer to seasonal changes in groundwater chemistry [69], and comparing bacterial community changes in epiphreatic karst pools following microbial colonization events associated with periodic flooding [70]. In all three studies listed above, T-RFLP was used in conjunction with cloning. rRNA gene cloning facilitated the identification of microbial taxa, and T-RFLP was used to quantify communities across a larger number of samples.

2.3.4 Denaturing gradient gel electrophoresis (DGGE)

DGGE is analogous to T-RFLP, in that environmental gene sequences are separated and analyzed by using electrophoresis. However, unlike T-RFLP, DGGE is performed on genetic sequences of the same length. Sequences are separated by their susceptibility to denaturing chemicals. When done with 16S rRNA genes, DGGE is performed by (i) extracting DNA from an environmental sample; (ii) amplifying short regions of the 16S rRNA gene; and (iii) loading and running the PCR product on a gel that includes a gradient of a denaturing agent, such as urea, formamide, or a mixture of the two. During amplification (step (ii)), a "GC-clamp," which is an approximately 40 bp region rich in guanine and cytosine (G and C) residues, is added to the end of the 16S rRNA sequence [71]. As fragments move through the gel (step (ii)), they are exposed to stronger and stronger concentrations of the denaturing agent. This eventually disrupts the hydrogen bonds that hold the DNA strands together. The GC-clamp prevents fragments from becoming fully denatured, so they form "Y"-shaped structures that remain stationary in the gel. Sequences with higher G+C content will move farther in the gel before denaturing and their final gel position is related ultimately to differences in base pair content. Like T-RFLP, the resulting gel pattern for DGGE represents community structure and diversity. Because no sequencing is required, a large number of samples can be simultaneously analyzed. Bands can be excised from the DGGE gel and sequenced to determine the identity of individual fragments.

DGGE has commonly been used in cave studies, often for comparative purposes. For example, using DGGE analysis of 16S rRNA genes, Portillo and Gonzalez [72] describe generally similar communities from white colonies on the cave walls from different parts of Altamira Cave, Spain. Other DGGE studies showed that bacterial and archaeal communities on the same speleothem surface in Kartchner Caverns are more similar to each other than to microbial communities on adjacent speleothems [73], and that fungal communities from show caves can be distinct from communities in caves with restricted access [74]. In another application, DGGE of 23S rRNA sequences was used to reveal the biogeography of *Epsilonproteobacteria* from different sulfidic caves and springs [75].

T-RFLP, DGGE, and related methods are collectively known as community fingerprinting. All community fingerprinting techniques employ electrophoretic separation of genetic variants in a PCR product, and the resulting electrophoresis pattern (the "fingerprint") represents a snapshot of the community structure. Other commonly used fingerprinting techniques include automated ribosomal intergenic spacer analysis, amplified ribosomal DNA restriction analysis, temperature gradient gel electrophoresis, and others. For more information, readers are referred to reviews on community fingerprinting by Nocker et al. [76] and Marzorati et al. [77].

2.3.5 Fluorescence *in situ* hybridization (FISH)

FISH is a technique by which specific microbial taxa are fluorescently labeled in an intact environmental sample and then are directly observed and quantified via epifluorescence microscopy or confocal laser scanning microscopy [78]. Environmental samples for FISH are first fixed to preserve cellular structures, usually using paraformaldehyde. Then, during a hybridization phase, fluorescently labeled DNA probes (typically 14–20 bp in length) are bound to the rRNA inside intact cells that can be imaged. Multiple FISH probes with different fluorophores can be applied to the same sample so that more than one population can be simultaneously observed. Because FISH targets rRNA transcripts and not genes, only active microbial populations with adequate ribosome numbers will produce a strong fluorescent signal. Furthermore, because rRNA genes are the basis for probe sequences, FISH probes can be designed to target specific species or entire microbial phyla, so long as appropriately conserved rRNA regions can be identified. Refer to Hugenholtz et al. [78], Wagner et al. [79], and Behrens et al. [80] for additional information on FISH.

The combined applications of cloning and FISH for microbial community characterization are commonly known as the "full-cycle rRNA approach." Cloning is first used to identify the microorganisms in a sample, and then FISH is applied to quantify those populations or to describe spatial associations among the populations. The full-cycle rRNA approach is powerful. Although FISH results can be biased by nonspecific probe binding or issues with probe accessibility [80], FISH is not subject to biases associated with DNA extraction, PCR, or *E. coli* transformation. Therefore, FISH is especially well suited to complement rRNA gene-based analyses.

In a study of unusual "sprout-like" microbial structures from Vjetrenica Cave, Bosnia and Herzegovina, Kostanjšek et al. [81] used FISH to map the position of different microbial populations *in situ*. From the careful preservation and examination of cross-sections of the structures, the authors found that certain bacterial populations are restricted to the exterior of the structures, while other populations occur in the interior. FISH has also been applied in other cave studies, such as to show that certain groups of Epsilonproteobacteria dominate stream communities in Lower Kane Cave, Wyoming, USA [82], to detect novel Acidobacteria in biofilms from the same cave [83], and to quantify populations of different sulfur-oxidizing microorganisms in sulfidic cave streams in the Frasassi and Acquasanta cave systems in Italy [84–86]. In another study in the Frasassi cave system, a FISH probe was designed to label attached *Thiothrix* epibionts that are symbiotically associated with a cave amphipod [87].

Many other methodological variations of FISH are possible. For example, catalyzed reporter deposition FISH (CARD-FISH) produces a more intense fluorescent signal than traditional FISH [88] and can improve detection of cells with low ribosomal numbers. For example, CARD-FISH was applied for the enumeration of microorganisms in cold oligotrophic karst aquifers where traditional FISH did not produce sufficient fluorescent signal to quantify bacterial populations [89]. Other variants combine FISH with radiolabeling and microradiography (MAR-FISH [90]), incorporate gold labeling for identification via electron microscopy (GOLD-FISH [91]), and even allow for the detection of nuclear genes [92]. In an application of one of these FISH variants to karst microbial processes, Wilhartitz et al. [93] used MAR-FISH to quantify the abundance of heterotrophic microorganisms and measure heterotrophic production rates in an oligotrophic karst aquifer.

2.4 PCR-Based Functional Gene Analysis

rRNA gene-based methods do not directly provide information on the metabolic capabilities of microorganisms in a sample. To identify microbes associated with a particular energy metabolism, it is often desirable to analyze functional genes. Many of the techniques described above, including cloning, T-RFLP, and DGGE, can also be applied to functional genes. For example, ammonia-oxidizing, sulfur-oxidizing, and autotrophic microbial communities from Movile Cave were described by cloning *amoA*, *soxB* and RubisCO gene sequences [6], and methanotrophic microbes were identified by cloning *mxaF*, pmoA, mmoX sequences [94]. DGGE analysis of *amoA* and *nifH* gene sequences was used to describe the diversity of ammonia-oxidizing and nitrogen-fixing organisms in lava tube caves in the Azores, Portugal [95], and sequencing *shc* genes from a sulfidic cave snottite community revealed evidence for hopanoid synthesis [42]. A significant challenge in analyzing functional gene content using PCR-based approaches is in the primer design step. Protein-coding genes have much faster mutation rates compared with rRNA genes, so developing universal primers is challenging and sometimes impossible.

2.5 Other Methods

In addition to the methods described above, multiple other culture-independent techniques have been applied to characterize the microbial diversity of caves, including DNA stable isotope probing [6, 94], enzyme activity assays [96], *in situ* microsensor profiling [85], synchrotron-based X-ray absorption spectroscopy [97], and radiolabel-based measures of autotrophy and heterotrophy [32, 98]. Electron microscopy has also been a valuable source of information on microbial morphology [81, 99] and potential geomicrobiological interactions [100, 101].

2.6 Metagenomics

All of the molecular techniques described above involve the direct amplification and analysis of specific genomic regions of interest, such as rRNA gene sequences or genes for particular enzymes. However, metagenomics is the analysis of genomic material directly from a mixed microbial community, which circumvents some of the biases and pitfalls of the other culture- and PCR-based methods. To construct a metagenomic dataset, DNA is extracted from an environmental microbial community and directly sequenced (▸ Fig. 2.1). The resulting metagenome contains genomic DNA sequences from multiple organisms in the community and includes sequences of both phylogenetic marker genes (e.g. 16S rRNA genes) and functional genes (e.g. genes involved in ammonia oxidation or nitrogen fixation). To effectively examine this mixed bag of microbial genomic information, the taxonomic affiliation and function must be determined for as many metagenomic sequences as possible. Generally, this is accomplished by first assembling short metagenomic sequences ("reads") into longer genome fragments, referred to as contiguous sequences. Assembly is followed by binning and annotation of the fragments. For a more in-depth introduction to metagenomics, consult recent reviews by Thomas et al. [102] and Teeling and Glöckner [103].

At the time of writing, metagenomics applications to caves and karst communities are currently limited to just a few studies from recent years. Tetu et al. [7] identified genes for nitrogen cycling and partially reconstructed the genome of a novel Thaumarchaeota from Weebubbie Cave, Australia. Ortiz et al. [3] studied community metabolic pathways from stalactite surfaces in oligotrophic Kartchner Caverns. Jones et al. [42, 104] determined microbial sulfur oxidation pathways from sulfidic cave snottites, which is highlighted in Section 2.8, Case Study, below. As both computational and sequencing tools advance, metagenomics will likely become more and more widely applied in cave studies.

2.7 RNA-Based Analyses and Other "-Omics" Approaches

DNA-based analyses of functional genes, such as metagenomics or PCR-based approaches, provide important metabolic information on environmental microorganisms. However, analysis of DNA alone only indicates the potential metabolic capabilities of an organism, and not which genes the organism is actively using. Gene expression can be determined by analyzing mRNA by either reverse transcription PCR (RT-PCR) or metatranscriptomics. In RT-PCR, a reverse transcriptase enzyme is used to convert single-stranded RNA into single-stranded DNA, and then that single-stranded DNA is replicated by standard PCR protocols. In contrast, metatranscriptomics is to mRNA what metagenomics is to DNA because the entire RNA content of a community is simultaneously extracted and sequenced. RNA-based approaches are powerful and provide direct information on the active microbial populations and ongoing metabolic processes.

RNA-based approaches have not been widely applied in cave studies. Desai et al. [96] used RT-PCR of *nifH* homologs to study active nitrogen fixation in the Frasassi cave system, and Gonzalez et al. [105] used 16S rRNA transcript analysis to show that Crenarchaeota are an active component of microbial communities on the walls of Altamira Cave. In the aforementioned study, Portillo and Gonzalez [72] used DGGE analysis of both rRNA transcripts and genes. However, like metagenomics, metatranscriptomics will certainly be applied to cave communities in the future because of its utility to provide information on active microbial responses to natural or artificial perturbations and because of its complementary nature to metagenomics. In addition, other "-omics" techniques are available to microbial ecologists. Metaproteomics and environmental metabolomics, the analysis of proteins and metabolites from environmental communities, respectively, also offer culture-independent insight into microbial processes in the environment [106, 107].

2.8 Case Study: Sulfidic Cave Snottites

To illustrate how some of the methods described in this chapter can be applied to cave communities, I conclude with a case study of unusual microbial formations known as snottites (► Fig. 2.2). Snottites are highly acidic (pH 0–1), pendulous biofilms that form on overhanging surfaces in $H_2S(g)$-rich caves. The earliest analyses of sulfidic cave snottites [4, 52] utilized 16S rRNA gene cloning to explore snottite microbial communities from Cueva de Villa Luz, Mexico, and the Frasassi cave system, Italy. In subsequent work in the Frasassi caves, a full-cycle rRNA approach was used to characterize snottites [108]. All three studies identified *Acidithiobacillus* spp. as the dominant bacteria in snottites, with smaller populations of *Acidimicrobium*- and *Sulfobacillus*-like bacteria and other rare taxa. Archaea from the Thermoplasmatales group were cloned from Frasassi snottites using universal primers (► Fig. 2.3) [108].

Fig. 2.2. Field image of snottite biofilms from the Frasassi Caves, Italy. Yellow scale bar is 2 cm.

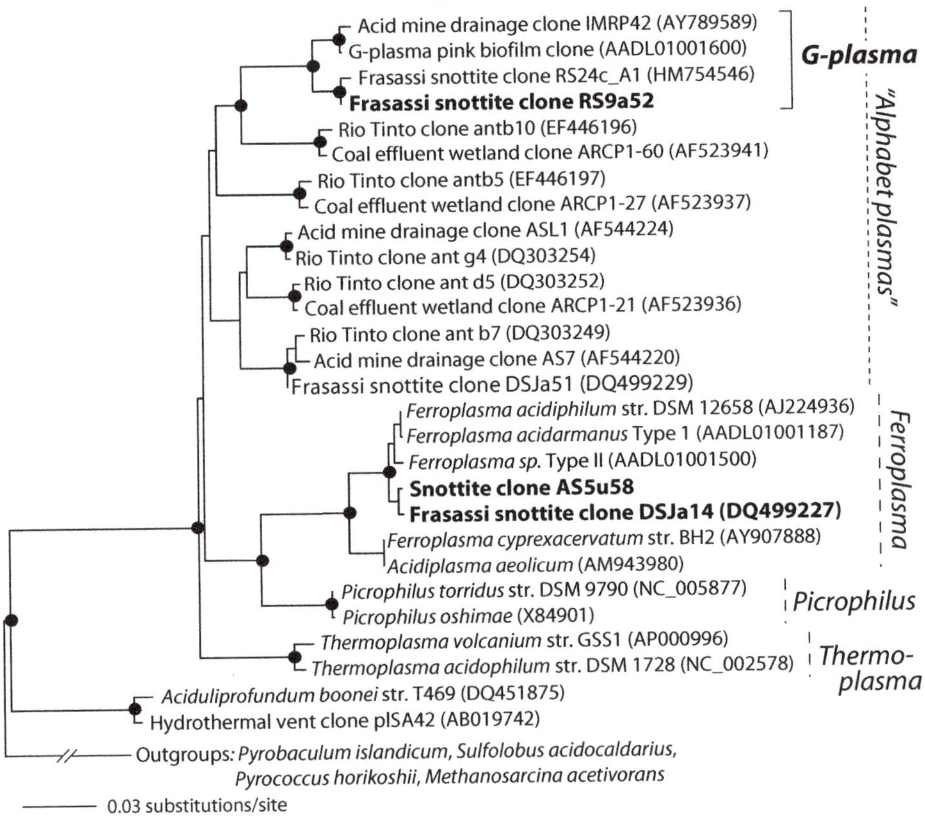

Fig. 2.3. Phylogenetic analysis of archaeal 16S rRNA gene clones identified populations of *Ferroplasma* spp. and a "G-plasma"-like organism in Frasassi Cave snottites. The tree shown here was constructed using neighbor-joining analysis and dots indicate nodes supported by > 95% bootstrap support. rRNA gene cloning of Frasassi snottites is described in [42, 104, 108].

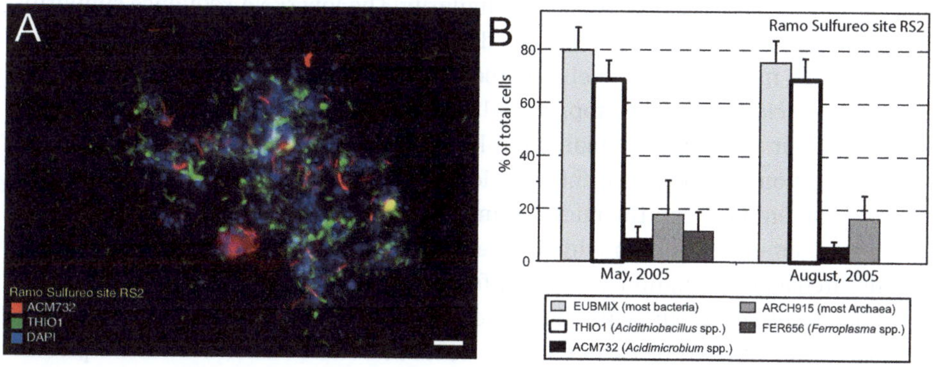

Fig. 2.4. FISH analyses of snottites from the Frasassi cave system. (a) A representative FISH photomicrograph of a snottite sample. Specificity of different FISH probes is given in the legend. Based on other FISH analyses not shown here, the blue cells in (a) are archaeal populations and the majority of green cells are *Acidithiobacillus* spp. (b) Cell counts based on FISH analyses were used to quantify snottite microbial populations. FISH analyses of Frasassi snottites are reported in [42, 104, 108].

Fig. 2.5. *Acidithiobacillus thiooxidans* strains isolated from snottites were found to produce biofilm material in culture (photo by A. Diefendorf).

FISH analyses [108] confirmed the cloning results, and were used to analyze a larger number of snottite samples from multiple locations in the cave system (► Fig. 2.4). FISH analysis revealed that archaeal populations constituted a large component of the snottite community, up to 40% in some cases, and that the *Acidimicrobium*-like organisms varied from 0% to 10% relative abundance. FISH analyses also showed that *Acidithiobacillus* spp. were perennially dominant in snottites. Strains of *Acidithiobacillus* spp. were isolated from snottites, and were shown to oxidize sulfur, fix carbon dioxide, and form biofilm (► Fig. 2.5).

However, rRNA methods and cultivation could not answer all questions. Some, but not all of the snottite Archaea hybridized with a FISH probe for the genus *Ferroplasma* [108]. Despite general agreement between FISH and cloning results [103], the other Archaea could not be identified. Furthermore, snottite Archaea and the *Acidimicrobium*-like bacteria were only distantly related to cultivated microorganisms based on 16S rRNA gene sequence similarity. The low rRNA sequence similarity, coupled with the fact that these groups defied all attempts at culturing, meant that little could be inferred about their metabolism. Finally, despite the identification of the snottite *Acidithiobacillus* spp. as sulfur-oxidizing autotrophs, other important aspects of their metabolism remained unknown. Therefore, Jones et al. [42, 104] applied metagenomic sequencing and additional full-cycle rRNA analyses to explore the metabolic potential of snottite microorganisms further.

Metagenomic analysis provided several important insights. First, the missing snottite archaeal population was identified as "G-plasma," an archaeon from the *Thermoplasmatales* group with no close culture representatives (▸ Fig. 2.3) [42]. "G-plasma" was missed by earlier studies because of a mismatch between G-plasma 16S rRNA genes and widely used archaeal PCR primers. However, metagenomics avoids PCR bias, and by using the new 16S rRNA gene sequence information recovered from metagenomics, primer sequences were modified and "G-plasma" sequences were successfully cloned [42]. Second, the energy metabolism of other snottite populations was inferred from metagenomic data (▸ Fig. 2.6). A lack of any known C-fixation pathways among the "G-plasma," *Ferroplasma*, and *Acidimicrobium*-like populations suggested that those organisms were heterotrophic. With deeper metagenome sequencing, snottite "G-plasma," *Acidimicrobium*, and *Ferroplasma* were each found to have an *sqr* gene that encodes the sulfide-oxidizing enzyme sulfide:quinoneoxidoreductase [104]. Third, metagenomics was used to characterize the sulfur oxidation pathway of the snottite *Acidithiobacillus*, which includes the SQR system, a partial SOX system, and four structurally distinct SQR enzymes [42, 104].

The combined use of cloning, FISH, culturing, and metagenomics has been essential to characterize the Frasassi snottite microorganisms. Cloning was used initially to identify the microbial inhabitants of snottites, and FISH allowed for a quantitative accounting of different microbial populations across multiple samples. Metagenomics identified an important primer bias against "G-plasma," which led to a more complete community description. General agreement between FISH and metagenomics confirmed the reliability of metagenomic-based community analysis. Metagenomics also provided functional information beyond what could be predicted from rRNA sequence analysis alone, including the identification of metabolic sulfur oxidation pathways in snottite microorganisms and development of a conceptual model of snottite biogeochemistry [42, 104]. Moreover, metagenomics generated hypotheses about the mechanism of sulfur oxidation by *Acidithiobacillus* that will be tested with culture-based manipulations in the future.

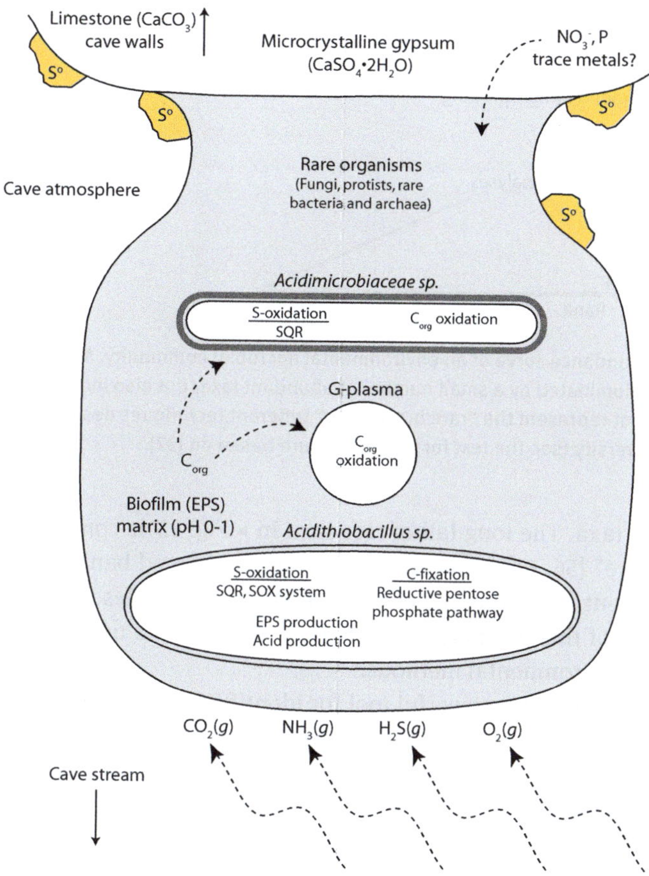

Fig. 2.6. Conceptual model of snottite biogeochemistry based in part on metagenomic analyses described in [42, 104]. Modified from [42].

2.9 Conclusions

Numerous methods are available to environmental microbiologists seeking to study microbial processes in caves and karst. The techniques reviewed here represent currently used approaches, as well as some that will become more widely employed in the future. When considering different techniques, it is important to be aware not only of the biases inherent to each, but also the extent to which each is capable of resolving the true microbial diversity of the sample. ▸ Fig. 2.7 depicts a rank abundance curve of a representative environmental community. Most environmental communities are dominated by a relatively small number of abundant microbial populations and larger

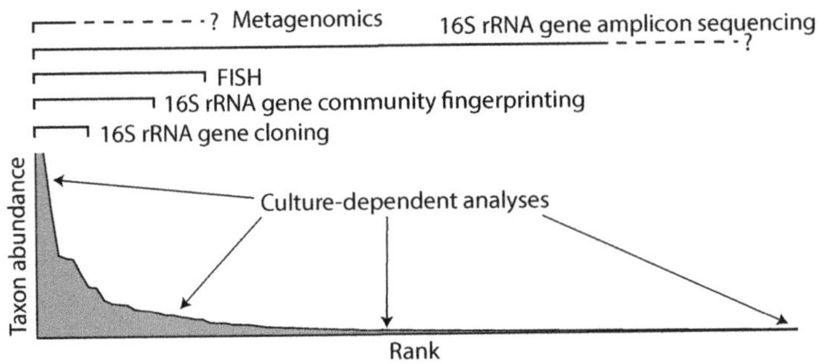

Fig. 2.7. A representative rank abundance curve of an environmental microbial community. Most environmental communities are dominated by a small number of abundant taxa, but also include a long tail of less abundant taxa that represent the "rare biosphere." Different techniques describe different portions of that total diversity (see the text for details). Figure based on [37].

numbers of low abundance taxa. The long tail of rare taxa in ▸ Fig. 2.7 is commonly known as the "rare biosphere" [56, 61], and can be thought of as a seed bank of microbial diversity. Each different technique reviewed here is capable of describing a slightly different component of this diversity, and ▸ Fig. 2.7 represents a useful context in which to summarize environmental methods.

Culture-based analyses represent a powerful tool for identifying microbial metabolic capabilities and fully characterizing isolates. However, they often represent a biased view of microbial abundance and diversity in the environment because the organisms that grow most readily in the lab might simply be "weeds" from the rare biosphere (▸ Fig. 2.7). Culture-independent methods use direct amplification of rRNA genes from environmental samples to produce a more accurate picture of the true microbial diversity. However, different techniques have different limitations. For example, rRNA amplicon sequencing produces large datasets that can begin to approach the true microbial diversity (▸ Fig. 2.7), but each sequence has low taxonomic resolution and amplicons libraries can have high error rates. Cloning, in contrast, produces full- or nearly full-length sequences but only relatively small libraries. Community fingerprinting techniques are currently the most cost-effective rRNA-based tools, but have neither the phylogenetic resolution of cloning nor the depth of amplicon libraries (▸ Fig. 2.7). Direct characterization of functional genes via PCR-based techniques or metagenomics provides culture-independent information on metabolic potential, but currently, those techniques are only effectively applied to the most abundant members of the community. However, DNA sequencing technology is rapidly advancing, and as costs decrease and throughput increases, metagenomics may even supplant certain rRNA-based approaches [109]. Together, environmental DNA sequencing, novel cultivation techniques, -omics approaches, and other microbiological methods offer innovative ways to probe environmental microbial processes and represent an exciting new toolbox for future cave and karst microbiology researchers.

Acknowledgments

Thanks to all of those who have advised and assisted me in the lab and field, especially J. Macalady, I. Schaperdoth, E. Lyon, T. Jones, S. Dattagupta, K. Dawson, and H. Albrecht. I extend sincere thanks to L. Hose, L. Rosales-Lagarde, A. Montanari, F. Baldoni, S. Carnevali, S. Cerioni, S. Galdenzi, M. Mainiero, S. Mariani, and the Gruppo Speleologico C. A. I. di Fabriano for wonderful guidance on caving and cave research. I also extend special thanks to A. Engel for organizing and editing this volume.

References

[1] Pace NR. Mapping the tree of life: progress and prospects. Microbiol Mol Biol Rev 2009, 73, 565–76.
[2] Barton HA, Taylor MR, Pace NR. Molecular phylogenetic analysis of a bacterial community in an oligotrophic cave environment. Geomicrobiol J 2004, 21, 11–20.
[3] Ortiz M, Legatzki A, Neilson JW, et al. Making a living while starving in the dark: metagenomic insights into the energy dynamics of a carbonate cave. ISME J 2014, 8, 478–91.
[4] Hose LD, Palmer AN, Palmer MV, Northup DE, Boston PJ, DuChene HR. Microbiology and geochemistry in a hydrogen-sulfide-rich karst environment. Chem Geol 2000, 169, 399–423.
[5] Engel AS. Microbial diversity of cave ecosystems. In: Barton LL, Mandl M, Loy A, eds. Geomicrobiology: Molecular and Environmental Perspective, Netherlands, Springer, 2010, 219–38.
[6] Chen Y, Wu L, Boden R, et al. Life without light: microbial diversity and evidence of sulfur-and ammonium-based chemolithotrophy in Movile Cave. ISME J 2009, 3, 1093–104.
[7] Tetu SG, Breakwell K, Elbourne LD, Holmes AJ, Gillings MR, Paulsen IT. Life in the dark: metagenomic evidence that a microbial slime community is driven by inorganic nitrogen metabolism. ISME J 2013, 7, 1227–36.
[8] Sarbu SM, Kane TC, Kinkle BK. A chemoautotrophically based cave ecosystem. Science 1996, 272, 1953–5.
[9] Engel AS, Stern LA, Bennett PC. Microbial contributions to cave formation: New insights into sulfuric acid speleogenesis. Geology 2004, 32, 369–72.
[10] Melim LA, Shinglman KM, Boston PJ, Northup DE, Spilde MN, Queen JM. Evidence for microbial involvement in pool finger precipitation, Hidden Cave, New Mexico. Geomicrobiol J 2001, 18, 311–29.
[11] Boston P, Spilde M, Northup D, et al. Cave biosignature suites: microbes, minerals, and Mars. Astrobiology 2001, 1, 25–55.
[12] Carmichael MJ, Carmichael SK, Santelli CM, Strom A, Bräuer SL. Mn (II)-oxidizing bacteria are abundant and environmentally relevant members of ferromanganese deposits in caves of the upper Tennessee River Basin. Geomicrobiol J 2013, 30, 779–800.
[13] Amend JP, Teske A. Expanding frontiers in deep subsurface microbiology. Palaeogeogr Palaeoclimatol Palaeoecol 2005, 219, 131–55.
[14] Kallmeyer J, Pockalny R, Adhikari RR, Smith DC, D'Hondt S. Global distribution of microbial abundance and biomass in subseafloor sediment. Proc Nat Acad Sci USA 2012, 109, 16213–6.
[15] Lin L-H, Wang P-L, Rumble D, et al. Long-term sustainability of a high-energy, low-diversity crustal biome. Science 2006, 314, 479–82.
[16] Palmer AN. Cave Geology. Dayton, OH, Cave Books, 2007.

[17] Engel AS, Northup DE. Caves and karst as model systems for advancing the microbial sciences. In: Martin JB, White WB, eds. Frontiers of Karst Research. Leesburg, Virginia, Karst Waters Institute, 2008, 90–95.
[18] Hugenholtz P, Goebel BM, Pace NR. Impact of culture-independent studies on the emerging phylogenetic view of bacterial diversity. J Bacteriol 1998, 180, 4765–74.
[19] Macalady JL, Hamilton TL, Grettenberger CL, Jones DS, Tsao LE, Burgos WD. Energy, ecology and the distribution of microbial life. Phil Trans R Soc B 2013, 368, 20120383.
[20] Madigan MT, Martinko JM, Stahl D, Clark DP. Brock Biology of Microorganisms (13th Edition). Boston, MA, Benjamin Cummings, 2010.
[21] Leadbetter JR. Cultivation of recalcitrant microbes: cells are alive, well and revealing their secrets in the 21st century laboratory. Curr Opin Microbiol 2003, 6, 274–81.
[22] Keller M, Zengler K. Tapping into microbial diversity. Nature Reviews Microbiology 2004, 2, 141–50.
[23] Epstein S. The phenomenon of microbial uncultivability. Curr Opin Microbiol 2013, 16, 636–42.
[24] Caumartin V. Review of the microbiology of underground environments. Nat Speleol Soc Bull 1963, 25, 1–14.
[25] Mikell Jr A, Smith C, Richardson J. Evaluation of media and techniques to enumerate heterotrophic microbes from karst and sand aquifer springs. Microb Ecol 1996, 31, 115–24.
[26] Vlasceanu L, Popa R, Kinkle BK. Characterization of *Thiobacillus thioparus* LV43 and its distribution in a chemoautotrophically based groundwater ecosystem. Appl Environ Microbiol 1997, 63, 3123–7.
[27] Brigmon R, Martin H, Morris T, Bitton G, Zam S. Biogeochemical ecology of *Thiothrix* spp. in underwater limestone caves. Geomicrobiol J 1994, 12, 141–59.
[28] Ikner LA, Toomey RS, Nolan G, Neilson JW, Pryor BM, Maier RM. Culturable microbial diversity and the impact of tourism in Kartchner Caverns, Arizona. Microb Ecol 2007, 53, 30–42.
[29] Baskar S, Baskar R, Mauclaire L, McKenzie J. Microbially induced calcite precipitation in culture experiments: Possible origin for stalactites in Sahastradhara caves, Dehradun, India. Curr Sci 2006, 90, 58–64.
[30] Cañaveras J, Hoyos M, Sanchez-Moral S, et al. Microbial communities associated with hydromagnesite and needle-fiber aragonite deposits in a karstic cave (Altamira, Northern Spain). Geomicrobiol J 1999, 16, 9–25.
[31] Rusznyák A, Akob DM, Nietzsche S, et al. Calcite biomineralization by bacterial isolates from the recently discovered pristine karstic Herrenberg cave. Appl Environ Microbiol 2012, 78, 1157–67.
[32] Engel AS, Porter ML, Kinkle BK, Kane TC. Ecological assessment and geological significance of microbial communities from Cesspool Cave, Virginia. Geomicrobiol J 2001, 18, 259–74.
[33] Northup DE, Barns SM, Yu LE, et al. Diverse microbial communities inhabiting ferromanganese deposits in Lechuguilla and Spider Caves. Environ Microbiol 2003, 5, 1071–86.
[34] Northup DE, Dahm CN, Melim LA, et al. Evidence for geomicrobiological interactions in Guadalupe caves. J Cave Karst Stud 2000, 62, 80–90.
[35] Popa R, Smith AR, Popa R, Boone J, Fisk M. Olivine-respiring bacteria isolated from the rock-ice interface in a lava-tube cave, a Mars analog environment. Astrobiology 2012, 12, 9–18.
[36] Staley JT, Konopka A. Measurement of in situ activities of nonphotosynthetic microorganisms in aquatic and terrestrial habitats. Annu Rev Microbiol 1985, 39, 321–46.
[37] Pedrós-Alió C. Marine microbial diversity: can it be determined? Trends Microbiol 2006, 14, 257–63.
[38] Woese CR, Fox GE. Phylogenetic structure of the prokaryotic domain: the primary kingdoms. Proc Nat Acad Sci USA 1977, 74, 5088–90.

[39] Sambrook J, Russell DW, Russell DW. Molecular cloning: a laboratory manual, third edition. Cold Spring Harbor, New York, Cold Spring Harbor Laboratory Press, 2001.
[40] Senko JM, Wanjugi P, Lucas M, Bruns MA, Burgos WD. Characterization of Fe (II) oxidizing bacterial activities and communities at two acidic Appalachian coalmine drainage-impacted sites. ISME J 2008, 2, 1134–45.
[41] Lakay F, Botha A, Prior B. Comparative analysis of environmental DNA extraction and purification methods from different humic acid-rich soils. J Appl Microbiol 2007, 102, 265–73.
[42] Jones D, Albrecht H, Dawson K, et al. Community genomic analysis of an extremely acidophilic sulfur-oxidizing biofilm. ISME J 2012, 6, 158–70.
[43] Dovichi NJ, Zhang J. How capillary electrophoresis sequenced the human genome. Angew Chem Int Ed 2000, 39, 4463–8.
[44] Sanger F, Nicklen S, Coulson AR. DNA sequencing with chain-terminating inhibitors. Proc Nat Acad Sci USA 1977, 74, 5463–7.
[45] Pruesse E, Quast C, Knittel K, et al. SILVA: a comprehensive online resource for quality checked and aligned ribosomal RNA sequence data compatible with ARB. Nucleic Acids Res 2007, 35, 7188–96.
[46] Konstantinidis KT, Ramette A, Tiedje JM. The bacterial species definition in the genomic era. Phil Trans R Soc B 2006, 361, 1929–40.
[47] Acinas SG, Marcelino LA, Klepac-Ceraj V, Polz MF. Divergence and redundancy of 16S rRNA sequences in genomes with multiple *rrn* operons. J Bacteriol 2004, 186, 2629–35.
[48] Kanagawa T. Bias and artifacts in multitemplate polymerase chain reactions (PCR). J Biosci Bioeng 2003, 96, 317–23.
[49] Polz MF, Cavanaugh CM. Bias in template-to-product ratios in multitemplate PCR. Appl Environ Microbiol 1998, 64, 3724–30.
[50] DeSantis TZ, Stone CE, Murray SR, Moberg JP, Andersen GL. Rapid quantification and taxonomic classification of environmental DNA from both prokaryotic and eukaryotic origins using a microarray. FEMS Microbiol Lett 2005, 245, 271–8.
[51] Angert ER, Northup DE, Reysenbach A-L, Peek AS, Goebel BM, Pace NR. Molecular phylogenetic analysis of a bacterial community in Sulphur River, Parker Cave, Kentucky. Am Mineral 1998, 83, 1583–92.
[52] Vlasceanu L, Sarbu SM, Engel AS, Kinkle BK. Acidic cave-wall biofilms located in the Frasassi Gorge, Italy. Geomicrobiol J 2000, 17, 125–39.
[53] Holmes AJ, Tujula NA, Holley M, et al. Phylogenetic structure of unusual aquatic microbial formations in Nullarbor caves, Australia. Environ Microbiol 2001, 3, 256–64.
[54] Margulies M, Egholm M, Altman WE, et al. Genome sequencing in microfabricated high-density picolitre reactors. Nature 2005, 437, 376–80.
[55] Neefs J-M, Van de Peer Y, De Rijk P, Chapelle S, De Wachter R. Compilation of small ribosomal subunit RNA structures. Nucleic Acids Res 1993, 21, 3025–49.
[56] Sogin ML, Morrison HG, Huber JA, et al. Microbial diversity in the deep sea and the underexplored “rare biosphere”. Proc Nat Acad Sci USA 2006, 103, 12115–20.
[57] Sahl JW, Fairfield N, Harris JK, Wettergreen D, Stone WC, Spear JR. Novel microbial diversity retrieved by autonomous robotic exploration of the world’s deepest vertical phreatic sinkhole. Astrobiology 2010, 10, 201–13.
[58] Sahl JW, Gary MO, Harris JK, Spear JR. A comparative molecular analysis of water-filled limestone sinkholes in north-eastern Mexico. Environ Microbiol 2011, 13, 226–40.
[59] Ortiz M, Neilson JW, Nelson WM, et al. Profiling bacterial diversity and taxonomic composition on speleothem surfaces in Kartchner Caverns, AZ. Microb Ecol 2013, 65, 371–83.
[60] Gray CJ, Engel AS. Microbial diversity and impact on carbonate geochemistry across a changing geochemical gradient in a karst aquifer. ISME J 2013, 7, 325–37.

[61] Kunin V, Engelbrektson A, Ochman H, Hugenholtz P. Wrinkles in the rare biosphere: pyrosequencing errors can lead to artificial inflation of diversity estimates. Environ Microbiol 2010, 12, 118–23.
[62] Bartram AK, Lynch MD, Stearns JC, Moreno-Hagelsieb G, Neufeld JD. Generation of multimillion-sequence 16S rRNA gene libraries from complex microbial communities by assembling paired-end Illumina reads. Appl Environ Microbiol 2011, 77, 3846–52.
[63] Faith JJ, Guruge JL, Charbonneau M, et al. The long-term stability of the human gut microbiota. Science 2013, 341, DOI:10.1126/science.1237439.
[64] Lundberg DS, Yourstone S, Mieczkowski P, Jones CD, Dangl JL. Practical innovations for high-throughput amplicon sequencing. Nat Methods 2013, 10, 999–1002.
[65] Harrison BK, Orphan VJ. Method for assessing mineral composition-dependent patterns in microbial diversity using magnetic and density separation. Geomicrobiol J 2012, 29, 435–49.
[66] Liu W-T, Marsh TL, Cheng H, Forney LJ. Characterization of microbial diversity by determining terminal restriction fragment length polymorphisms of genes encoding 16S rRNA. Appl Environ Microbiol 1997, 63, 4516–22.
[67] Osborn AM, Moore ER, Timmis KN. An evaluation of terminal-restriction fragment length polymorphism (T-RFLP) analysis for the study of microbial community structure and dynamics. Environ Microbiol 2000, 2, 39–50.
[68] Chelius MK, Beresford G, Horton H, et al. Impacts of alterations of organic inputs on the bacterial community within the sediments of Wind Cave, South Dakota, USA. Int J Speleol 2012, 38, 1–10.
[69] Moss JA, Nocker A, Snyder RA. Microbial characteristics of a submerged karst cave system in Northern Florida. Geomicrobiol J 2011, 28, 719–31.
[70] Shabarova T, Widmer F, Pernthaler J. Mass effects meet species sorting: transformations of microbial assemblages in epiphreatic subsurface karst water pools. Environ Microbiol 2013, 15, 2476–88.
[71] Muyzer G, De Waal EC, Uitterlinden AG. Profiling of complex microbial populations by denaturing gradient gel electrophoresis analysis of polymerase chain reaction-amplified genes coding for 16S rRNA. Appl Environ Microbiol 1993, 59, 695–700.
[72] Portillo MC, Saiz-Jimenez C, Gonzalez JM. Molecular characterization of total and metabolically active bacterial communities of "white colonizations" in the Altamira Cave, Spain. Res Microbiol 2009, 160, 41–7.
[73] Legatzki A, Ortiz M, Neilson JW, et al. Bacterial and archaeal community structure of two adjacent calcite speleothems in Kartchner Caverns, Arizona, USA. Geomicrobiol J 2011, 28, 99–117.
[74] Adetutu EM, Thorpe K, Bourne S, et al. Phylogenetic diversity of fungal communities in areas accessible and not accessible to tourists in Naracoorte Caves. Mycologia 2011, 103, 959–68.
[75] Rossmassler K, Engel AS, Twing KI, Hanson TE, Campbell BJ. Drivers of epsilonproteobacterial community composition in sulfidic caves and springs. FEMS Microbiol Ecol 2012, 79, 421–32.
[76] Nocker A, Burr M, Camper AK. Genotypic microbial community profiling: a critical technical review. Microb Ecol 2007, 54, 276–89.
[77] Marzorati M, Wittebolle L, Boon N, Daffonchio D, Verstraete W. How to get more out of molecular fingerprints: practical tools for microbial ecology. Environ Microbiol 2008, 10, 1571–81.
[78] Hugenholtz P, Tyson GW, Blackall LL. Design and evaluation of 16S rRNA-targeted oligonucleotide probes for fluorescence *in situ* hybridization. In: Aquino de Muro M, Rapley R, eds. Methods in Molecular Biology, vol 179, Gene Probes: Principles and Protocols. Totowa, NJ, USA, Humana, 2002, 29–42.
[79] Wagner M, Horn M, Daims H. Fluorescence *in situ* hybridisation for the identification and characterisation of prokaryotes. Curr Opin Microbiol 2003, 6, 302–9.

[80] Behrens S, Fuchs BM, Mueller F, Amann R. Is the in situ accessibility of the 16S rRNA of Escherichia coli for Cy3-labeled oligonucleotide probes predicted by a three-dimensional structure model of the 30S ribosomal subunit? Appl Environ Microbiol 2003, 69, 4935–41.

[81] Kostanjšek R, Pašić L, Daims H, Sket B. Structure and community composition of sprout-like bacterial aggregates in a Dinaric Karst subterranean stream. Microb Ecol 2013, 66, 5–18.

[82] Engel AS, Lee N, Porter ML, Stern LA, Bennett PC, Wagner M. Filamentous "Epsilonproteobacteria" dominate microbial mats from sulfidic cave springs. Appl Environ Microbiol 2003, 69, 5503–11.

[83] Meisinger DB, Zimmermann J, Ludwig W, et al. In situ detection of novel Acidobacteria in microbial mats from a chemolithoautotrophically based cave ecosystem (Lower Kane Cave, WY, USA). Environ Microbiol 2007, 9, 1523–34.

[84] Jones D, Tobler D, Schaperdoth I, Mainiero M, Macalady J. Community structure of subsurface biofilms in the thermal sulfidic caves of Acquasanta Terme, Italy. Appl Environ Microbiol 2010, 76, 5902–10.

[85] Macalady JL, Dattagupta S, Schaperdoth I, Jones DS, Druschel GK, Eastman D. Niche differentiation among sulfur-oxidizing bacterial populations in cave waters. ISME J 2008, 2, 590–601.

[86] Macalady JL, Lyon EH, Koffman B, et al. Dominant microbial populations in limestone-corroding stream biofilms, Frasassi cave system, Italy. Appl Environ Microbiol 2006, 72, 5596–609.

[87] Dattagupta S, Schaperdoth I, Montanari A, et al. A novel symbiosis between chemoautotrophic bacteria and a freshwater cave amphipod. ISME J 2009, 3, 935–43.

[88] Pernthaler A, Pernthaler J, Amann R. Fluorescence *in situ* hybridization and catalyzed reporter deposition for the identification of marine bacteria. Appl Environ Microbiol 2002, 68, 3094–101.

[89] Wilhartitz I, Mach RL, Teira E, Reinthaler T, Herndl GJ, Farnleitner AH. Prokaryotic community analysis with CARD-FISH in comparison with FISH in ultra-oligotrophic groundand drinking water. J Appl Microbiol 2007, 103, 871–81.

[90] Lee N, Nielsen PH, Andreasen KH, et al. Combination of fluorescent *in situ* hybridization and microautoradiography—a new tool for structure-function analyses in microbial ecology. Appl Environ Microbiol 1999, 65, 1289–97.

[91] Schmidt H, Eickhorst T, Mußmann M. Gold-FISH: A new approach for the *in situ* detection of single microbial cells combining fluorescence and scanning electron microscopy. Syst Appl Microbiol 2012, 35, 518–25.

[92] Wagner M, Haider S. New trends in fluorescence *in situ* hybridization for identification and functional analyses of microbes. Curr Opin Biotechnol 2012, 23, 96–102.

[93] Wilhartitz IC, Kirschner AK, Stadler H, et al. Heterotrophic prokaryotic production in ultra-oligotrophic alpine karst aquifers and ecological implications. FEMS Microbiol Ecol 2009, 68, 287–99.

[94] Hutchens E, Radajewski S, Dumont MG, McDonald IR, Murrell JC. Analysis of methanotrophic bacteria in Movile Cave by stable isotope probing. Environ Microbiol 2004, 6, 111–20.

[95] Marshall Hathaway JJ, Sinsabaugh RL, Dapkevicius MdLN, Northup DE. Diversity of ammonia oxidation (*amoA*) and nitrogen fixation (*nifH*) genes in lava caves of Terceira, Azores, Portugal. Geomicrobiol J 2013, 31, 221–35.

[96] Desai MS, Assig K, Dattagupta S. Nitrogen fixation in distinct microbial niches within a chemoautotrophy-driven cave ecosystem. ISME J 2013, 7, 2411–23.

[97] Engel AS, Lichtenberg H, Prange A, Hormes J. Speciation of sulfur from filamentous microbial mats from sulfidic cave springs using X-ray absorption near-edge spectroscopy. FEMS Microbiol Lett 2007, 269, 54–62.

[98] Porter ML, Engel AS, Kane TC, Kinkle BK. Productivity-diversity relationships from chemolithoautotrophically based sulfidic karst systems. Int J Speleol 2009, 38, 27–40.

[99] Melim LA, Northup DE, Spilde MN, Jones B, Boston PJ, Bixby RJ. Reticulated filaments in cave pool speleothems: microbe or mineral? J Cave Karst Stud 2008, 70, 135–41.

[100] Cunningham K, Northup D, Pollastro R, Wright W, LaRock E. Bacteria, fungi and biokarst in Lechuguilla Cave, Carlsbad Caverns National Park, New Mexico. Environ Geol 1995, 25, 2–8.

[101] Spilde MN, Northup DE, Boston PJ, et al. Geomicrobiology of cave ferromanganese deposits: A field and laboratory investigation. Geomicrobiol J 2005, 22, 99–116.

[102] Thomas T, Gilbert J, Meyer F. Metagenomics – a guide from sampling to data analysis. Microb Inform Exp 2012, 2, 3. DOI:10.1186/2042-5783-2-3.

[103] Teeling H, Glöckner FO. Current opportunities and challenges in microbial metagenome analysis—a bioinformatic perspective. Brief Bioinform 2012, 13, 728–42.

[104] Jones DS, Schaperdoth I, Macalady JL. Metagenomic evidence for sulfide oxidation in extremely acidic cave biofilms. Geomicrobiol J 2014, 31, 194–204.

[105] Gonzalez JM, Portillo MC, Saiz-Jimenez C. Metabolically active Crenarchaeota in Altamira cave. Naturwissenschaften 2006, 93, 42–5.

[106] Bundy JG, Davey MP, Viant MR. Environmental metabolomics: a critical review and future perspectives. Metabolomics 2009, 5, 3–21.

[107] Ram RJ, VerBerkmoes NC, Thelen MP, et al. Community proteomics of a natural microbial biofilm. Science 2005, 308, 1915–20.

[108] Macalady JL, Jones DS, Lyon EH. Extremely acidic, pendulous microbial biofilms from the Frasassi cave system, Italy. Environ Microbiol 2007, 9, 1402–14.

[109] Logares R, Sunagawa S, Salazar G, et al. Metagenomic 16S rDNA Illumina tags are a powerful alternative to amplicon sequencing to explore diversity and structure of microbial communities. Environ Microbiol 2014, 16, 2659–71.

Kathleen H. Lavoie

3 "A Grand, Gloomy, and Peculiar Place": Microbiology in the Mammoth Cave Region

Abstract: Mammoth Cave is the longest cave in the world, with over 640 km of mapped passage on five levels in South Central Kentucky, USA. Stephen Bishop, an early guide at Mammoth Cave, described it as "A grand, gloomy, and peculiar place." The Mammoth Cave region covers 800 km^2 of the Interior Low Plateaus. The Pennyroyal Plateau is a sinkhole plain catchment area where water sinks underground and flows toward the master stream in the area, the Green River. Mammoth Cave is in the Chester Upland where a layer of sandstone caprock protects the underlying limestone and the cave. Mammoth Cave is an important hotspot of subterranean biodiversity, with 41 species of troglobites and stygobites across nine higher groups. The food base is typically detrital, supporting a limited number of trophic levels. The earliest work on bacteria in Mammoth Cave was done to study the formation of saltpeter, an important mining product during the War of 1812. One of the earliest studies using molecular methods from any cave was performed in Parker Cave on the sinkhole plain and identified a microbially based chemolithoautotrophic system. Mammoth Cave holds a record of human diseases with its tuberculosis huts and paleofeces left by native peoples. White-nose syndrome, a catastrophic fungal disease of bats, arrived in the Mammoth Cave National Park in the winter of 2012–2013. Despite its size and complexity, we are just beginning to learn about microbial distribution, diversity, and activity, and how to protect microbial communities in Mammoth Cave and the region.

3.1 Introduction to Mammoth Cave and the Region

Stephen Bishop, an early guide at Mammoth Cave and former slave, described Mammoth Cave as "A grand, gloomy, and peculiar place" [1]. He arrived at Mammoth Cave in 1835 and died in 1857, only one year after being freed. Bishop was a guide and the greatest explorer of the unknown passages of Mammoth Cave at the time [1]. Mammoth Cave, technically the Mammoth Cave–Flint Ridge System, is the longest cave in the world, with more than 640 km of mapped passage in 2013 [2]. Speculations of an imminent connection with Fisher Ridge Cave have been around for a long time, and may be exaggerated [3], but this connection would push the system to over 800 km. Fisher Ridge Cave is currently the 10th longest cave in the world, with nearly 200 km of mapped passage [4].

Mammoth Cave is known as a hotspot of subterranean biodiversity with a total of 41 species of troglobites and stygobites across nine higher groups, and 6th of the 20 most biologically diverse caves and wells in the world [5]. Subterranean biodiversity is associated with sites that are large, have high primary productivity or rich organic inputs, or with permanent groundwater. Mammoth Cave has a rich geologic history with extensive area above the cave, permanent groundwater, and its great size allows for a range of habitats [6, 7]. In Mammoth Cave, one-third of the animal groups are aquatic and two-thirds are terrestrial. Most of the species listed by Culver and Sket [5] are easily observed and collected; there is a relative lack of study of smaller organisms, particularly microorganisms.

3.1.1 The Mammoth Cave region

The Mammoth Cave region is part of the Interior Low Plateaus that extend west from the Appalachian Mountains to the southern boundary of continental glacial sediments, west to the Mississippi Valley and north to the Ohio River [8]. The Mammoth Cave region is an area of low-relief karst that covers an area of about 800 km^2 in South Central Kentucky. A cross-section of the Mammoth Cave region shows distinct geologic provinces (▸ Fig. 3.1). The Pennyroyal Plateau is to the southeast where St. Louis and Ste. Genevieve limestones are exposed. Surface drainage sinks into the sinkhole plain and water moves underground along the 0.3° stratigraphic dip toward the northwest and the Green River. The Chester Upland to the northwest (▸ Fig. 3.1) is comprised of limestone protected by overlying Big Clifty caprock. The Big Clifty is comprised of Late Mississippian sandstones and shales that overlay Pennsylvanian sandstones and conglomerates. The Girkin Formation and other limestones are about 120 m thick below the caprock [8, 9]. Erosion through the caprock results in irregular, flat-topped ridges separated by karst valleys. There are many cracks and breaks in the caprock that allow for infiltration of water. Vertical shafts form where water enters along the eroded edges of the caprock. Palmer [8] states that the geologic setting is

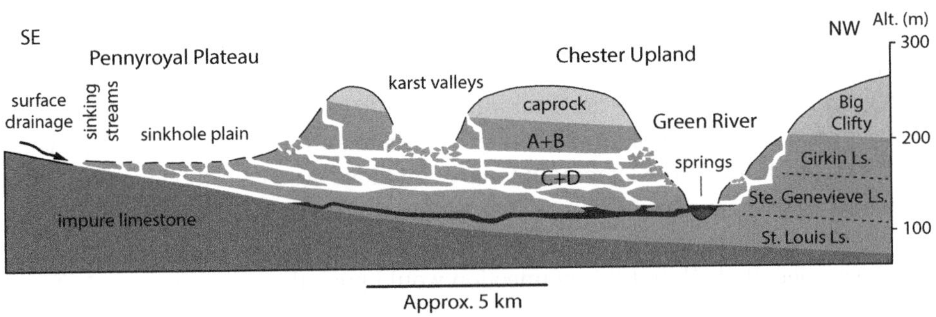

Fig. 3.1. Geologic cross-section through the Mammoth Cave region. A, B = large upper level canyons and tubes. C, D = tubes. The lowest passages are flooded by a late Pleistocene rise in base level [8]. © Arthur Palmer. Reproduced by permission of Arthur Palmer. Permission to reuse must be obtained from the rights holder.

ideal for the development of long caves due to extensive drainage basins, low relief, and many varied recharge points.

Regional base level is controlled by the master stream in the area, the Green River [8, 9]. There are four passage levels within Mammoth Cave that are independent of the limestone stratigraphy that cluster at elevations of 210, 183–189, 168, and 152 m and reflect the geomorphology of the area. Each level represents a base level change and pauses in stream downcutting. The oldest upper level passages (Late Tertiary, A and B in ▸ Fig. 3.1) are large canyons and tubes. The lower level passages (Pleistocene, C and D in ▸ Fig. 3.1) are smaller tubes formed after fragmentation of surface topology [8]. The lowest passages are flooded by a late Pleistocene rise in base level. Level A and B passage sediments date to 2.6–3.5 million years, and level C and D sediments date to 1.5–1.2 million years ago [8]. Sediments are younger than the passages.

3.1.2 Mammoth Cave National Park

Most of Mammoth Cave is located under Mammoth Cave National Park (MCNP) [10], but portions extend well beyond the park boundaries. MCNP is the second oldest tourist attraction in the United States (after Niagara Falls) and began offering guided tours in 1816. Currently, over 400,000 tourists visit the cave each year. In addition to Mammoth Cave, there are over 400 known caves within MCNP, and more than 21,000 hectares of land around the Green River Valley. The primary cover is mostly secondary growth of oak and hickory forest to the west and mixed Mesophytic forest of beech, maple, and poplar to the east and north [11].

Mammoth Cave became a National Park in 1941, and was recognized as a World Heritage Site in 1981 and by UNESCO as an International Biosphere Reserve in 1990. Because biosphere reserves balance conservation of plants, animals, and microorganisms with sustainable use, the Mammoth Cave Area International Biosphere Reserve [12] legally protects the core area in MCNP. To the north and south of MCNP, there is a zone of cooperation that is managed to benefit local residents and the local environment. The outermost zone in the Reserve is a transition zone that extends into surrounding counties where conservation and economic development are linked.

3.2 Microorganisms in Caves

We are most familiar with the abundant life we observe on the Earth's surface, but Gold [13] speculates that there is as much microbial biomass below the surface as all of the biomass above ground. Microbes are of central importance to the biosphere and to biogeochemical cycling and allow us to explore the strategies and limits of life. Caves serve as important laboratories that are simplified natural ecosystems [6], and provide relatively easy access for the study of underground environments. We can use microbes to monitor environmental change, water pollution, the quality of an environment, and the recovery of a system to a particular stress. Microbes play a major role in conservation and restoration biology, and provide important models for understanding principles of ecology and evolution. Caves also serve as model systems for us to study what life might be like on other planets [14].

We originally thought that microbes in caves were just a subset of organisms found on the surface that had washed into caves and could eke out a minimal existence in the low-nutrient cave environment. The first review of the microbiology of underground environments was published by Caumartin in 1963 [15]. Early microbiological studies used traditional culture media designed to resemble the nutrient-rich environments of the human body. Incubation was usually done at human body temperature of 37 °C. Since the 1990s, much has changed in terms of methods used to study environmental microorganisms. Our original ideas about cave microbes were wrong. New strategies involve using low-nutrient media and media specifically designed to isolate microbes with particular characteristics from a habitat, including caves [16]. Other improved methods include microscopy and staining, electron microscopy, measures of physiological characteristics, and the incredibly powerful and versatile tools of molecular biology [16, 17]. Microbial activity can be monitored through analysis of atmospheric gases and water chemistry [17, 18]. Molecular phylogenic advances show that the main diversity of life is microbial [19] with many unexpected evolutionary lineages.

Molecular techniques can detect the presence of many novel organisms without having to culture them or even see them, although old and new techniques should be seen as complimentary, not clashing, since each strategy reveals different diversity [20]. Visual evidence of microbes and microbial activities in caves [17] include colonies seen as dots on surfaces; discoloration of rock surfaces caused by corrosion residues; banded mineralization and mineral precipitation; structural changes to surfaces; and coatings and biofilms, such as white filaments in streams with inputs of sulfur.

3.2.1 Bacteria and Archaea

Bacteria and Archaea are both domains of single-celled organisms that can metabolically utilize any chemical reaction that potentially has energy. Archaea were first found in extreme environments, like thermal springs and salt marshes, but have a cosmopolitan distribution [19]. Very little work has been done on Archaea in caves, and I know of no studies of Archaea in Mammoth Cave. Jarrell et al. [21] speculate that Archaea are adapted to chronic energy stress, which might be a factor in differentiating bacterial and archaeal ecology. Archaea may be important in nutrient-limited cave ecosystems by contributing to nutrient cycling [21], through sulfur oxidation, methanogenesis, nitrogen fixation, nitrification, and ammonia oxidation. Archaea compete successfully in all mainstream environments and are dominant in soils with low concentrations of nitrogen and low nitrification rates. As in most environments, the focus of study in caves is on the diversity of Bacteria.

Microbiology has traditionally explained patterns of microbial biogeography using the ecological concept that "everything is everywhere: but the environment selects" [22, 23]. Recent discoveries of geographical patterns in microbial distribution are changing our fundamental assumptions and show that microbial biogeography is much more complicated than we thought. Metagenomics will show how microorganisms make their living in caves where both chemolithoautotrophy and chemoorganotrophy are viable growth strategies. Comparisons of bacterial phylogeny are showing the presence of a worldwide level of diversity of the microbiota from limestone and lava caves [24]. Many of the bacteria identified are unique, and may represent cave-adapted microorganisms. Much more study is needed.

3.2.2 Early microbiological studies from Mammoth Cave

Studies of the biomass and activity of bacteria in limestone caves in MCNP were conducted by Feldhake [25]. Measurements of metabolic rates were done on substrates at 12 sites in four caves, with comparisons to overlying forest soils. Except for a site rich in cricket guano, Feldhake found that organic matter content, metabolic activity, and biomass were much lower in the cave than in forest soil, with significant variation among sample sites within the cave. He encountered methodological problems when samples were removed from the cave, and then transported to the lab for analysis 24–48 h after collection. He measured chemoorganotrophic activity by monitoring incorporation of radiolabeled acetate into lipids and autotrophic activity by incorporation of radiolabeled bicarbonate into lipids. Heterotrophs responded to disturbance with an increase in metabolism but no increase in biomass within six hours, while addition of glucose caused a large increase in metabolic activity and biomass after eight days. Addition of water alone caused no increase in biomass after eight days. Autotrophic activity was very low at two of twelve sites, and absent at the remaining ten.

In general, studies show low numbers and activity of bacteria in caves. The exception is a study performed to compare the bacterial activity, density, and diversity of two aquatic sediment sites in Mammoth Cave by Rusterholtz and Mallory [26]. The study included counts of cells in the sediment, staining to determine metabolic activity of soil microbes, plate counts using both high- and low-nutrient media, followed by extensive physiological testing of isolates from the plate counts. They recovered between 11% and 58% of the total cell count on culture medium. The recovery rate for most surface soils is typically 0.4–1.7%. Active metabolism was detected in 53–58% of the population, despite very low levels of total organic carbon per liter of water. The diversity of populations was extremely high, with 42% of the isolated species similar to surface organisms, with the remainder unidentified. There were no dominant species, and the type of growth medium used strongly influenced the species isolated. Although they were one of the first to use low-nutrient media, it is very likely that their unusual results were due to the methods used. We know that cultural studies and molecular approaches show very different findings when applied to the same sample [20].

3.2.3 Recent microbiological studies from Mammoth Cave

Early studies [25, 26] mostly identified heterotrophs and chemoorganotrophs in contrast with more recent investigations that combine cultural methods with sequencing of the 16S rRNA genes. Pemberton et al. [27] compared microbial diversity using both approaches from an upper level dry passage, as a low nutrient location, with a lower level stream passage, as a high nutrient location, in Jack Bradley Cave in southern Kentucky, located in the middle of the Interior Low Plateaus. Based on colony type, the greatest diversity resulted from growing microbes on the lowest nutrient medium, but the species that grew in culture were not major components of the species detected using genetic techniques [20]. These are very different findings from previous culture-based work. Many species retrieved from both approaches were not previously described. There was also little overlap in species diversity between the two environments, suggesting greater metabolic flexibility in isolates from extremely starved environments.

3.2.4 Actinobacteria

Descriptions of cave wall slime, wall fungus, and lava wall slime all refer to the growth of Actinobacteria and associated microbes (▸ Fig. 3.2). Actinobacteria are widespread in caves and may make up 10–33% of total soil microbes [28]. These bacteria produce geosmin, the characteristic earthy odor associated with caves and soils. Their main role in nature is in decomposition of organic matter. Many Actinobacteria fix atmospheric nitrogen either in association with plant roots or as free-living cells. The organic nitrogen produced may represent a significant amount of soil nitrogen, particularly in extreme environments. The role of Actinobacteria in nitrogen fixation in caves has not been studied.

Actinobacteria are a highly varied group of heterotrophic and chemoorganotrophic bacteria that have the unusual characteristics of filamentous growth and external spore production, which macroscopically makes them resemble fungi. On close inspection, individual colonies have a branched appearance, and are commonly white to yellow in color, but there are also tan, pink, and orange colonies. Actinobacterial colonies are hydrophobic, with water or secreted fluids beading up on the surface. This water often reflects cavers' lights and cause the colonies to appear reflective, sometimes described as "cave silver" (▸ Fig. 3.3). Actinobacteria are blamed for the biodeterioration of cave paintings studied extensively in Altamira and other Spanish caves [29]. Refer to Chapter 13 in this book for more details about Altamira Cave.

Fig. 3.2. Individual colonies and dense growth of Actinobacteria beside an adult female cave cricket (*Hadenoecus subterraneus*). Photograph credit: thomaslavoiephotography.com.

Fig. 3.3. "Cave silver" forms on actinobacterial colonies and microbial mats when water beads reflect light. Photograph credit: thomaslavoiephotography.com.

Actinobacteria produce about 75% of all antibiotic compounds, including streptomycin and tetracycline. Antibiotics may have important ecological roles in reducing competition for scarce resources, but antimicrobials may also serve as toxins, ionophores, bioregulators, and signal molecules produced by "metabolically talented" microbes [30]. Caves are an extreme habitat and the cave microbiome has great potential as a novel resource for the development of new bioactive microbial metabolites [30, 31].

Frisch et al. [32] reported that they isolated bacteria from Mammoth Cave that produced potential drugs that blocked cancer, tuberculosis, and angiogenesis, but it takes many years before such a discovery can be brought to actual treatment. “Biological deserts” in Great Onyx Cave in MCNP are areas high in gypsum with no animal life in passages extending under the caprock. Poulson et al. [33], in a preliminary study, reported on a higher proportion of antibiotic producing bacteria in the gypsum area compared to an area low in gypsum and high in biology only 25 m away. Presumably the need to outcompete other microbes is higher in the gypsum environment with reduced input of organics.

Surveys of Actinobacteria distribution were made by Lavoie and Northup (unpublished, 1994) in White Cave, a small limestone cave in MCNP and in Four Windows Cave, a lava tube cave in El Malpais National Monument, New Mexico. Actinobacterial density was strongly related to areas of water seepage on ceilings and walls. Cover ranged from isolated, individual colonies to complete coverage of surfaces, with greatest density in lava tube caves. Growth rate data collected from White Cave show no increase in the size of individual, isolated colonies over a 12-year period, which suggests either episodic growth or growth in marginal habitats (Lavoie, unpublished).

3.3 Cave Ecosystem Energy

To understand any ecosystem, you have to “follow the energy.” All ecosystems are based on two sources of energy: light or chemicals. Mammoth Cave is a typical karst cave dissolved out of carbonate rocks with most of the ecosystem energy coming indirectly from photosynthesis based on inputs of detrital organic matter [6, 7]. Although photosynthesis is not possible in the dark zone of caves, the lighted sections of Mammoth Cave open to tourists have growth of algae, cyanobacteria, and other phototrophic microbes, known as lampenflora. Refer to Chapter 12 for more information about this topic. Microbial utilization of inorganic chemicals for energy sustains chemosynthetic ecosystems, but there are few examples of these ecosystems in the Mammoth Cave region.

Because of this low-energy detrital food base in most caves, communities tend to have very simplified trophic structure compared with surface ecosystems, with only one or two trophic levels in a food pyramid (▸ Fig. 3.4) and much reduced total biomass. Compared to nearly any surface ecosystem, the size of the cave pyramid would be miniscule. The pyramid shape also reflects the lower numbers and biomass found at higher trophic levels; carnivores are much rarer than their herbivore prey. The energy base of the detrital ecosystem supports decomposer microorganisms, mostly bacteria and fungi, and some small invertebrates. In Mammoth Cave, the decomposers are then eaten by a guild of primary consumers with apex predators (secondary con-

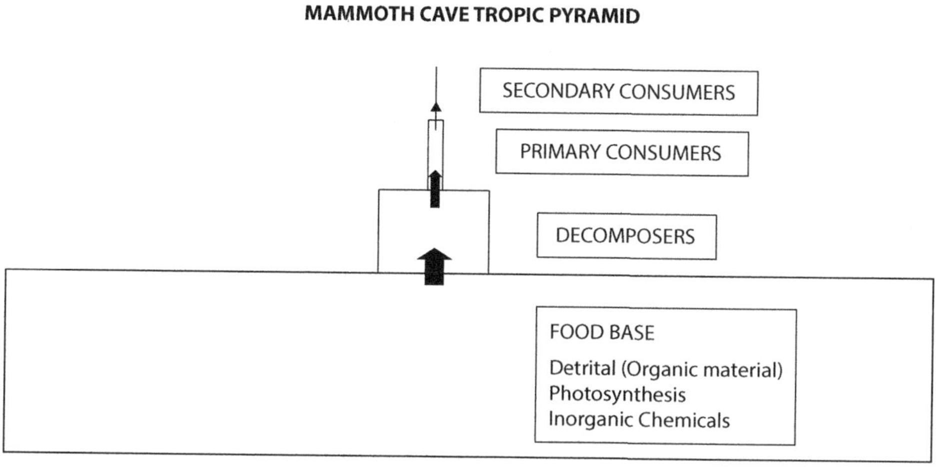

TROPICHIC LEVEL	TERRESTRIAL EXAMPLES	AQUATIC EXAMPLES
Secondary Consumers Predators	*Meta* cave spiders, Rove beetles, salamanders, rats	Cavefish, cave crayfish, salamanders
Primary Consumers Herbivores & Omnivores	Colembola, mites, crickets, millipedes, beetles	Amphipods, isopods, copepods, planaria
Decomposers Microorganisms & Decomposers	Bacteria, fungi, protozoa, nematodes and other small invertebrates	Bacteria, fungi, protozoa Small invertebrates
Food Base	Organic debris Seeping water and drips Guano from bats and crickets Animal droppings (rat, raccoon) Carcasses	Organic debris Floods Seeping water and drips Guano from bats and crickets H_2S, Oil seeps

Fig. 3.4. Cave food pyramid for terrestrial and aquatic communities at Mammoth Cave.

sumers) at the top of the food pyramid. Simon et al. [34] followed the movement of dissolved organic matter in cave streams to primary consumers and predators, and found organic matter decomposition and uptake rates through the food web to be similar to those they measured in surface streams.

Troglobites and troglophiles respond differently to a bonanza of food; horse excrement that Tom Poulson brought into caves in MCNP [35] resulted in increased reproduction by troglophiles, but not by troglobites. Troglobites are adapted to a steady level of low food and energy input, and maintain their low levels of growth and reproduction. These responses are typical of how ecologists describe r-adapted and k-adapted reproductive strategies.

3.3.1 Detrital-based ecosystems

Detrital ecosystems are dependent on secondary inputs of organic matter made through photosynthesis. Detrital ecosystems are actually very common, and include soils, deep rivers, lakes, oceans, the rumens of cows, and most caves. In detrital-based cave ecosystems, surface-derived organic material like dead leaves, wood, and dissolved organic matter enters a cave through entrances, is washed in during floods, or seeps in through cracks in the rock from the surface. Dissolved organic carbon and microbial biofilms growing on rock substrates in cave streams are important energy sources for stream communities even when coarse particulate organic matter is an abundant resource [34]. Once in caves organic matter decomposition is successional where nutrients get leached out or used up by organisms, usually microbes, until all that remains is material highly resistant to decomposition, like wood and leaves. Microbial communities also change as a function of organic matter succession.

Byl et al. [36] studied substrate utilization from water samples collected at levels B, C, and D in the Styx River drainage in Mammoth Cave. They quantified bacterial communities from each level and measured substrate utilization using Biolog's EcoPlates™ with 31 different substrates. Samples from the surface and the upper levels of the cave had the greatest initial substrate richness values. Charlotte's Dome from level C was initially also high, but substrate use was slower than for other samples. After 72 h, the richness of all of the sites was similar and indicated an initial preference for particular substrates. But, over time, the bacterial communities were able to utilize almost all of the tested substrates to some extent. After 120 hours, microbial communities from the surface and higher level passages of the cave used the substrates more evenly than communities from deeper in the cave. Sampling was in summer of 2012,

Fig. 3.5. *Mucor* spp. fungus growing on fresh rat droppings in White Cave, MCNP. Photograph credit: thomaslavoiephotography.com.

during a period of hard drought, suggesting that the distinct community patterns observed may vary by season or with different amounts of rainfall [36].

Detrital energy input to caves can also be from transient organic matter contributed by trogloxenes and troglophiles that leave caves to forage, and then import energy in the form of guano, organic material, carcasses, and eggs. In Mammoth Cave, cave crickets, bats, and cave rats are the most important contributors of transient organic matter which can support specialized cave communities [6, 7, 35] (▸ Fig. 3.5). Cave crickets support two specialized communities based on eggs [37] and guano [38]. Compared to bat guano, cave cricket guano is very low in energy (2.4 kcal/g) [39], but does support its own specialized community of guanobites, such as *Scoterpes* millipedes.

3.3.2 Phototrophy due to tourism

Show caves are in a constant battle to remove algae from artificial lighting. Refer to Chapter 11 in this book for more information about show caves, and Chapter 12 about microbial communities associated with artificial lighting in caves. Algae are an important component of lamp flora or lampenflora around artificial light sources. Lamp flora displaces natural microbiota and can damage speleothems. Smith and Olson [41] identified phototrophic species around artificial lights in Mammoth Cave. Half of the isolates were cyanobacteria (50%), with eight species of green algae (29%), and six species of diatoms (21%). They found no correlation between species diversity and temperature, although there was a general trend of increasing diversity with warmer temperatures.

Olson [42] evaluated lampenflora control by using different lighting in the Frozen Niagara Entrance of Mammoth Cave. This passage is the most heavily visited in the park, and it is also the best lit. A test area was photographed, cleaned with bleach, and both gold-phosphor fluorescent and yellow LED lamps installed. Existing white lamps were used as a control. After 2.5 years of daily tours, the greatest growth was with the traditional white light, the gold-phosphor supported limited growth, and there was no growth with the gold LED lighting. Other research indicates that bright light encourages more graffiti left by visitors and dries out surfaces that can alter mineral composition [42]. Bright light may also decrease relative humidity that could kill cave-adapted animals or cause them to move into different areas of a cave.

3.3.3 Chemolithoautotrophically based cave ecosystems

Primary production of food in caves away from entrances can be from chemolithoautotrophic microbial activity. While it is usually a minor component of the food base in most caves, it can sustain entire ecosystems.

One of the earliest studies of microbes in caves using molecular techniques was performed by Angert et al. [43] from Sulphur River in Parker Cave, Kentucky. Parker Cave is located in the middle of the sinkhole plain about 7.5 km south of the main entrance to MCNP [44]. The cave has a mapped extent of 6.9 km of five parallel stream passages formed in the St. Louis limestone. The streams all sump downstream and are blocked upstream by breakdown or constrictions and connect by short crossover passages. Each stream has its own unique chemistry. The first is Brown River, fed by sinking streams with water chemistry typical of the area; a slightly basic pH, specific conductance of 450–550 μS/cm with 80 mg/L calcium and 10 mg/L chloride. The fourth stream is Sulphur River. The air smells strongly of hydrogen sulfide (H_2S) gas; at concentrations of 50 ppmv, that is enough to cause eye damage [44]. Water chemistry in the Sulphur River is strikingly different than the other passages, with near neutral pH, specific conductance of 4000–9000 μS/cm, 270 mg/L calcium, and 1000–2600 mg/L chloride. Biofilms include gelatinous snottites on the walls and white filaments in the stream. Upstream in Sulphur River is the Phantom Waterfall, a soft cascade of gypsum and bacterial filaments that is about 2.5 m tall and 1m wide. The water source for Phantom Waterfall comes from a narrow canyon that has water with slightly basic pH, specific conductance >26,000 μS/cm, 2750 mg/L sulfate, 80 mg/L calcium, and 8000 mg/L chloride. The water source is probably oil-field brines that seep up into the cave. Precipitated sulfur deposits coat the sediments and make sulfur stalactites that hang from the ceiling. The mud floor in the upper room of the cave is coated with elemental sulfur and has a highly acidic pH of 0.13, while the ceiling has an acidic layer of microbial biofilm. Sulfuric acid appears to have significantly enlarged the room from the original small crawl.

The terrestrial community in the Sulphur River area is more diverse than other areas in Parker Cave probably due to increased food availability. Lisowski et al. [45] reported two types of annelids, a cave snail, two types of beetles, a spider, and several mite species. Thompson and Olson [46] found at least 13 genera of protozoa in eight orders from the stream. They also did a microscopic study of the white filaments, finding *Beggiatoa* and *Thiothrix* bacteria that deposit elemental sulfur granules inside their cells or in sheaths. Both of these bacteria are known sulfur oxidizers, leading the authors to speculate on a community based on chemolithoautotrophic production of organic material because of the complete isolation from the usual indirect photosynthetic input of energy in detrital-based ecosystems.

Comparing the 16S rRNA genes from the microbial mat with databases of known species, Angert et al. [43] showed that the Sulphur River bacterial community in Parker Cave had the greatest similarity to sulfur-oxidizing bacteria known from sulfur-based deep-sea hydrothermal vents communities. Several are related to an epibiont of a vent tubeworm. Others were related to species that have the ability to fix CO_2 as a source of carbon. This chemolithoautotrophic cave community is much more accessible for study than geothermal vents at the bottom of the ocean. They also speculate on contributions of chemolithoautotrophic microorganisms on the dissolution and precipitation of minerals in caves.

Olson [47] recently investigated a potential hydrogen sulfide seep in the area of Marianne's Pass in Historic Mammoth Cave. A sulfur spring in this area was first described in 1845 [48], and is marked as a sulfur spring on the 1908 Kaemper Mammoth Cave map. The seep is deeply weathered into the rock, and there are thousands of springtails feeding on the microbial mats, with crickets and cave beetles nearby. Other sulfur inputs are rare in Mammoth Cave, although Olson [47] points out that the vast length of Mammoth Cave and the unknown number of smaller voids and passages means we just do not know enough about the extent of chemolithoautotrophy in ecosystem development.

3.4 Geomicrobiology

The field of geomicrobiology is at the intersection of geology and microbiology. Microbes are important agents either actively or passively in chemical reactions that influence geological formations on localized scales to landscape scales (biokarst). The development of new tools and techniques in both biology and geology are contributing to a better understanding of the intersection of the two fields. The first conference on the geomicrobiology of caves was sponsored by the Karst Waters Institute [49]. We see geomicrobiology at work in caves [50–52] in the formation of some speleothems, mineral deposits, and the formation of karst caves, including Mammoth Cave. Karst caves form primarily by dissolution of carbonate rock by acidic water. Rainwater is slightly acidic (pH 5.6), and additional acid comes from microbial activities as water moves through soil overlying the caves. Karst caves also form by sulfuric acid speleogenesis, where nonbiological chemical reactions and chemolithoautotrophic microbes produce sulfuric acid as a waste product that dissolves the carbonate.

There are many types of speleothems and other secondary minerals found in caves [53]; speleothems are a secondary mineral deposit formed by physical and chemical reactions from a primary mineral. Many of the reactions resulting in speleothem formation or destruction are both biotic and abiotic, but microbes are probably responsible for all or most low-energy reactions.

Most microorganisms in subsurface environments grow as biofilms or groups of cells attached to rock surfaces. Rocks and minerals are not homogeneous and can influence microbial diversity at the microniche level [54, 55]. Components of the stratum may be toxic or beneficial to microbial growth by providing mineral nutrients, pH buffering, or other advantageous conditions. Given the long evolutionary history of microbes in geologic time, Jones and Bennett [55] hypothesize that each mineral surface in caves is specifically altered by the best adapted and most comprehensive microbial community that can use the mineral surface to the greatest advantage. Their experimental study under oligotrophic conditions concluded that community structure can be differentiated at the level of individual mineral grains supporting the unique microbial consortium best adapted to those conditions.

3.4.1 Saltpeter formation

One of the best examples of geomicrobiology in caves is saltpeter, also known as nitre or niter (KNO_3). The earliest work on the microbiology of saltpeter formation in Mammoth Cave was published by Fliermans and Schmidt [56, 57]. They used species-specific fluorescent antibodies to study the presence, distribution, and population densities of *Nitrobacter* in Mammoth Cave soils. *Nitrobacter* is a chemolithoautotrophic nitrifier associated with production of saltpeter. *Nitrobacter* was isolated from saltpeter deposits in Mammoth Cave with densities more than a hundred times higher than in surface soils. *Nitrobacter agillis* predominated in cave sediments and *N. winogradsky* in surface soils.

Saltpeter is the principle component of gunpowder, making up about 75% of the total with varying amount of charcoal and sulfur [58]. Historically, caves were mined for saltpeter throughout the southeastern United States for many years. In addition to producing gunpowder for personal use, mining was of strategic importance during the War of 1812 when coastal harbors in the United States were blockaded by the British and gunpowder could not be imported. Kentucky produced the greatest amount of saltpeter in the early 19th century from both large-scale mining, like from Mammoth Cave, to small-scale mining in rockshelters [59]. Saltpeter leaching vats and piping made of tulip poplar logs are a major exhibit on the historic tour of Mammoth Cave (► Fig. 3.6) [58].

Saltpeter earth appears dry and fluffy. The first step in saltpeter mining is leaching nitrates with water [60]. At Mammoth Cave, the leachate was pumped up to the surface where it was boiled down to evaporate water and concentrate the calcium and magnesium from the nitrate solution [60]. During the lixivation process, wood ash (potash lye with 5.4% K) was added to the leachate to crystallize KNO_3 [60]. Further purification steps are needed to get a better grade of saltpeter.

Fliermans and Schmidt [56, 57] found no strong correlation between *Nitrobacter* populations and nitrate concentrations or sediment moisture. Other research supports the role of nitrifying bacteria in the origin of saltpeter earth (reviewed in [60–62]) including stable isotope analysis that shows saltpeter is selectively enriched in the lighter isotope of nitrogen [63], indicative of a bacterial source. Through nitrification, *Nitrosomonas* bacteria convert ammonium ions (or ammonia) to nitrite at a maximum rate of 1–30 million μg N/d/g dry cells by

$$2NH_3 + 3O_2 \rightarrow 2NO_2^- + 2H^+ + 2H_2O \tag{3.1}$$

Fig. 3.6. Saltpeter leaching vat on the Historic Tour in Mammoth Cave. These square vats were used in large-scale production of saltpeter [59, 60].

Nitrite is then converted into nitrate by *Nitrobacter* at a maximum rate of 5–70 million µg N/d/g dry cells by

$$2NO_2^- + O_2 \rightarrow 2NO_3^- \tag{3.2}$$

Both processes can accumulate nitrate as the end product to a maximum of 2000–4000 µg N/mL [64]. Nitrification is widespread in soils and water associated with decomposition of organic compounds.

Nitrogen content in surface soils varies with climate, vegetation type, topography, soil type and porosity, and microbial activity. There is a strong correlation between the distribution of saltpeter caves in the southeastern United States and dense overlying oak or oak-hickory forests [61]. Saltpeter earth in caves may be limited to the northern latitudes by lower temperatures and higher moisture content, to the southern latitudes by higher temperatures and less organic deposits or by nitrogen-retaining grasslands. Drill cores 30 cm in depth from surface limestone show very low levels of nitrate, but drill cores from all areas of limestone from caves have high concentrations of nitrate (a few ppm vs. hundreds to thousands of ppm) [60, 61]. Nitrate concentration generally drop off in the first few centimeters, can be were within an order of magnitude in concentration at deeper points. Evaporation of water in dry underlying passages would result in a buildup of nitrate in the saltpeter earth.

One reported aspect to the production of saltpeter in caves is the ability of cave soils to regenerate nitrate if allowed to rest undisturbed in contact with the walls and floor of the cave. Olson and Krapac [62] investigated nitrate regeneration over a six year period. The highest niter levels were obtained with leached soil placed in contact with bedrock or sediment. There was some production with samples in contact with cave air alone, and no production in soils exposed to surface air. After two and a half years, no sample had reached the preleaching level of nitrate concentration and only one sample had reached a level consistent with conversion of residual nitrogen. Essentially, observed regeneration rates were so slow that groundwater percolation could not account for the original high nitrate concentrations. These times also do not support the cited time span for regeneration of 3–10 years. During the saltpeter leaching process, *Nitrobacter* populations did not change, but total bacteria decreased 57% [62], possibly accounting for the regeneration of niter in cave soils. Olson and Krapac [62] concluded that guano from large former populations of bats, woodrats, and raccoons was the dominant and original source of ammonia in Mammoth Cave, although many other suggestions have been made [65].

3.4.2 Ferromanganese deposits

Dark colored ferromanganese deposits (FMD) are common in caves. Black coatings exposed to flowing water in limestone caves are deposits of manganese oxides that are produced by microbial action (▸ Fig. 3.7). Microbes oxidize soluble Mn^{2+} to trivalent or tetravalent manganese, which results in the deposition of poorly crystallized coatings. Refer to Chapter 7 in this book for more details about manganese cycling. White et al. [66] did a systematic study of manganese oxide deposits from caves in central and eastern United States, including Mammoth Cave. They found that all samples con-

Fig. 3.7. Black manganese oxide deposits on a speleothem and coating the floor.

tained both manganese and iron, but in different ratios. Samples from Mammoth Cave had a ratio of about 20 Mn:5 Fe. They also reported enrichment of the black deposits in transition metals (e.g. copper, cobalt, nickel, vanadium, and zinc) at the fractional percent level, which was on the order of a million times greater than found in surrounding rock and water. FMD from Tennessee caves had Mn:Fe ratios approaching 1.0 and have a unique consortia of Mn(II)-oxidizing bacteria based on 16S rRNA gene sequences, including *Pseudomonas*, *Leptothrix*, *Flavobacterium*, and *Janthinobacterium* spp. [67]. Boston et al. [18] identified characteristic biosignature suites in caves that could be applied to detection of life, past and present, on other planets. FMD could be studied as biomarkers in Mammoth Cave. FMD can be recognized by microscopic and naked-eye observation and unusual concentrations of minerals.

3.5 Eukaryotic Microorganisms

3.5.1 Protozoa and algae

Protozoa and other protists are important consumers that feed largely on bacteria and can be an important food source for smaller invertebrates. Most protozoa live in water, although their numbers may be high in any moist environment like soils and guano. The first paper on cave protozoa appeared in 1845 and since that time hundreds of species have been identified from cave aquatic habitats and as parasites of cave biota. Most species are the same as those found in forest litter, but there may be some true troglobites.

Algae are single-celled, largely photoautotropic microorganisms. They are not of importance in caves except around entrances and artificial light sources. Barr and Kuehne [40] found some evidence of algae reproducing in Mammoth Cave streams as part of their plankton studies. Peaks in algal growth inside caves can be associated with heavy rains and spring snow melt that wash algae into the cave, and chemoheterotrophic growth.

Protozoa are important in community interactions in several aquatic environments in Mammoth Cave. Barr and Kuehne [40] found most biological activity located on or in bottom sediments. Water with a direct connection to the surface, such as Echo River, had higher densities of plankton and some seasonal changes compared to Crystal Lake, an isolated body of water perched above the current water table. Shelia Seale (unpublished data) quantified protozoa numbers using the dilution most probable number method [68]. She found protozoa in all aquatic pools and streams sampled in Mammoth Cave, but none entering the cave from drip water.

3.5.2 Fungi

Fungi are important decomposers and serve as a food source for many organisms. All fungi depend on sources of organic material, as shown in ▸ Fig. 3.5, for their growth, which they absorb through their cell walls after excreting digestive exoenzymes. Many types of fungi living in caves have been identified, as reviewed by Vanderwolf et al. [69]. All fungi known from caves are identical to surface forms, and may have grown from spores brought in with organic material or by colonizing organic matter from spores found in the cave. Food spoilage by microbial growth is a way microbes can monopolize a food resource and keep it away from much larger consumers. Fungi can produce dangerous compounds during growth, such as mycotoxins. Most animals reject moldy food if they have a choice; the abundance of cedar in cave rat middens and nests may be brought in by the cave rats as a way to decrease mold growth on stored materials. Caumartin [15] felt that we are not likely to find fungi unique to caves because of their high energy demands, but Vanderwolf et al. [69] say that the low-nutrient, stable, low temperatures in caves favors communities of oligotrophic, cold-tolerant fungi. Much more study needs to be performed on fungi in caves.

3.5.2.1 Fungal diversity

Lavoie [70] studied the factors controlling fungal succession on cave rat fecal pellets, and interactions of fungi with higher organisms at two sites in MCNP: a deep cave site inside the Austin entrance and in Little Beauty Cave, a small isolated cave. Like all organic matter, fecal material goes through a secondary succession of fungi, with simple, filamentous forms appearing first and utilizing simple sugars and protein. Next to appear are the Deuteromycetes and Ascomycetes growing on more complex compounds, and last to appear are Basidiomycetes (mushrooms) growing on more recalcitrant materials. Is the succession due to time to develop more complex structures, nutritional availability, or physical form of the dung? The same amount of ground cave rat droppings was mixed with water and shaped to resemble other naturally deposited feces in caves as a single piece (raccoon scat), a pile of pellets (cave rat droppings), and spread on the soil in a thin veneer (cave cricket guano). All forms were similar in the timing of appearance of fungal species and the type of fungi found, until the last stage of mushroom formation. Mushrooms developed on the single piece and the pile of pellets, but not on the thin veneer. Apparently the fungal hyphae are able to penetrate and grow through the larger forms, but the thin veneer did not permit movement of nutrients to one location to allow for formation of a mushroom. The numbers of beetle and fly larvae that grew on the raccoon and cave rat feces were similar, while numbers were reduced on the cricket guano because the larvae had no refuge inside the veneer and were accessible to predators, like staphylinid rove beetles. Lab experiments with *Ptomaphagus* beetles and sciarid fl ies, which naturally colonize fecal deposits in Mammoth Cave, showed that if early fungi start to degrade organic matter before the invertebrates, then their hyphae are quite effective at blocking access.

Shapiro and Pringle [71] investigated possible human impacts on fungal diversity in caves, including Dogwood Cave and Diamond Caverns that are hydrologically connected to the Mammoth Cave system. They sampled soils from a range of areas in terms of human impact, including two sites that may never have had human contact. The particular cave did not affect fungal diversity if the samples were incubated at room temperature, but did when cultures were incubated at 10 °C. The fungi that grew at 25 °C were unlikely to grow under cave conditions. They did not isolate any fungi from the area that had never been visited, as predicted by Caumartin [15]. The highest diversity was associated with low human impact. Disturbance allows faster growing weedy species to take over, and results in decreased species diversity. In these caves, fungal diversity rises with moderate levels of disturbance, and peaks in minimally disturbed sites. They concluded that impacts of human disturbance are highly localized [71].

3.5.2.2 Mushroom farming

An 1881 issue of *Scientific American* briefly reports on a proposal by a Frenchman to build a commercial mushroom farm in Mammoth Cave based on the success of farms in France where there are often miles of mushroom beds in caves. A proposal was made to the Trustees of Mammoth Cave to rent a portion of Audubon Avenue for a mushroom farm. The special advantage of subterranean culturing was the uniform temperature. It was speculated that the extensive deposits of bat guano from the swarms of bats in the Great Bat Room could be mixed with other fertilizers to form a substrate for growing the fungus. The report notes some concern with dryness of the soil but ventured

Fig. 3.8. Mushroom beds in Audubon Avenue of Mammoth Cave. Photograph by Elizabeth Lavoie.

that it could easily be moistened. Algeo [79] reviewed the history of the venture from original company records and family letters, and identified sources of tension between the mushroom business and tourism. She reports that the Mammoth Cave mushroom farm may have been the first such operation in the United States. The mushroom beds were sabotaged with coal oil in December 1881, which led to the failure of the company before it had produced a single mushroom. The few beds that were built are still at the far end of Audubon Avenue and identified on the 1908 Kaemper map of Mammoth Cave (► Fig. 3.8).

3.5.2.3 White-nose syndrome

Currently, the best-known fungus in caves is *Pseudogymnoascus destructans* (formerly *Geomyces destructans*), the fungus that causes white-nose syndrome (WNS) in hibernating bats, killing them by the millions across the United States and Canada. First detected in the winter of 2006–07 in a cave near Albany, New York, WNS has been confirmed in 25 USA states and five Canadian provinces [72]. One control effort has been decontamination and clean caving to prevent possible human spread to uncontaminated sites.

Pseudogymnoascus destructans is an introduced species from Europe where it does not cause the high mortality seen in naïve North American bats [73]. The fungus co-evolved with European bats over thousands of years, but it is an invasive species in North American bats that have not developed any resistance to the fungus, although some bats are now surviving the infection. Zukal et al. [74] characterized the fungus as a generalist parasite of bats. They found WNS in a wide range of ecologically diverse hibernating bats in the Czech Republic, and suggested that all hibernating bats in the range of the fungus may be at risk.

WNS was detected at Mammoth Cave in the winter of 2012–2013, where it was found in a northern long-eared bat (*Myotis septentrionalis*) from Long Cave, the largest hibernacula in MCNP [75]. Toomey et al. [75] reviewed actions taken at MCNP before and after WNS detection. Beginning in 2009, the park increased surveillance and monitoring of hibernacula and summer bat roosts to establish baseline data and to document potential population changes once WNS arrived. Visitor education provides an opportunity to increase understanding of the value of bats and awareness of WNS, and was done by public announcements, pre-tour briefings by guides, and posters. At the time of writing, WNS continues to spread and is following the major flight routes of infected bats [72]. A modeling study by Alves et al. [76] predicts that WNS will not spread throughout North America, but could cause population declines in 32% of hibernating bat species.

At MCNP, the high use of quaternary ammonia compounds (QAC) during decontamination of caving gear and visitor shoes raised concerns about the potential for QAC to be washed into the cave via storm runoff. Byl et al. [36] studied possible effects of QAC contamination on the bacteria in the River Styx watershed that drains waters from the campground and Visitor's Center into Mammoth Cave. Dye tracing [77] allowed for selection of water from levels B, C, and D of the system [8]. They also tested antibiotic resistance, since Chapman [78] reported that bacteria resistant to QACs were likely to be resistant to other antibiotics. Bacteria from cave waters were sensitive to increasing QAC concentrations, and to four of five antibiotics tested. Erythromycin showed a slight stimulation of growth at low concentrations (< 1 mg/L), but was inhibitory at higher concentrations. Slightly greater resistance was observed with bacteria from water from higher level passages.

3.6 Infections and Parasites

3.6.1 Tuberculosis

Tuberculosis (TB) is a serious lung infection largely caused by the bacterium *Mycobacterium tuberculosis*. Even now, TB infects (active and inactive) one third of the population of the world. In 1900, an estimated 110,000 Americans died each year from TB [80]. TB was the second leading cause of death after pneumonia and influenza, heart disease was fourth, and cancer eighth. Historically, TB was known as consumption, wasting disease, or the white plague. The prevailing ideas for treatment of TB were that patients should have good food, lots of fresh air, and inactivity, which led to the establishment of sanatoriums [81].

People were willing to live in stone TB huts deep in Mammoth Cave in hope of a cure. In 1839, Mammoth Cave was purchased by Louisville physician and entrepreneur, Dr. John Croghan [82]. He believed the constant temperature and high humidity in caves would benefit TB patients. In 1841, he allowed 16 TB patients to live in wooden and stone huts located near the Star Chamber beyond Giant's Coffin in the cave; huts were also built in other locations. Cool temperatures necessitated open fires and the patients wanted light, which led to sooty conditions. Bushes were brought in to cheer up the patients and tours passed by the huts regularly. The deaths of some patients and worsening condition of others ended the experiment by 1843. Dr. Croghan died of TB in 1849 [82].

3.6.2 Parasites

Giardia is a protozoan that can cause diarrhea 2–4 weeks after drinking polluted waters. This microbe has been reported from numerous caves and springs around the world. Filters and boiling water are effective at removing it. Human paleofeces from Salts Cave in MCNP show evidence of infestation with *Giardia* [83]. Humans were known to use the cave over 2000 years ago. Parasitism of similar prehistoric human populations was found in Big Bone Cave in Tennessee. Faulkner et al. [84] found fleas, *Giardia*, and worms in paleofeces. Five samples were positive for *Enterobius*, a tiny roundworm 2–10 mm in length, that causes pinworm or threadworm, the most common worm infection in the United States today. They also found eggs of *Ascaris* worms, which is still the most common roundworm intestinal infection for humans worldwide.

3.7 Human Impact

Caves are conduits into the subterranean world, and surface actions impact water quality and cave life. For microbiologists, water pollution is defined as the presence of feces in water because many serious human diseases, such as cholera and dysentery, are transmitted by water. However, it is impossible to look for every kind of disease-causing microbe since there are so many possibilities. Instead, indicator bacteria are used because they are easier to test for, live longer than disease-causing bacteria, and are always present proportional to the amount of fecal contamination. Common indicator bacteria are coliforms, like *Escherichia coli*, and fecal streptococci. Water is considered to be nonpotable if any coliform bacteria are found.

At MCNP, large areas of the Green River watershed and groundwater basins that flow through Mammoth Cave are outside park boundaries. Activities outside MCNP that influence water quality inside MCNP include sewage treatment and solid waste disposal from private homes, municipalities, and industries in the area, as well as agricultural and forestry management activities. Other factors that contribute to point source and nonpoint contamination are recreational activities and oil and gas production. The flow of water into and through MCNP has been determined through dye tracing.

In October 2006, elevated levels of fecal coliforms and *E. coli* were identified in MCNP in waterfalls and seeps fed by water from the developed area of the park that includes the visitor center, parking lots, and campground [85]. All of these areas overlay the historic portion of Mammoth Cave around the entrance. Rapid testing determined which areas showed fecal coliform contamination. Monitoring sites were established at waterfalls at the Historic Entrance and Mammoth Dome. The standards for secondary contact were not exceeded; people could safely come in contact with the water but they should not drink it. In addition, inspections of the sewer lines in MCNP showed some breaks that were repaired. Elevated levels of fecal coliforms are an on-going issue of concern and are continuously monitored at MCNP.

Hidden River Cave in Horse Cave, Kentucky, located about 15 km due east from MCNP, is a success story [86]. The cave was a popular tourist attraction and was once the source of drinking water and electric power for Horse Cave. But the cave had to be closed in 1944 because it stank due to contamination from municipal wastes, wastes from a local creamery, and runoff from a chrome-plating plant. The once abundant cave fish and cave crayfish were gone. On a visit to Horse Cave in the 1980s, Tom Poulson and I accidently ignited methane pooled at the entrance and watched in awe as a bubble of blue flame raced up the sinkhole before fizzling out (it was dramatic, but no harm to us). A new regional wastewater treatment plant was built in the 1980s and many of the cave animals have since returned. The American Cave Conservation Association is located in Horse Cave, and the organization has worked to restore the cave and offers regular tours [85].

A study of soil communities from Mammoth Cave and Lechuguilla Cave in New Mexico (USA) showed the harmful effects of urine [87]. Urine addition caused the growth of a layer of microbes on the surface, changed the population number and type of microbes in the soil, and it looked and smelled bad. Cavers are advised to pack out all wastes.

Lavoie and Northup [88] looked for evidence of microbes associated with humans by comparing areas of Mammoth Cave and Carlsbad Cavern in New Mexico that had high versus low impacts from visitation. They used swabs and cultures to look for human indicator microbes (*E. coli*, *Staphylococcus aureus*) and bacteria that could be tracked in from the surface (high-temperature *Bacillus*). They found some trends related to impact. Interpreting the results is complicated because we do not know how long these microbes actually survive in the cave environment.

Lint accumulations of clothing fibers and dust in heavily visited caves, besides being unaesthetic, can serve as a source of nutrients for the growth of microbes with the potential to damage and discolor underlying formations. Jablonsky [89] began organizing "lint camps" in 1992 where volunteers work to remove lint from caves, including Mammoth Cave.

3.8 Microbes and Cave Crickets

Cave crickets are keystone species in caves in the Mammoth Cave region because they import fixed energy in the form of guano, eggs, and carcasses. Monitoring the health of cricket populations is a good indicator of the health of the cave ecosystem [38].

3.8.1 Cricket crop microbes

I sometimes describe *Hadenoecus subterraneus* cave crickets (▸ Fig. 3.2) as little "cave cows." Some orthopterans, including crickets, grasshoppers, and cockroaches, consume plant detritus, decaying fruit, and herbivore dung, and ingest a variety of bacteria, protozoa, and fungi along with their food. The crop of cave crickets is a very thin-walled structure that lies between the esophagus and hindgut. The inner wall of the crop contains chitinous "teeth" (▸ Fig. 3.9) that aid in mixing and movement of food through the digestive system. Crickets can eat up to 300× their body weight in food to the point of physical distortion. If microbes that are ingested survive and pro-

Fig. 3.9. Scanning electron photomicrograph of chitinous "teeth" surrounded by bacterial cells from the crop of a cave cricket (*H. subterraneus*).

liferate in the digestive tract or excrete enzymes, then ingested microbes can augment or extend the digestive and metabolic capabilities of the organism that consumes and harbors them.

Studier and Lavoie [90] found that cave crickets from Mammoth Cave rapidly lost weight in water-saturated air only 2 °C above cave ambient temperatures. Crickets die in a few hours if held at temperatures above room temperature (23 °C). Crickets may be partially restricted to a narrow temperature range to keep their crop microbes under control. Most bacteria isolated from cave cricket crops grew best above ambient cave temperatures of 13.5 °C with crop enzyme activity optimal at 23 °C [91]. When cave crickets were fed diets rich in either carbohydrates or protein compared to the natural diet, the activities of specific enzymes respond rapidly to the different diets, as expected if microbes were producing the digestive enzymes. Some experimental crickets, as well as an occasional field-collected specimen, had crops visibly distended with gas, occasionally to the point of rupture. We think that the crickets were killed by unregulated growth of crop microbes that produce excessive gases or toxic metabolites at elevated temperatures.

Whatever the reasons for the extreme thermal sensitivity of *H. subterraneus*, even a small increase in cave ambient temperature could have profound negative effects on cave crickets. If global climate change causes an increase of 2–6 °C over the next 50 years, it would greatly increase metabolic demands and evaporative water loss in cave crickets, which would force more frequent foraging bouts and exposure to surface conditions and predators. These changes would probably result in extinction of cave crickets and the concomitant loss of the major source of fixed carbon energy inputs into caves in central Kentucky and other areas.

3.8.2 Crickets and fungi

Packer [92], in his 1881 review of the cave fauna of North America, includes a sketch of a *H. subterraneus* cricket with fungi growing out of its back end (Plate XVII, ▸ Fig. 3.3 in reference [92]) from infection with a *Cordyceps* fungus, a highly specialized parasite of particular insects.

Finding what looks like a marshmallow with legs is a not uncommon occurrence in caves (▸ Fig. 3.10). The white bloom is caused by the fungus *Beauveria bassiana*. It is an aggressive parasite of many different insect host species and is used in insect pest management [93]. Once eaten by the insect, spores germinate and produce the toxin beauvericin that weakens the host's immune system. The fungal hyphae secrete enzymes that allow it to penetrate and grow into the insect's body. After the insect dies, oosporein antibiotic (seen as pink droplets) is produced that enables the fungus to outcompete intestinal bacteria. Eventually, the entire body cavity is filled with fungal hyphae. When conditions are favorable (e.g. relative humidity >92%), the fungus will grow out through the softer parts of the insect's body, resulting in the characteristic

Fig. 3.10. Cave cricket "marshmallow" in the Frozen Niagara Entrance of Mammoth Cave. The cricket has been killed by the growth of a parasitic fungus. © Kenneth Ingham. Reproduced by permission of Kenneth Ingham. Permission to reuse must be obtained from the rights holder.

white bloom appearance. These external hyphae produce conidial spores that ripen and are released into the environment to infect the next insect on contact, thereby completing the cycle.

The mycoflora of parthenogenic *Hadenoecus cumberlandicus* cave crickets from caves in Carter Caves State Resort Park in Carter County, Kentucky, in the Bluegrass region of the Interior Low Plateaus [8] show mostly saprophytic fungi with two crickets carrying *Beauveria* [94]. Yoder et al. [95] reported finding *Beauveria* on a *Meta ovalis* cave spider. *Meta* were not susceptible to infection by the fungus, but could transfer the infection to the crickets.

3.8.3 Cricket parasites

Crickets are parasitized by horsehair worms. The worm eggs are taken up as the cricket drinks from contaminated pools. The juvenile worm leaves the digestive tract and enters the body cavity of the cricket. The worm grows to adult size and bursts through the side of the cricket to drop into water pools below the cricket roost. The worms mate and lay eggs in the pool to complete their life cycle. Studier et al. [86] found an infection rate of 9.6% among *Ceuthophilus stygius* camel crickets and 0.5% in *H. subterraneus* cave crickets from several locations within MCNP. Infected *C. stygius* showed a reduction in egg production from an average of 34 eggs per adult female to an average of 2.2 eggs per parasitized female (7 of the 9 had no eggs). The difference is because *C. stygius* must drink water so is more likely to ingest eggs, and *H. subterraneus* gets nearly all of their water from their food.

3.9 Conservation of Microbes

Because we do not usually see microorganisms, conservation of microbes and their habitats have not received much attention. Microbes are clearly impacted by human activity. Saiz-Jimenez [29] studied impacts of mass tourism on show caves in Spain that contain significant archaeological remains including paintings. Visitation has led to a deterioration of the cave environment and the paintings and areas remote from entrances. He stresses the importance of understanding colonization patterns and dispersal mechanisms, and potential effects on human health when studying microorganisms in caves.

Northup [97] has advocated setting aside areas in caves as Microbial Preserves; one advantage is that they would not have to be very large. Humans continually shed hair, skin cells, and lint from clothing, and drop crumbs of food, which are all great food resources for microbes. Cavers are advised to pack out all wastes and to practice clean caving to limit human impact. To reduce collateral damage to microbes, higher organisms, and formations, consideration of microbes should be a factor when choosing cleaning methods and techniques to remove contamination [98]. Reducing the addition of organic carbon is also important in the low-nutrient cave environment to prevent overgrowth of non-native microbes. Boston et al. [98] and Northup [97] have offered suggestions on how to preserve native cave microbes while removing or reducing contaminants, and how to tell the difference between native and contaminating microbes.

There are currently no microbial preserves anywhere in MCNP, but the cave is protected because of restricted access to certain locations beyond tourist trails and control of researchers through an application process.

3.10 Conclusions

Mammoth Cave may be grand and gloomy, but it is not at all peculiar and is a very typical karst cave in terms of its microbiology. Microorganisms are critically important components of every ecosystem, including caves, but we still have much to learn about their activities. Geomicrobiological effects on cave formation, enlargement, and mineral precipitation have clear microbial contributions [5–7]. New techniques have caused great changes in our understanding of microbial ecology and diversity from Mammoth Cave and other caves in the region. Despite the great size and complexity of the Mammoth Cave system, surprisingly little microbiological research has been conducted in the system. There are many opportunities for future study in these caves, particularly associated with fungal microbiology and WNS, animal-microbe interactions, and the geomicrobiology.

Caves do not exist in isolation from the surface [99]. Caves, mineral formations, archaeological resources, organisms, and microorganisms all can be damaged by direct visitation and any surface activity that alters quality and quantity of inputs of water, nutrients, and air exchange. Comparisons of bacterial phylogeny are showing the presence of an indigenous subsurface microbiota from limestone and lava caves around the world. Many of the bacteria identified are unique, and may represent cave-adapted microorganisms. There is still much to be learned about microorganisms in caves in the Mammoth Cave Region.

Acknowledgments

Thanks are extended to the many colleagues and friends for the evolution and revolution in our understanding of microorganisms in caves, especially Tom Poulson, Diana Northup, the late Gene Studier, Penny Boston, Louise Hose, and Kurt Helf. I am always appreciative of the many students, colleagues, and family who have participated in field work over the years, especially Elizabeth Lavoie and Jim Lavoie. I thank Art Palmer for use of ▸ Fig. 3.1, Kenneth Ingham for the marshmallow, Diana

Northup for image processing, and Thomas Lavoie for the indicated photos and image processing. Support from the University of Michigan-Flint, the National Park Service, and the State University of New York College at Plattsburgh has been greatly appreciated. Thanks are also extended to the editor and an anonymous reviewer for their insightful edits, questions, and comments.

References

[1] Brucker RW, Watson RA. The Longest Cave. 2nd ed; first published in 1976. Southern Illinois University Press, 1987.
[2] Mammoth Cave reaches 400 miles in 2013. Accessed 13 November 2013 at www.nps.gov/maca/parknews/mammoth-cave-400-miles.htm.
[3] Quick P. The Fisher Ridge Cave system. In: Palmer AN, Palmer MV, eds. Caves and Karst of the USA. Huntsville, Alabama, National Speleological Society, Inc., 2009, 114–115.
[4] NSS Geo2 Long & Deep Caves. Accessed 13 November 2014 at www.caverbob.com.
[5] Culver DC, Sket B. Hotspots of subterranean biodiversity in caves and wells. J Cave Karst Stud 2000, 62, 11–17.
[6] Poulson TL. The Mammoth Cave ecosystem, In: Camacho A, ed. The Natural History of Biospeleogy. Madrid, Museo Nacional de Ciencias Nataurales, 1992, 569–611.
[7] Poulson TL, Lavoie KH. The trophic basis of subsurface ecosystems. In: Wilkens H, Culver DC, Humphreys WF, eds. Ecosystems of the World 30: Subterranean Ecosystems. Elsevier, 2000, 231–49.
[8] Palmer AN. The Mammoth Cave region, Kentucky. In: Palmer AN, Palmer MV, eds. Caves and Karst of the USA. Huntsville, Alabama, National Speleological Society, Inc., 2009, 108–113.
[9] Palmer AN. A Geologic Guide to Mammoth Cave National Park. Zephyrus Press, Teaneck, NJ. 1981.
[10] Mammoth Cave National Park. Accessed 25 Nov 2014 at www.nps.gov/maca.
[11] Olson R, Scoggins L, Toomey RS III, Burton J. 2011 vegetation map for Mammoth Cave National Park. In: Trimboli S. (ed.) Mammoth Cave National Park's 10th Research Symposium: Celebrating the Diversity of Research in the Mammoth Cave Region, 2013, 4–8.
[12] International Biosphere Reserve. Accessed 15 Nov 2014 at www.nature.nps.gov/biology/globalconservation/biosphere.cfm
[13] Gold T. The deep, hot biosphere. Proc Natl Acad Sci USA 1992, 89, 6045–9.
[14] Boston PJ, Ivanov MV, McKay CP. On the possibility of chemosynthetic ecosystems in subsurface habitats on Mars. Icarus 1992, 95, 300–8.
[15] Caumartin V. Review of the microbiology of underground environments. Natl Speleol Soc Bull 1963, 25, 1–14.
[16] Stewart EJ. Growing unculturable bacteria. J Bacteriol 2012, 194, 4151.
[17] Barton HA. Introduction to cave microbiology: A review for the non-specialist. J Cave Karst Stud 2006, 68, 43–56.
[18] Boston PJ, Spilde MN, Northup DE, et al. Cave biosignature suites: Microbes, minerals and Mars. Astrobiology J 2001, 1, 57–70.
[19] Pace NR. A molecular view of microbial diversity and the biosphere. Science 1997, 276, 734–40.
[20] Donachie SP, Foster JS, Brown MV. Culture clash: challenging the dogma of microbial diversity. ISME J 2007, 1, 97–102.

[21] Jarrell KF, Walters AD, Bochiwal C, Borgia JM, Dickinson T, Chong JPJ. Major players on the microbial stage: why archaea are important. Microbiol 2011, 157, 919–36.
[22] Baas-Becking LGM. Geobiologie of Inleidingtot de Milieukunde. The Hague, Netherlands, W. P. Van Stockum& Zoon, 1934.
[23] O'Malley MA. 'Everything is everywhere: but the environment selects': ubiquitous distribution and ecological determinism in microbial biogeography. Stud Hist Phil Bio Biomed Sci 2008, 39, 314–25
[24] Northup DE, Melim LA, Spilde MN, et al. Lava cave microbial communities within mats and secondary mineral deposits: Implications for life detection on other planets. Astrobiology 2011, 11, 1–18.
[25] Feldhake DJ. Microbial activity and biomass of limestone caves. Cincinnati: University of Cincinnati, 1986, 204 pp.
[26] Rusterholtz KM, Mallory LM. Density, activity, and diversity of bacteriaindigenous to a karstic aquifer. Microbial Ecol 1994, 28, 79–99.
[27] Pemberton A, Millette J, Barton HA. Comparative study of oligotrophic bacterial species cultivated from Jack Bradley Cave, KY. Proceedings of the 14th International Congress of Speleology, Kalmas, Hellas, 2005, 1, 245–6.
[28] Janssen PH. Identifying the dominant soil bacterial taxa in libraries of 16S rRNA and 16S rRNA genes. Appl Environ Microbiol 2006, 72, 1719–28.
[29] Saiz-Jimenez C. A review: Microbiological and environmental issues in show caves. World J Microbiol Biotech 2012, 28, 2453–64.
[30] Zhang L. Integrated approach for discovery of novel drugs from microbial natural products. In Zhang L, Demain AL, eds. Natural Products. Drug Discovery and Therapeutic Medicine. Totowa, New Jersey, Humana, 2005, 30–55.
[31] Gabriel CR, Northup DE. Microbial Ecology: Caves as an extreme habitat. In: Cheeptham N, ed. Cave Microbiomes: A Novel Resource for Drug Discovery. Springer, 2013, 85–108.
[32] Frisch R, Schaffer R, Haak P, Hacker M, Morgan, R. Identification of three bacteria from Mammoth Cave, Mammoth Cave National Park, Kentucky, USA that are potential anti-angiogenesis, anti-tuberculosis, or anti-microbial producing organisms. Presented at the 103rd Gen Mtg Am Soc Microbiol, Washington D.C., 2003.
[33] Poulson TL, Lavoie K H, Keith JH. Biological desert under the cap rock. Cedar Falls, Iowa, Cave Research Foundation 1985 Ann Rep, 1986, 24–5.
[34] Simon KS, Benfield EF, Macko SA. Food web structure and the role of epilithic biofilms in cave streams. Ecology 2003, 84, 2395–2406.
[35] Poulson TL, Kane TC. How food type determines community organization in caves. Proceedings of the 8th International Congress of Speleology 1981, 1, 56–9.
[36] Byl P, Trimboli S, Toomey R, Byl J, Solomon D. Antibiotic resistance and substrate utilization by bacteria affiliated with cave streams at different levels of Mammoth Cave. In: Trimboli S, ed. Mammoth Cave National Park's 10th Research Symposium: Celebrating the Diversity of Research in the Mammoth Cave Region, 2013, 33–9.
[37] Studier EH. Composition of bodies of cave crickets (*Hadenoecus subterraneus*), their eggs, and their egg predator, *Neaphanop stellkampfii*. Am Midl Nat 1996, 136, 101–9.
[38] Poulson TL, Lavoie KH, Kelf K. Long-term effects of weather on the cricket guano community in Mammoth Cave National Park. Am Midl Nat 1995, 134, 226–36.
[39] Studier EH, Lavoie KH, Wares II WD, Linn JA-M. Bioenergetics of the cave cricket, *Hadenoecus subterraneus*. Comp Biochem Physiol Part A: Physiology 1986, 84, 431–6.
[40] Barr TC, Kuehne RA. Ecological studies in the Mammoth Cave system of Kentucky. II. The ecosystem. Annal Speleol 1971, 26, 4–96
[41] Smith T, Olson R. A taxonomic survey of Lamp Flora (Algae and Cyanobacteria) in electrically lit passages within Mammoth Cave National Park, KY. Int J Speleol 2007, 36, 105–114.

[42] Olson R. Control of lamp flora in developed caves. In: Hildreth-Werker V, Werker JC. eds. Cave Conservation and Restoration. Huntsville, Alabama, National Speleological Society, 2006, 343–348.
[43] Angert ER, Northup DE, Reysenbach A-L, Peek AS, Goebel BM, Pace NR. Molecular phylogenetic analysis of a bacterial community in Sulphur River, Parker Cave, Kentucky. Am Mineralo 1998, 83, 1583–92.
[44] Meiman J, Palmer A. The Pennyroyal Plateau, Kentucky. In: Palmer AN, Palmer MV, eds. Caves and Kart of the USA. National Speleological Society, Inc., Huntsville, AL. 2009, 116–118.
[45] Lisowski E, Olson R, Thompson B. Geochemistry and biology of Sulphur River, Parker Cave, Kentucky. Cave Res Found 1985 Ann Rep, 1985, 85–86.
[46] Thompson DB, Olson R. A preliminary survey of the protozoa and bacteria from Sulphur River in Parker's Cave, KY USA. Natl Speleol Soc Bull 1988, 50, 42–46.
[47] Olson R. Potential effects of hydrogen sulfide and hydrocarbon seeps on Mammoth Cave ecosystems. In: Trimboli S, ed. Mammoth Cave National Park's 10th Research Symposium: Celebrating the Diversity of Research in the Mammoth Cave Region, 2013, 25–31.
[48] Bullitt A. Rambles in the Mammoth Cave during the year 1844 by a visitor. Morton and Griswold, Louisville, KY 1845, 88.
[49] Sasowsky ID, Palmer MV, eds. Breakthroughs in Karst Geomicrobiology and Redox Chemistry. Leesburg, Virginia, Karst Waters Institute Special Publications 1, 1994.
[50] Northup DE. Lavoie KH. Geomicrobiology of caves: A review. Geomicrobiol J 2001, 18, 199–222.
[51] Barton HA, Northup DE. Geomicrobiology in cave environments: Past, current, and future perspectives. J Cave Karst Sci 2007, 69, 163–178.
[52] Engel AS. Microbial diversity of cave ecosystems. Chapter 10. In: Barton LL, Mandi M, Loy A. eds. Geomicrobiology: Molecular and Environmental Perspective. Springer Science+Business Media BV, 2010, 219–258.
[53] Hill CA, Forti P. Cave Minerals of the World. 2nd edn. Huntsville, Alabama, National Speleological Society, 1997.
[54] Barton HA, Taylor NM, Kreate MP, Springer AC, Oehrle SA, Bertog JL. The impact of host rock geochemistry on bacterial community structure in oligotrophic cave environments. Int J Speleol 2007, 36, 93–104.
[55] Jones AA, Bennett PC. Mineral microniches control the diversity of subsurface microbial populations. Geomicrobiol J 2014, 31, 246–261.
[56] Fliermans CB, Schmidt EL. *Nitrobacter* in Mammoth Cave. Int J Speleol 1977, 9, 1–19.
[57] Fliermans CB, Schmidt EL. Autoradiography and immunofluorescence combined for autecological study of single cell activity with *Nitrobacter* as a model system. Appl Microbiol 1978, 30, 676–684.
[58] George AI. Mammoth Cave Saltpeter Works. HMI Press, 2005.
[59] Duncan MS. Examining early nineteenth century saltpeter caves: An archaeological perspective. J Cave Karst Stud 1997, 59, 91–94.
[60] Hill CA. Origin of cave saltpeter. Nat Speleol Soc Bull 1981, 43, 110–126.
[61] Hill CA, Eller PG, Fliermans CB, Hauer PM. Saltpeter conversion and the origin of nitrates. Nat Geogr Soc Res Report 1983, 15, 295–309.
[62] Olson R, Krapac IG. Origin and regeneration of nitrates in Mammoth Cave sediment. J Cave Karst Stud 2001, 63, 114.
[63] Jameson RA, Boyer DG, Alexander EC. Nitrogen isotope analysis of high nitrate and other karst waters and leached sediments at Friar's Hole Cave, West Virginia (abs). In: Sasowsky ID, Palmer MV, eds. Breakthroughs in Karst Geomicrobiology and Redox Geochemistry, Karst Waters Institute, Special Publication 1, 1994, 1, 36–7.

[64] Focht DD, Verstraete W. Biochemical ecology of nitrification and denitrification. Adv Microb Ecol 1977, 1, 135–214.
[65] Olson R. This Old Cave: The ecological restoration of the historic entrance ecotone of Mammoth Cave, and mitigation of visitor impact. In: Proc Fifth Ann Mammoth Cave Nat Park Sci Conf 1996, 5, 87–95.
[66] White WB, Vito C, Scheetz BE. The mineralogy and trace element chemistry of black manganese oxide deposits from caves. J Cave Karst Stu 2009, 71, 136–143.
[67] Carmichael MJ, Carmichael SK, Santelli CM, Strom A, Bräuer SL. Mn(II)-oxidizing bacteria are abundant and environmentally relevant members of ferromanganese deposits in caves of the upper Tennessee River basin. Geomicrobiol J 2013, 30, 779–800.
[68] Singh BN. A method of estimating the numbers of soil protozoa, especially amoebae, based on their differential feeding on bacteria. Ann Appl Biol 1946, 33, 112–120.
[69] Vanderwolf K, David Malloch JD, McAlpine DF, Forbes GJ. A world review of fungi, yeasts, and slime molds in caves. Int J Spel 2013, 42, 77–96.
[70] Lavoie KH. The influence of biotic and abiotic factors on the successional decomposition of cave rat dung. PhD Dissertation, Univ Illinois Chicago, 1982.
[71] Shapiro J, Pringle A. Anthropogenic influences on the diversity of fungi isolated from caves in Kentucky and Tennessee. Am Midl Nat 2010, 163, 76–86.
[72] White Nose Syndrome. Accessed 24 February 2015 at www.whitenosesyndrome.org.
[73] Puechmaille SJ, Wibbelt G, Korn V, et al. Pan-European distribution of white-nose syndrome fungus (*Geomycesdestructans*) not associated with mass mortality. PLoS ONE 2011, 6, e19167.
[74] Zukal J, Bandouchova H, Bartonicka T, et al. White-Nose Syndrome fungus: A generalist pathogen of hibernating bats. 2014, PLoS ONE, 9, e97224.
[75] Toomey RS, Thomas SC, Gillespie J, Carson V, Trimboli SR. White-nose Syndrome at Mammoth Cave National Park: Actions before and after its detection. In: Trimboli S, ed. Mammoth Cave National Park's 10th Research Symposium: Celebrating the Diversity of Research in the Mammoth Cave Region. 2013, 206–7.
[76] Alves DMCC, Terrible LC, Brito D. The potential impact of White-Nose Syndrome on the conservation status of North American bats. PLoS ONE 2014, 9, e1o7395.
[77] Embry L, Roland V, Painter R, Toomey R, Sharpe Jr L. Quantitative dye tracing: Development of a new interpretive method. In: Proc 22nd TN Water Res Symp. Burns, TN. 2012, 1C-6–16.
[78] Chapman JS. Disinfection resistance mechanisms, cross-resistance, and co-resistance. Int Biodeterior Biodegr 2003, 51, 271–276.
[79] Algeo K. The Mammoth Cave Mushroom Company: A brief history of a short-lived venture. In: Trimboli S. ed. Mammoth Cave National Park's 10th Research Symposium: Celebrating the Diversity of Research in the Mammoth Cave Region. 2013, 133.
[80] CDC. Achievements in public health, 1900–1999: Control of infectious diseases. Morb Mortal Weekly Rep 1999, 48, 621–629.
[81] Sucre R. The Great White Plaque: The culture of death and the TB Sanatorium. Accessed December 6, 2013 at www.faculty.virginia.edu/blueridgesanatorium/death.htm
[82] NPS History. Accessed February 26, 2015 at www.nps.gov/maca/historyculture/physicianhealthyself.htm
[83] Fry GF, Watson PJ. Ovum and parasite examination of Salts Cave human paleofeces. In: Watson PJ, ed. Archeology of the Mammoth Cave Area. Academic Press. 1974, 61.
[84] Faulkner CT, Patton S, Strawdridge-Johnson S. Prehistoric parasitism in Tennessee: Evidence from the analysis of desiccated fecal material collected from Big Bone Cave, Van Buren County, Tennessee. J Parasitol 1989, 75, 461–463.

[85] Streater S. Threat down below: Polluted caves endanger water supplies, wildlife. Env Health News 2006. Accessed 12 January 2015 at www.environmentalhealthnews.org/ehs/news/polluted-caves

[86] American Cave Conservation Association. Accessed December 1, 2014 at http://caveconservation.com/Projects.html

[87] Lavoie KH. The effects of urine deposition on microbes in cave soils. In: Pate DL, ed. Proc 1993 Nat Cave Mgt Symp, Carlsbad, New Mexico, 1995, 302–11.

[88] Lavoie KH, Northup DE. Bacteria as indictors of human impact in caves. Proc 2005 Nat Cave Karst Mgt Symp 30th Anniversary Issue. 2006, 119–24.

[89] Jablonsky PL. Lint in caves. Proc 1993 Nat Cave Mgt Symp, Carlsbad, New Mexico, 1995, 73–81.

[90] Studier EH, Lavoie KH. Biology of cave crickets, *Hadenoecus subterraneus*, and camel crickets, *Ceuthophilus stygius* (Insecta: Orthoptera): Metabolism and water economies related to size and temperature. Comp Biochem Physiol A: Physiol 1990, 95, 15–161.

[91] White CR. 1989. Digestive enzymes of the cave cricket, *Hadenoecus subterraneus*. In: Lindsey K, ed. 1988 Cave Res Found Ann Rep. 1989, 61–63.

[92] Packer AS. The cave fauna of North America with remarks on the anatomy of the brains and origin of blind species. Mem Nat Acad Sci 1881, IV(1), 3–156.

[93] Glare TR, Milner RJ. Ecology of entomopathogenic fungi. In: Aroroa DK, Ajello L, Mukerji KG, eds. Handbook of Applied Mycology. Human, Animals, and Insects. CRC Press, 1991, 2, 613–64.

[94] Benoit JB, Yoder JA, Zettler, Hobbs HH III. Mycoflora of a trogloxenic cave cricket, *Hadenoecus cumberlandicus* (Orthoptera: Rhaphidophoridae), from two small caves in northeastern Kentucky. Annal Entomo Soc Am 2004, 97(5), 989–93.

[95] Yoder JA, Benoit JB, Christensen BS, Croxall TJ, Hobbs III HH. Entomopathogenic fungi carried by the cave orb weaver spider, *Meta ovalis* (Araneae, Tetragnathidae), with implications for mycoflora transfer to cave crickets. J Cave Karst Stud 2009, 71, 116–20.

[96] Studier EH, Lavoie KH, Chandler CM. Biology of cave crickets, *Hadenoecus subterraneus*, and camel crickets, *Ceuthophilus stygius* (Rhaphidaphoriae); Parasitism by hairworms. J Helminth Soc Washington 1991, 58, 248–50.

[97] Northup DE. Managing microbial communities in caves. In: van Beynen PE, ed. Karst Management. Berlin, Springer, 2011, 225–240.

[98] Boston PJ, Northup DE, Lavoie KH. Protecting microbial habitats: Preserving the unseen. In: Hildreth-Werker V, Werker JC, eds. Cave Conservation and Restoration. Nat Speleol Soc 2006, 66–83.

[99] Jones WK, William K, Hobbs HH III, et al. Recommendations and Guidelines for Managing Caves on Protected Lands. Special Publication 8, Leesburg, Virginia, Karst Waters Institute, 2003.

Hazel A. Barton

4 Starving Artists: Bacterial Oligotrophic Heterotrophy in Caves

Abstract: Past microbiology in caves often focused on environments that contain significant energy input, such as detritus from trogloxene activity or from geochemical inputs support autotrophic activity. In the 1990s, the discovery of Lechuguilla Cave (USA) and interest in its unusual speleothems provided microscopic evidence of microbial life from the most nutrient and energy-poor sites, although many of the microbial species identified remained unculturable. By combining the tools of modern molecular microbiology, geochemistry and geology with more sophisticated cultivation approaches, we can now probe many of the geochemical activities and physiological adaptations that allow microbial species to survive within nutrient-limited cave environments.

4.1 Introduction

Past research on the biology of caves has often focused on environments that contain significant energy input, either detritus from trogloxene activity and flooding events [1–3], or from geochemical inputs that drive autotrophic activity [4–6]. Outside these zones, cave systems were presumed to be sparsely populated with surface-derived microbial life transported into the cave system either by air movement, percolating water, or human activity [7–10]. In the 1990s, the discovery of Lechuguilla Cave, New Mexico (USA) and interest in the unusual speleothems and minerals of the cave, provided microscopic evidence that there was microbial life in mineral samples from even in the most nutrient and energy-poor sites [11–13]. Rather than contaminants, these microorganisms appeared to be residents that alter the chemical and structural geology of their surroundings [13, 14]. Nonetheless, many of the microbial species identified within such cave sites either remained unculturable or were insufficiently numerous to be studied using traditional techniques [14, 15]. Today, with advances in molecular biology and DNA extraction, the emergence of high-throughput DNA sequencing, and our ability to assemble and read microbial genomes rapidly, it is now possible to identify and characterize these microbial communities more properly [12, 16–25]. By combining the tools of modern molecular microbiology, geochemistry, and geology with more sophisticated cultivation approaches, we can probe many of the geochemical activities and physiological adaptations that allow microbial species to survive within cave environments, and firmly establish cave geomicrobiology as an emergent and exciting discipline within microbiology [26–28].

4.1.1 Oligotrophy

The majority of microbiologists receive their training on the growth and physiological characteristics of human pathogens and their near-relatives, including such genera as *Escherichia*, *Staphylococcus*, or *Pseudomonas*. Evolving in the kill-or-be-killed environment of the human body, where nutrients are plentiful and success is dependent upon the ability to grow as quickly as possible, these bacteria have little in common physiologically with the slow-growing species found in nutrient-limited caves [29]. Oligotrophy is operationally defined as any environment containing < 2 mg/L of total organic carbon (TOC), with oligotrophs variously described as any organism that has adapted to grow under such nutrient limitation [29]. Dark and geologically isolated from the surface, caves receive varying allochthonous input of dissolved organic carbon (DOC) that percolates in from the photosynthetic and soil-derived ecosystems of the surface. The amount and type of available DOC entering the cave depends on the cave depth and its relationship to the surface, both in terms of geology and regional hydrology [30, 31]. A vast majority of caves thus qualify as oligotrophic and their endemic microbial populations qualify as oligotrophs. This chapter will focus on the oligotrophic heterotrophic bacteria found in caves, with metabolisms that depend on surface-derived allochthonous organic carbon that leaches into the cave system. The reader is directed to Chapter 6 in this book for more information about nitrogen-based chemolithoautotrophic microorganisms, to Chapter 7 for manganese and iron chemolithoautotrophic processes, and Chapters 9 and 10 for sulfur-based chemolithoautotrophic metabolism within cave environments.

4.2 To Grow or Not to Grow

Prior to the advent of molecular techniques, microbiology in caves relied on cultivation and/or microscopy [8, 11, 32–34]. Studies that used general, nontargeted cultivation methods suggested the dominance of the Proteobacteria, Actinobacteria, and Firmicutes in these environments, represented by members of the genera *Pseudomonas*, *Arthrobacter*, *Streptomyces*, *Micrococcus*, *Nocardia*, and *Bacillus* [8, 32–35]. Such species are now recognized as culture "weeds": fast-growing and well-adapted to laboratory culture conditions. These generalists often dominate agar plates and obscure the growth of rarer or seldom identified species [36, 37]. These species are also well suited to make the transition from the nutrient-poor conditions of the cave environment to the nutrient-rich conditions of an agar plate, through flexible metabolism (e.g. pseudomonads) or the production of spores (e.g. *Bacillus* and *Streptomyces*). Such transitions are generally lethal to oligotrophic species [29, 36, 38]. The fact that these genera share significant similarity with those isolated from soils reinforced the notion in earlier studies that microbial cave communities could be populated by contaminating species from the surface [9]. Modern molecular microbiology, fluorescent staining, and cultivation techniques demonstrate the bias of cultivation toward "weed" organisms. We now recognize that in virtually all environments, the vast majority of microorganisms remain unculturable and traditional cultivation approaches in caves provide limited informational gain [39, 40].

4.3 The Culture-independent View of Heterotrophy in Caves

The first culture-independent, molecular study of an oligotrophic cave environment was carried out by Schabereiter-Gurtner et al. [41] in Tito Cave, Spain. This was not the first molecular investigation of a microbial community in a cave, but the preceding studies examined chemolithoautotrophic communities and will not be discussed in the scope of this review [4, 42]. The Tito Cave study examined the microbial biofilms threatening the Paleolithic paintings of the cave by using denaturing gradient gel electrophoresis (DGGE) of the bacterial 16S ribosomal RNA gene sequence (16S rDNA) to demonstrate the presence of a more diverse microbial ecosystem than had been suggested by cultivation [8, 33, 34, 41, 43]. These investigators demonstrated that, in addition to the previously recognized Proteobacteria, Actinobacteria, and Firmicutes, Tito Cave also contained members of the generally anaerobic Bacteroidetes, the phototrophic Chloroflexi (at the time referred to as green nonsulfur bacteria), the structurally and metabolically unusual Planctomycetes (at the time called the BD group), and the enigmatic Acidobacteria (▸ Fig. 4.1) [41]. In rapid succession, other molecular investigations followed, and samples were collected from caves throughout North America, Europe, and Asia, including those from oligotrophic surfaces and biofilms, iron-rich precipitates, sediments, calcite formations, and wall colonies (▸ Fig. 4.1) [12, 16–18, 22–24, 41, 44–46].

These studies from oligotrophic cave environments revealed a surprisingly broad but consistent range of phyla, in addition to the functionally well-known Proteobacteria, Actinobacteria, Firmicutes, and Bacteroidetes. Caves also routinely contain representatives of the Chloroflexi, Planctomycetes, Acidobacteria, Verrucomicrobia, Gemmatimonadetes, and Nitrospira (▸ Fig. 4.1) [47–50]. The similarity in community structure across geographical and geochemical cave environments led to the suggestion of a core set of carbonate cave phyla representative of metabolisms best suited to the physiochemical and nutritional conditions of the cave [23]. The drivers of this core composition can be hinted at when comparing cave environments to oligotrophic soils and noncarbonate caves (e.g. quartzite), or environments with microbial communities that display structural similarity yet contain no overtly similar physiochemical conditions, such as spacecraft assembly facilities [51, 52]. Such comparisons suggest that aphotic conditions, pH, and the presence and type of organic carbon all play an important role in shaping microbial community structure in oligotrophic cave environments.

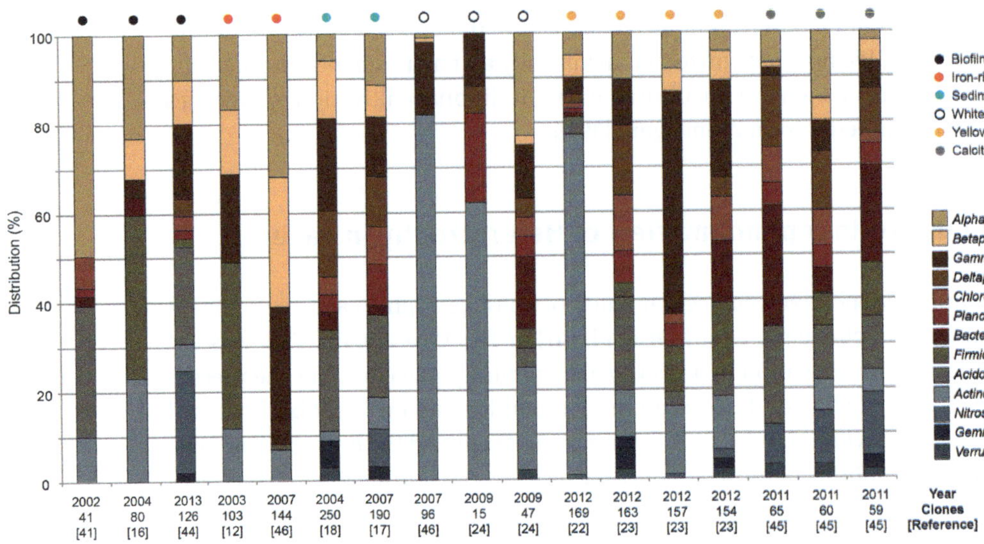

Fig. 4.1. The compiled distribution of bacterial phylotypes identified in caves (at the time of writing). The studies used were selected as those that target carbonate cave systems in zones that would be considered oligotrophic (no direct input of known energy source), indicated by the relevant reference number. The studies were further constrained by those that used the same analytical method – PCR amplification of the 16S rDNA gene sequence followed by cloning and sequencing (deep sequencing (454/Illumina) methods were not included). The phylogenetic profiles only include phyla that represented >1% of the community profile and are grouped based on the type of environment sampled (biofilms, sediments, etc.).

4.4 Diversity of Oligotrophic Microbes in Caves

4.4.1 Old friends: The Proteobacteria, Actinobacteria, Firmicutes, and Bacteroidetes

The most dominant phyla in oligotrophic cave environments are the Proteobacteria, Actinobacteria, Firmicutes, and Bacteroidetes (▸ Fig. 4.1). The Proteobacteria are primarily represented by members of the Alphaproteobacteria, Betaproteobacteria, Gammaproteobacteria, and Deltaproteobacteria (▸ Fig. 4.1), whereas members of the Epsilonproteobacteria are absent, likely due to their association with sulfidic habitats [53]. The samples that demonstrate the highest relative abundance (> 60%) of the Proteobacteria are often associated with some of the deepest and most nutrient-limited caves examined (▸ Fig. 4.1) [12, 46], and their role may be hinted at by their representation. Both phylogenetic and cultivation libraries indicate that members of the genera *Bosea*, *Bradyrhizobium*, *Burkholderia*, *Cupriavidus*, *Devosia*, *Methylobacter*, *Methylococcus*, *Mesorhizobium*, *Ochrobactrum*, *Rhizobium*, and *Sinorhizobium* are present, all of which are known diazotrophs with nitrogen-fixing capabilities [54].

To date, there are very few studies on the nitrogen cycle in cave environments and none have examined whether *in situ* nitrogen fixation occurs in deep caves [12, 55–57]. Given the absence of a geologic nitrogen source, it is hard to imagine that this nutrient is not a primary driver of microbial community structure in caves [58]. Yet nitrogen fixation is energetically costly and makes the dominance of nitrogen-fixing species in oligotrophic cave environments surprising. Even so, all representative members of the core cave phyla contain species that encode the nitrogenase gene, *nifH* (▸ Fig. 4.1) [54]. Although the energy necessary for nitrogen fixation could be provided by chemolithoautotrophic activity, and it may be possible for altruistic chemolithoautotrophs to power nitrogen fixation within a microbial community, it seems unlikely because most chemolithoautotrophic communities in caves demonstrate their own nitrogen-fixing capabilities [20, 55]. A high relative proportion of nitrogen-fixing species within cave environments may, therefore, suggest that successful heterotrophic growth in caves, even under severe oligotrophy, may be tied to nitrogen-fixing capabilities.

The order Actinomycetales (within the Actinobacteria), Firmicutes, and Bacteroidetes are among the best-studied heterotrophs outside human pathogens. Each is a generalist with a broad environmental distribution and impressive arsenal of biodegrading enzymes [59–61]. While the actinobacteria are best known for their secondary metabolic products, and the Firmicutes for their fermentative growth, all are able to degrade some of the most complex and recalcitrant polymers on earth, including cellulose, pectin, xylan, chitin, and lignin [59, 60, 62]. This capability gives these phyla an important role in the breakdown of macromolecular plant, animal, and even fungal matter entering the cave. In terms of successional carbon breakdown, these phyla may carry out principal heterotrophic breakdown that releases sugars and other monomers from these complex macromolecules to fuel community growth [19, 22–24, 41, 54].

Within caves, the Actinomycetales and Firmicutes appear to be the yin and yang of heterotrophy. When actinobacteria dominate, the Firmicutes are in the minority, and vice versa (▸ Fig. 4.1). A similar relationship is seen in the human intestine between the Firmicutes and Bacteroidetes: when fed a plant-rich diet, our intestinal flora is dominated by the Bacteroidetes, while a protein-rich diet leads to the dominance of the Firmicutes, with consequences for energy acquisition and digestive health [63, 64]. Within caves, the relationship between the Actinomycetales and Firmicutes appears to be dictated by depth. The Actinomycetales dominate in near-surface environments, but Firmicutes dominate deeper systems (▸ Fig. 4.1). Before placing a strong emphasis on this inverse correlation, it is important to note that these studies (represented in ▸ Fig. 4.1) vary in a number of factors, including sequencing depth (▸ Fig. 4.2), whether the studies targeted metabolically active cells, the availability of oxygen in the microenvironment, and the reliance on PCR amplification rather than direct counts [12, 16–18, 23, 24, 41, 44–46].

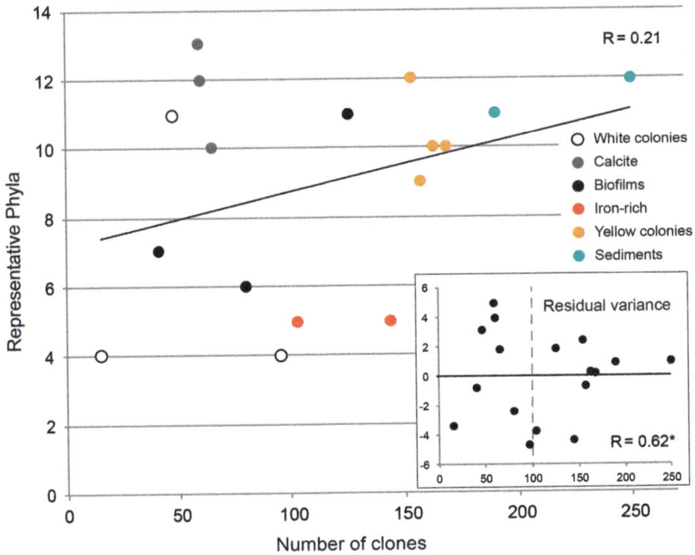

Fig. 4.2. Impact of sequencing depth on community profile structure. The 17 phylogenetic studies examined varied widely in the number of clones sequenced (15–250 clones). To determine the impact that clones number has on the number of phyla identified, a plot was created of these two variables (a). The Spearman rank coefficient demonstrated a poor correlation between library size and number of identified phyla ($R = 0.21$; 16 *df*); however, these data had a large variance around the expected values [E(*X*)] for small libraries [< 100 clones (b)]. If smaller data sets are removed from these analyses, then the correlation between library size and identified phyla increases dramatically ($R = 0.62$, $P < 0.02$; 10 *df*, data not shown). These data suggested that a deeper phylogenetic analysis of these environments would reveal a much broader diversity than the current studies show, as demonstrated by Ortiz et al. [19]. Nonetheless, even in relatively small clone libraries, the representative core phyla could clearly be observed (► Fig. 4.1). The environmental location of each library is shown (colored circles).

The Actinomycetales and Firmicutes are known to sporulate upon nutrient limitation, which could lead to the accumulation of spores rather than metabolically active cells within these environments. To determine whether the detection of species using molecular techniques correlates with metabolic activity requires the isolation of ribosomal RNA (rRNA), which is rapidly degraded outside of a living host cell. Although very few studies have attempted to identify the bacterial populations in caves using rRNA, due to the difficulty in extracting RNA from low biomass, geologic samples [65], Stomeo et al. [25] examined actinobacterial colonies in a cave using this approach. Their work demonstrated that, in these samples at least, the actinobacteria present are indeed metabolically active. Whether an oxygen depleted microenvironment could be responsible for the apparent dominance of the Firmicutes is ruled out by their common association with the family Flavobacteriaceae (Bacteroidetes), which are strict aerobes [60], and may be attributed to a more degraded or reduced level of organic carbon. Although little more than a superficial assessment, comparing the results of these cave studies to date (▸ Fig. 4.1) suggests that untangling the physical and geochemical relationships that drive the dominance of the Proteobacteria, Actinomycetales, Firmicutes, and Bacteroidetes may require a better understanding of nitrogen acquisition, the amount and type of organic carbon for growth, and even how competition for available resources drives the relative abundance of these generalists [66].

4.4.2 ... and new: The Planctomycetes, Chloroflexi, Acidobacteria, and Verrucomicrobia

Justifying a role for well-described phyla in cave environments is relatively simple. Justifying the roles for less well understood taxa is much more challenging, such as for Planctomycetes, Chloroflexi, Acidobacteria, Nitrospirae, Gemmatimonadetes, and Verrucomicrobia. This difficultly is primarily due to a dearth of cultured representatives, with less than 0.3% of the >220,000 bacterial cultures identified in the Ribosomal Database Project representing these six phyla combined (http://rdp.cme.msu.edu). Nonetheless, environmental sequence analysis suggests that the Planctomycetes, Chloroflexi, and Acidobacteria have a phylogenetic (and presumably metabolic) diversity as broad as that observed in either the Proteobacteria or Actinomycetales. Although any predictive statements about the environmental physiology of these phyla made here remains presumptive at best, their repeated identification in carbonate caves suggests that they contain an exploitable niche consistent with the metabolic capabilities of these organisms [48, 67–71].

The Planctomycetes belong to the Planctomycetes–Verrucomicrobia–Chlamydiae (PVC) superphylum, along with Lentisphaerae, and the Candidate Divisions Poribacteria, OP3, and WWE2 [72]. To date, no Chlamydia, Poribacteria, or Lentisphaerae have been identified in caves (presumably due to the absence of suitable hosts or environmental conditions [72]), but members of the Planctomycetes, Verrucomicrobia, and Candidate Division OP3 appear to be common (although the OP3 appear to be restricted to aquatic cave environments) [20, 21]. For brevity, I will consider members of the PVC superphylum (including the Verrucomicrobia) as represented by members of the Planctomycetes. With dramatically diverse taxonomic (and metabolic) affiliations for planctomycete sequences isolated from cave environments, I created a phylogenetic tree of all the 16S rRNA gene sequences from caves in the NCBI database identified as Planctomycetes in the original studies (▸ Fig. 4.1) to understand their function in these environments (▸ Fig. 4.3), although it would be advantageous to carry out such an exhaustive phylogenetic analysis of the 16S rDNA sequences for each of the lesser well known core phyla.

In the past decade, there has been a surge of interest in the Planctomycetes due to their identification as the "missing" chemolithoautotroph, carrying out the anaerobic oxidation of ammonia (ANAMMOX) to N_2 gas [47]. Yet, the uniqueness of this phy-

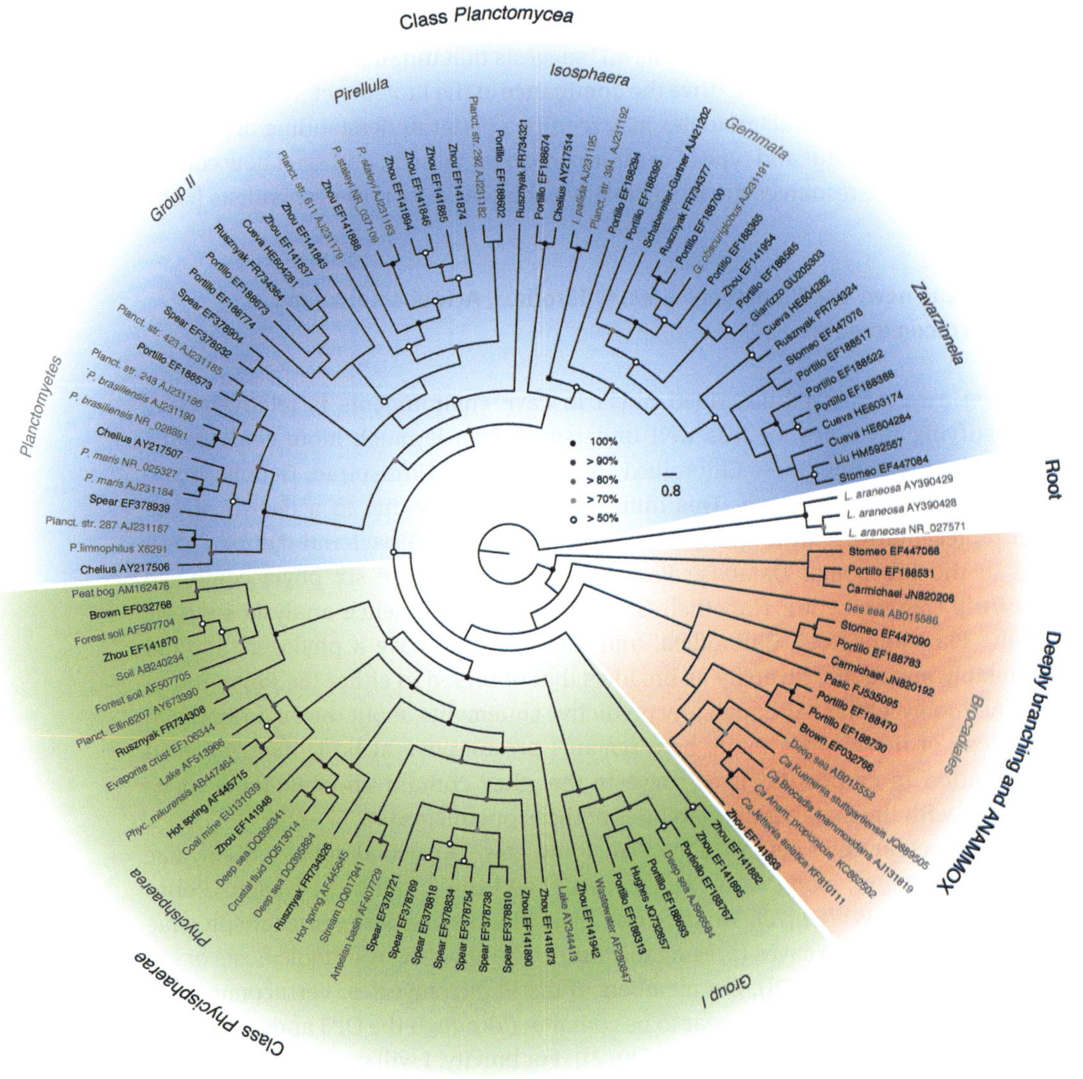

Fig. 4.3. Phylogeny of cave Planctomycetes. All 16S rRNA gene sequences identified as Planctomycetales from cave studies in the NCBI Genebank database (www.ncbi.nlm.nih.gov) and amended with curated Planctomycetes sequences from the Ribosomal Database Project (http://rdp.cme.msu.edu). All the sequences were aligned using MUSCLE version 3.7 on the CIPRES Portal (www.phylo.org) and trimmed in Mesquite version 2.75. Maximum likelihood (ML) trees were generated using RaxML on the CIPRES portal using RAxML-HPC BlackBox version 7.27. The best ML tree is shown, rooted with *Lentisphaerae araneosa*. The robustness of each branching topography was calculated by averaging 1000 bootstrapping analyses, with the percentage of branch topologies fitting the observed pattern are shown (circles).

lum is not restricted to chemolithoautotrophy [69]. Members of the Planctomycetes are unusual for a number of features, including the absence of peptidoglycan, the use of budding for cell division, and the presence of intracellular organelles, such as a membrane-bound compartment for DNA and the pirellulosome, both of which have profound implications for the evolution of eukaryotes [47]. Their proteinaceous cell wall also makes them resistant to a number of antibiotics and the hydrolytic enzymes produced by predatory species [47, 69], which would certainly give them an advantage in competitive microbial environments. They also use other unusual metabolic strategies, with members of the genus *Gemmata* spp. feeding on environmental proteins using endocytosis, a process that is otherwise only seen in eukaryotes [47, 73]

Although the identification of Planctomycetes in caves might lead microbiologists to assume that members of this phylum use ANAMMOX chemolithoautotrophy as an adaptive growth strategy, the ANAMMOX physiology is restricted to a few candidate species that are poorly represented within caves (▸ Fig. 4.3). The representative genome of "*Candidatus* Kuenenia stuttgartiensis" reveals that, in addition to genes involved in ANAMMOX, this organisms encodes degradative enzymes that would allow heterotrophic growth, along with pathways for Fe/Mn reduction and even Fe oxidation [47]. This led to the suggestion that, given the appropriate environmental conditions, these bacteria may be able to carry out autotrophic or mixotrophic growth utilizing reduced Fe and Mn. The identification of potential ANAMMOX taxa within iron-rich and ferromanganese deposits of caves certainly supports such a suggestion (▸ Fig. 4.1) [12, 46, 47], as does the absence of an obvious geological source of ammonia for ANAMMOX, along with inhibition of this physiology by carbonates [74].

The majority of representative 16S rDNA gene sequences of the Planctomycetes within caves appear to affiliate with the class Planctomycetia (▸ Fig. 4.3) [69]. Representative clones within the unculturable class Phycisphaerae were retrieved in the Spear et al. [75] study, although these samples were collected within a thermal (52 °C) system and unlikely reflect the majority of planctomycetes in temperate caves (▸ Fig. 4.3). The class Planctomycetia are characterized as obligate oligotrophs, with the growth of isolates such as *Isosphaera pallida* inhibited by the addition of as little as 0.02% of glucose, fructose, or maltose [73]. Indeed, while the use of 0.005% peptone as a sole carbon source is insufficient for the growth of other heterotrophs [73], it can actually enrich for members of the Planctomycetia. In addition, Planctomycetia cultures must be incubated in the dark, which is interesting in the scope of cave microbiology, although aphotic conditions are primarily considered to prevent the overgrowth of autotrophic Cyanobacteria [73].

With regard to the type of organic carbon available for growth, many Planctomycetia have been successfully isolated on the breakdown products of biogenic macromolecules, such as *N*-acetylglucosamine and glucuronate (produced from the hydrolysis of chitin and polysaccharides, respectively) [73]. Nonetheless, genomic screening has revealed the presence of methanopterin-linked enzymes that would allow for growth on C1 compounds like methane, methanol, methylamine, and formaldehyde [76], while the observed accumulation of Fe and Mn oxides during growth opens up the possibility of mixotrophic growth on these poor carbon substrates [47, 73, 77]. Unsurprisingly the observed, often obligate, oligotrophy within Planctomycetia isolates also correlates with slow growth, and even the fastest-growing members of this class have an average doubling time of 10 days [47, 73]. Slow growth on poorly-reduced substrates, in addition to the structural adaptations for competition avoidance and growth, make members of the Planctomycetia well adapted to an extremely oligotrophic niche within caves (► Fig. 4.3).

The Chloroflexi, originally isolated from thermal environments and referred to as green nonsulfur bacteria, are familiar to most microbiologists as anoxygenic phototrophs [68]. But the Chloroflexi are not strict phototrophs and many species quickly switch to heterotrophy in aphotic conditions [68] or lack chlorosomes all together and grow as obligate heterotrophs in a wide variety of habitats [68, 78]. Within caves, 16S rDNA analyses suggest the Chloroflexi are dominated by the Dehalococcoidia, which were originally identified due to their fermentative growth on organohalide compounds [79]. The sequenced genome of *Dehalococcoides* sp. RBG-2 (which shares the closest similarity to cave clones) suggests catabolic pathways for fermentative growth on a variety of halogenated and plant polymers [68, 78], while a complete β-oxidation pathway suggests that fatty acids can also be utilized [78]. The lack of a complete electron transport chain in *Dehalococcoides* sp. RBG-2 constrains this organism to fermentative growth, which appears to be a standard heterotrophic strategy in the Chloroflexi as many cultured members of the *Dehalococcoides* are unable to use an external electron acceptor for growth [78, 79]. Reliance on fermentative growth in a cave with an oxygenated atmosphere would seem an energetically poor choice; however, such an approach could provide a metabolic advantage where fermentation is sufficient to meet the energy demands of slow growth by decoupling the cell from the need to compete for available redox-active compounds [78].

As the name suggests, the Acidobacteria were first cultured from acidic environments [80]. But members of this phylum are now known to be widely distributed across a variety of habitats, including soils and subsurface sediments [48, 81–85]. The broad phylogenetic diversity of the Acidobacteria is grouped into 26 distinct clades or "subgroups" (Gp1–26) [82]. Although little is understood about the function of each subgroup, their relative abundance appears to be dependent on both the pH and type of carbon present [48, 82, 84, 86–88]: Acidobacterial subgroups Gp1 and Gp2 are found in acidic soils (pH 3.5–5.5), subgroups Gp4, Gp6, Gp7, and Gp16 are associated with alkali conditions (pH ~8.5), and the remaining subgroups appear to tolerate a variety of pH conditions [48]. The work of Zimmerman et al. [88–90] demonstrates that the Acidobacteria are notable contributors to microbial community structure in caves, with specific representation by subgroups Gp3–7 and Gp9–11 in carbonate caves, with dominance by members of subgroups Gp4 and Gp10 [89].

Again, the difficulty in assigning a functional role for Acidobacteria comes from the dearth of representative cultures. The majority of cultured Acidobacteria are from the fast-growing Gp1, and only a handful of representative cultures exist for the extremely slow-growing Gp2, Gp3, Gp4, Gp6, Gp8, Gp10, and Gp23 [84, 91–93]. Nevertheless, a combination of 16S rRNA gene screening and genomic approaches suggests that the Acidobacteria are important heterotrophs in oligotrophic environments, utilizing plant polymers and their breakdown products for growth, albeit with unusual catabolic strategies [86, 94, 95]. Of the Acidobacteria that have been cultured from cave-relevant subgroups (Gp4 and Gp10), aerobic heterotrophy appears to be the dominant physiology. Although members of Gp10 have been shown to reduce S^0 to S^{2-}, reduction did not stimulate growth and suggests that sulfur does not serve as an electron acceptor [96].

An interesting aspect of acidobacterial physiology in regard to this review is that Acidobacteria are more dominant in older soils (unamended for decades or longer) where the carbon has been heavily oxidized [48, 81, 86, 87, 97]. Such studies suggest a pattern of ecological succession that may mirror the processes occurring in caves when complex organic carbon, in the form of biopolymers (lignin, xylan, cellulose, starch, chitin, etc.), enters the ecosystem and faster growing heterotrophs dominate. However, as this organic carbon pool becomes increasingly oxidized and poor nutritionally, the populations of Acidobacteria are seen to increase [98–100]. In support of this successional role, the Acidobacteria have been observed to degrade more recalcitrant polymers and hydrocarbon substrates, suggesting that they have a competitive advantage for resources that are more difficult to breakdown and/or catabolize [83, 86, 93, 94, 101].

4.5 Something Wicked This Way Comes – Understanding Carbon in Caves

Based on the varying heterotrophic physiologies found in caves, it is readily apparent that understanding the amount, type, and route of organic carbon entering the system is critical to our understanding of the resultant microbial community dynamics. Soil-derived DOC, primarily in the form of plant polymers (e.g. celluloses, polysaccharides, and lignin) and their breakdown products (i.e. humic and fulvic acids) can enter the cave through the movement of water into the unsaturated zone and underlying groundwater system. This entering DOC can follow a variety of routes through the vadose zone, which is the portion of the subsurface that extends from the surface to directly above the saturated (phreatic) zone [31, 102]. Given the unsaturated nature of the vadose zone, solute transport, and particularly the movement of DOC and nutrients, is controlled by the movement of water and the sorption/desorption of molecules to geochemical surfaces [103]. As a result, even though DOC may be present, it may not be readily available to microbial species. Organic carbon may become limiting as a result of precipitation, co-precipitation, sorption, redox, complexation, and colloidal interactions *en route* to the subsurface [104]. To further complicate DOC transport, different parameters control the movement of water through the vadose zone, including bedrock structure (e.g. sedimentary or igneous), fracturing and preferential flow, porosity and geochemistry [31, 102, 103, 105]. Thus, a shallow (<20 m) cave that is overlain with an impermeable sandstone caprock may be more nutrient limited than a deep (>300 m) cave that routinely floods [21, 106]. Equally, the presence of groundwater in a cave may not necessarily be associated with high DOC, and deep hypogenic caves may contain ancient aquifers that are among the most nutrient-limited environments on Earth [107]. Under such varying scenarios, we measured the amount of DOC in caves from <0.02 to >50.00 mg/L, although the average concentration in the unsaturated zone appears to be around 0.5 mg/L [31, 46, 105]. In caves, it may be possible to relate microbial community dynamics to the geology of the cave, its geochemistry (i.e. related to sorption/desorption), and the route this organic carbon takes into the system. In near-surface caves or caves with numerous paths for preferential flow, microbial communities have greater access to allochthonous, soil-derived organic carbon and should be dominated by heterotrophs. In deeper caves or caves where geologic features restrict the movement of water to a slow, percolating route that allows ex-

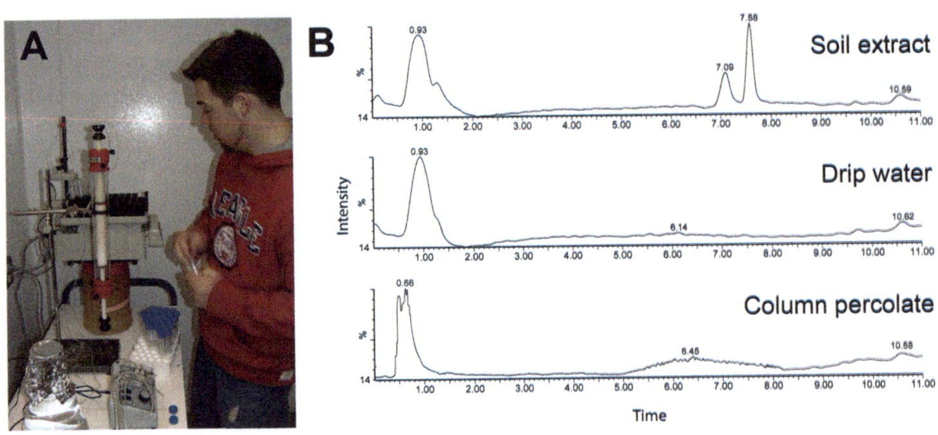

Fig. 4.4. Organic carbon within cave systems. (a) To determine the influence of the bedrock on the structure of organic carbon entering the cave system, we compared the chemistry of cave drip water with an extract of soil (prepared as described in [110]) from above the cave (soil extract) and by running the latter through an FPLC column packed with crushed limestone (column percolate) prior to analysis. (b) These samples were then examined using electron spray mass spectroscopy. The larger molecular weight (mwt) peaks (at 7.09 and 7.58 min) seen in the soil extract are not seen in the drip water, while the drip water and column percolate demonstrated similar organic carbon profiles. ESI/MS and GC/MS analysis of the column percolate demonstrated the presence of organic compounds that were chemically similar to those found in the dripping water. GC/MS analysis revealed that these small organic carbon compounds were comprised of small cyclic and aromatic carbon molecules, indicative of fractionated humic and fulvic acids. Together, these data highlight the way in which organic carbon can be shaped by its geochemical interactions en route into the cave.

tensive oxidation by heterotrophs *en route*, a poorer allochthonous carbon source is available for heterotrophy [108].

As the growth of any organism is determined by its genetic potential and whether it can express the catabolic pathways necessary to utilize available carbon and energy sources for growth, the type of organic carbon entering a cave can have a dramatic impact on microbial community structure [33, 41, 109]. Therefore, when attempting to predict the nature of heterotrophic microorganisms in any particular cave system, it is important to recognize how the geology and hydrology of the cave affects the influx and type of DOC. Accordingly, in addition to measuring DOC levels, an understanding of the nature and structure of the organic carbon (▸ Fig. 4.4) can substantively explain both the types of microorganisms found within the system and the potential methods for cultivation [110–112].

4.6 Whether 'Tis Nobler to Grow

Although genomics can reveal significant information about many uncultured species, such approaches are limited by the fact that the vast majority of environmentally identified genes have no known function [113]. Even in organisms that do contain a high proportion of functionally described genes, there is a disconnect in our understanding of genotype and resultant phenotype in the environment [114, 115]. Therefore, our ability to understand heterotrophic strategies in caves continues to rely on the identification of cultured representatives.

Numerous studies of oligotrophic cave environments have attempted to culture the heterotrophic communities found there, with variable success, and often reidentifying the same "weed" species [36]. To date no culture studies have managed to isolate cultured members of the Planctomycetes, Chloroflexi, Acidobacteria, Nitrospirae, Gemmatimonadetes, or Verrucomicrobia from caves. The lack of culturability in these oligotrophs may be due to a variety of factors within the media, such as the type and amount of carbon used for growth (▸ Fig. 4.5), the culture conditions (e.g. cave temperatures range from 0–20 °C with humidity >95%) and whether the media chemistry reflect the *in situ* cave chemistry. Water in carbonate caves is generally buffered to ~pH 8.3 and contains low total dissolved solids and numerous trace minerals (Fe, Mn, Mo, Ni, etc.). Other micronutrients, such as vitamins and cofactors produced from community turnover, can also be critical in the cultivation of difficult-to-isolate species.

Few studies have attempted to determine how geochemistry influences oligotroph culturability. We examined the impact that carbon sources, geochemistry, and micronutrients had on the culturability of bacteria from a near-surface carbonate cave using 10 different types of media (▸ Fig. 4.5) [116]. In carrying out this study, we assumed that the greatest culturability of cave species would be found on the plates that most closely resembled the geochemistry of the cave. However, we observed broad diversity on one of the simplest media used: distilled water agar (▸ Fig. 4.5). This finding

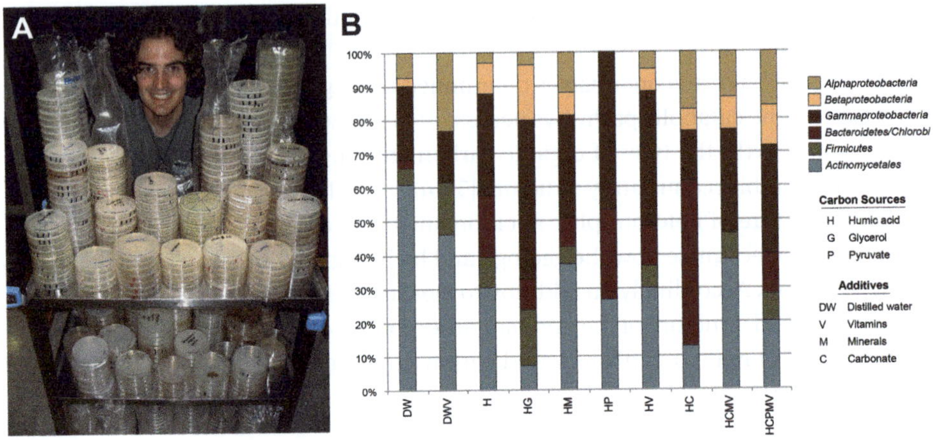

Fig. 4.5. The Jack Bradley Cave, Kentucky (USA), culture library. (a) A portion of the culture collection for a single site within Jack Bradley Cave. (b) Ten different types of media were inoculated from the same sample, generating a culture library of 501 isolates. Each isolate was purified by single colony isolation and identified using the 16S rDNA gene sequence. The relative distribution of cultured phyla identified on each media type is shown [116].

was presumably due to the contaminating organic carbon in our laboratory distilled water (measured at ~2.0 mg/L), which was close to the level of organic carbon found in the cave (0.5 mg/L) [116]. We were also surprised that the addition of calcium carbonate negatively affects the culturability of common species (► Fig. 4.5). Although Ca^{2+} ions are toxic to bacteria, we had assumed that life in a cave would have selected for microbial species more resistant to the effects of this ion [117]. But, this was not the case (► Fig. 4.5). Moreover, even though we also incubated our agar plates for up to two months to isolate the slow-growing strains, screening >500 cultures using 16S rDNA sequencing demonstrated that we were unable to culture representatives from 7 of the 13 core cave phyla (► Fig. 4.5).

Our inability to culture slow-growing oligotrophs in these experiments may reflect the nature of the oligotrophy itself. Microbiologists often assume that slow growth is simply due to a lack of available nutrients, but many obligate oligotrophs have fixed growth rates, as demonstrated by a significantly reduced enzymatic V_{max} rates, that cannot be further stimulated by excess nutrient supplies [38, 118]. Rather, when oligotrophs are exposed to excess nutrients, their cellular catabolic and reductive systems are simply overwhelmed [119, 120]. Without the ability to deal with immediate oxidative stress rapidly, cellular damage accumulates and may lead to cell death [38]. In an attempt to reduce the oxidative stress in our cultivation experiments, we used pyruvate, which can scavenge reactive oxygen species, and glycerol (i.e. a more oxidized carbon source). As should have been expected, the addition of excess carbon decreased diversity (► Fig. 4.5). The only factor that seemed to consistently increase culturability was soil-derived humic acid (▸ Fig. 4.5), which consists of the breakdown products of plant-derived polymers [121]. Despite our poor results, such cultivation attempts are not simply exercises in futility, as even our low diversity of cultured heterotrophs has allowed us to identify some important physiological strategies that would not be readily apparent by molecular approaches alone.

4.7 Out, Out, Brief Candle – Competition and Death in Cave Oligotrophs

In distinguishing the roles of different heterotrophic species in caves, we were surprised by the dominance of Proteobacteria in some of the most nutrient-poor cave environments (▸ Fig. 4.1). Initially, we suspected that such dominance was in some way related to nitrogen fixation, yet the organic carbon necessary to fuel such an energy-intensive process is increasingly scarce in deep cave environments. To investigate this discrepancy, we cultured representatives of the Proteobacteria from a deep site in Lechuguilla Cave and identified them using 16S rRNA gene sequencing. Many of the obtained isolates belonged to the genera *Ensifer* (Alphaproteobacteria) and *Myxococcus* (Deltaproteobacteria). Although these species are indeed known nitrogen fixers, they are better known for their predatory behavior [122]. After examining the cave isolates for a predatory phenotype, in every case they demonstrate active predatory behavior (▸ Fig. 4.6). Expanding this screen to include *Ensifer* strains cultured from caves in Kentucky and South Dakota, we found that all cave isolates demonstrated the same predatory behavior at a pH consistent with cave (pH 8.0) rather than soil (pH 6.0) environments [123]. Such results suggest that when energy or nutrients become scare, these environments harbor a population of *selfish-cheaters*; species that benefit from ecosystem processes without themselves contributing [123, 124]. Predation may be an important strategy for heterotrophic survival in caves, but not all bacteria hunt prey; others may use a more passive approach to selfish competition.

Bacteria produce numerous antimicrobial compounds, but not all of them function as antibiotics. For example, the production of siderophores can sequester iron and inhibit the growth of species that do not possess similar iron-scavenging systems. Thus, while growth inhibition can be observed on a plate, this does not necessarily confirm the presence of antibiotics. For instance, the Actinomycetales are known antibiotic producers and there is no debate as to their ability to either limit or kill competitors. But, the precise role of antibiotics within the environment continues to be debated [123, 124]. The relatively high abundance and culturability of actinobacteria in caves has driven significant interest in the potential for secondary metabolite production by these species. Nonetheless, screening for antibiotic production is a difficult task.

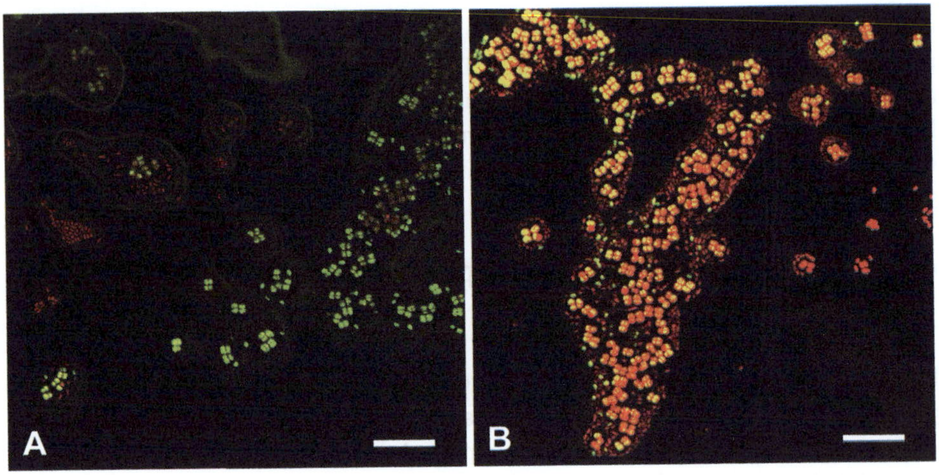

Fig. 4.6. Predation in cave communities. We have identified numerous strains from genera containing predatory species in cave cultures. To determine if these isolates demonstrate a predatory phenotype, we examined the ability of these species, such as this *Ensifer adhaerens* isolated from Lechuguilla Cave, to obtain nutrients by killing the prey species *Micrococcus luteus*. Using the LIVE/DEAD fluorescent stain to examine the fate of each species in a mixed culture, we saw that after 24 h (a) the *Micrococcus* (green tetrads) are surrounded by red *Ensifer* cells (*Ensifer* stain red when they come into contact with the prey, but remain viable throughout the assay); after 72 h (b) the majority of the *Micrococcus* cells stain red with propidium iodide, indicating cell death. The scale bar indicates 10 μm.

The best approach to identify presumptive antibiotic compounds is through analytical chemistry techniques (such as gas chromatography, high-pressure liquid chromatography, etc.), although even with these techniques it is difficult to identify antibiotics against the huge background of primary and secondary metabolites [125, 126]. An easier approach to identify whether antibiotics are present within cave environments is to screen for intrinsic antibiotic resistance. Using a culture library of 93 bacterial isolates from Lechuguilla Cave, we demonstrated resistance to 26 different antibiotic compounds, with the only exception being synthetic and semisynthetic derivatives [110]. While every cave isolate tested was resistant to at least one antibiotic, one *Streptomycete* was shown to be resistant to 14 different antibiotic compounds [110]. Given the wide, phylogenetic range of actinobacteria in caves, the antibiotics we tested possibly reflect a tiny percentage of the compounds that might play a role in interspecies competition. We selected the isolate *Streptomyces* LC38 for exhaustive screening for antimicrobial compounds. This isolate produced >30 different antimicrobial compounds, including a novel macrolide antibiotic, named Lechamycin [127]. Such results, even from a small culture population, indicates that antibiotics are numerous in caves and that, despite oligotrophy, these environments are likely highly competitive.

4.8 Heterotrophic Community Dynamics in Caves – If You Can Look into the Seeds of Time and Say Which Grain Will Grow and Which Will Not...

The traditional view in ecology of extremely nutrient-limited environments is that species diversity is limited through competitive exclusion for the scant resources available [128]. Nonetheless the observation of high diversity under oligotrophy, such as observed from caves, was first described by Hutchinson in 1961 after looking at planktonic lake communities [129]. He referred to his observation of high species diversity, in spite of limited resources as the *paradox of the plankton* [129]. In attempting to describe how such broad diversity can be seen under oligotrophic conditions, microbial ecologists have argued that local dispersion, nutrient flux, or competition account for high species richness, although none of these hypotheses adequately explain the diversity seen in the relative stability of a cave [130–132]. An alternate hypothesis is that such diversity is driven by the formation of novel niche space. Rather than directly competing for the same nutrients, bacterial species can exploit the products of the metabolic breakdown of allochthonous carbon to create discrete populations (▸ Fig. 4.7).

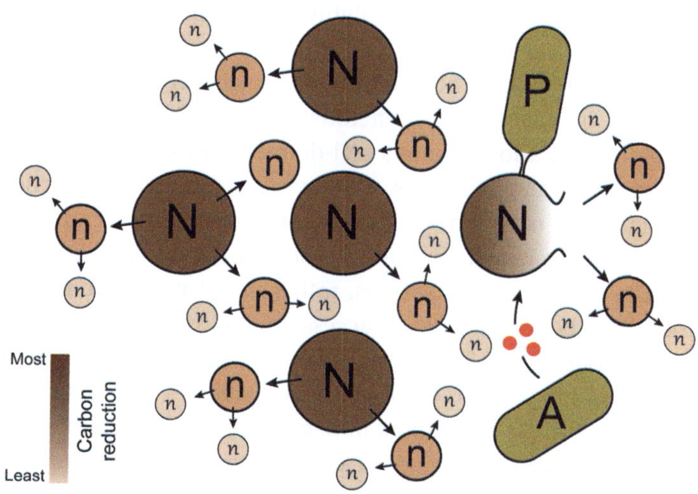

Fig. 4.7. Proposed model of successional carbon usage in caves. Highly reduced organic carbon enters the cave, such as soil-derived polymeric substances, creating a niche for the growth of the principle heterotrophic community (N). As these polymeric substances are oxidized into simpler compounds, such as monomeric sugars, alcohols, aldehydes, organohalides, and small cyclic molecules, they create a niche for the successional heterotrophs (n and n). Eventually the organic carbon pool becomes sufficiently poor that it only supports the growth of strict oligotrophs (n). The activities of selfish-cheaters, such as predatory bacteria and antibiotic producers, short-circuit this successional carbon utilization and release nutrients directly into the environment. The relative quality of the available carbon (brown), selfish-cheaters (green-gray), predatory bacteria (P), and antibiotic producers (A) are shown.

This strategy provides a metabolic parallel to *r*/*k* selection in animal ecology: fast growing and competitive (*r*-type) *principle heterotrophs* metabolize allochthonous organic carbon directly, while slow-growing oligotrophs serve as (*k*-type) *successional heterotrophs*, surviving on the breakdown products and poorly reduced carbon of this principle metabolic pathways (▸ Fig. 4.7) [133, 134]. Evidence for this two-tier metabolic approach has been seen in numerous bacterial populations and can be seen in caves, both in the structure of the core phyla and through cultivation studies (▸ Fig. 4.1 and 4.7).

The principle heterotrophs within caves include species within the Proteobacteria, Actinomycetales, Firmicutes, and Bacteroidetes – the weeds. Frequently cultivated on a variety of rich carbon sources, these fast-growing heterotrophs are able to directly outcompete other species for organic carbon and nutrients, either by more efficient scavenging mechanisms or maximizing growth and simply dominating other species by nutrient exclusion [110]. Yet the dominance of these principle populations also makes them vulnerable to attack, either *via* the release of antimicrobial products or outright predation [110]. The observed dominance of the Actinomycetales in near-surface caves, including the bloom of obvious white, yellow, and gray colonies associated with seeping surface water, suggests that these principle heterotrophs are well suited for the breakdown of this plant-derived, highly reduced polymeric carbon pool [23–25].

The catabolism of complex organic carbon sources, especially in rapidly growing organisms, is surprisingly inefficient and numerous metabolic byproducts are generated [119, 120]. In the case of allochthonous soil-derived carbon, this can include the release of monomeric sugars, alcohols, aldehydes, organohalides, cyclic and aromatic organic molecules, all of which have been shown to support the growth of successional heterotrophs like the Planctomycetes, Chloroflexi, Acidobacteria, and Verrucomicrobia [47, 62]. These successional species, rather than trying to compete for resources, use alternate metabolic strategies to access a more oxidized and less-desirable organic carbon pool (▸ Fig. 4.7). The slow-growth of these species prevents the generation of large pools of offspring to be targeted by predators, even as they express alternate cell architectures that make them impervious to attack from predators or antibiotic compounds [47]. While in periods of famine, the *r*-type heterotrophs either sporulate or revert to cheating or order to ensure survival, the slow-and-steady catabolism of *k*-type heterotrophs ensures that even trace amounts of carbon can provide sufficient energy to maintain the cell integrity [118].

Despite what I hope is a persuasive argument, the assumption that it is possible to predict particular microbial metabolic strategies in any environment at the phylum-level is naïve. It is likely that numerous factors play a role in shaping heterotrophic community dynamics under nutrient limitation in these oligotrophic cave environments. To understand these drivers will require experimental techniques beyond simply descriptive and phylogenetic approaches. The use of analytical techniques in geology and chemistry may allow us to better understand the fate and turnover of organic compounds entering the cave, and whether these products could support successional heterotrophy [108]. The *in situ* analysis of cell growth and nitrogen-fixing rates *via* stable isotope probing would help us understand the rate at which microbial populations mineralize organic carbon, and the impact that feast and famine has on such processes. Most importantly, the continued culture of microbial species from caves can provide valuable information as to the actual physiological adaptations of species subjected to such nutrient limitation [135]. Together, these results would not only provide important information regarding the structure and nature of cave microbial ecosystems, but would help us to better understand global themes in bacterial survival during periods of extreme famine.

Acknowledgments

I would like to thank the numerous collaborators who's insight and contributions, particularly in the fields of chemistry and geology, have helped me better understand the complex nature of microbial ecosystems in caves. Much of the relevant research to this review was supported by a grant from the National Science Foundation Microbial Interactions and Processes Program (NSF# 0643462).

References

[1] Culver DC. Cave Life. Cambridge, MA, Harvard University Press, 1982.
[2] Gibert J, Deharveng L. Subterranean ecosystems, a truncated functional diversity. BioScience 2002, 52, 473–81.
[3] Simon KS, Benfield EF, Macko SA. Food web structure and the role of epilithic biofilms in cave streams. Ecology 2003, 84, 2395–406.
[4] Angert ER, Northup DE, Reysenbach A-L, Peek AS, Goebel BM, Pace NR. Molecular phylogenetic analysis of a bacterial community in Sulphur River, Parker Cave, Kentucky. Am Min 1998, 83, 1583–92.
[5] Engel AS, Lee N, Porter ML, Stern LA, Bennett PC, Wagner M. Filamentous "*Epsilonproteobacteria*" dominate microbial mats from sulfidic cave springs. Appl Environ Microbiol 2003, 69, 5503–11.
[6] Sarbu SM, Kane TC, Kinkle BK. A chemoautotrophically based cave ecosystem. Science 1996, 272, 1953–5.
[7] Hess WH. The origin of nitrates in cavern earths. J Geol 1900, 8, 129–34.
[8] Williams MAM, Benson-Evans K. Summary of results obtained during a preliminary investigation into the bacterial and botanical flora of caves in South Wales. Int J Speleol 1967, 2, 397–402.
[9] Caumartin V. Review of the microbiology of underground environments. Nat Speleol Soc Bull 1963, 25, 1–14.
[10] Høeg OA. Cyanophyceae and bacteria in calcareous sediments in the interior of limestone caves in Nord-Rana, Norway. Nytt Magazin for Naturvidenskapene 1946, 85, 99–104.

[11] Cunningham KI, Northup DE, Pollastro RM, Wright WG, LaRock EJ. Bacteria, fungi and biokarst in Lechuguilla Cave, Carlsbad Caverns National Park, New Mexico. Environ Geol 1995, 25, 2–8.
[12] Northup DE, Barnes SM, Yu LE, et al. Diverse microbial communitiens inhabiting ferromanganese deposits in Lechuguilla and Spider Caves. Environ Microbiol 2003, 5, 1071–86.
[13] Provencio PP, Polyak VJ. Iron oxide-rich gilmanets: Possible fossil bacteria in Lechuguilla Cave, New Mexico. Geomicrobiol J 2001, 18, 297–309.
[14] Spilde MN, Northup DE, Boston PJ, et al. Geomicrobiology of cave ferromanganese deposits: A field and laboratory investigation. Geomicrobiol J 2005, 22, 99–116.
[15] Barton HA. Introduction to cave microbiology: A review for the non-specialist. J Cave Karst Stud 2006, 68, 43–54.
[16] Barton HA, Taylor MR, Pace NR. Molecular phylogenetic analysis of a bacterial community in an oligotrophic cave environment. Geomicrobiol J 2004, 21, 11–20.
[17] Zhou J, Gu Y, Zou C, Mo M. Phylogenetic diversity of bacteria in an earth-cave in Guizhou Province, southwest of China. J Microbiol 2007, 45, 105–12.
[18] Chelius MK, Moore JC. Molecular phylogenetic analysis of Archaea and Bacteria in Wind Cave, South Dakota. Geomicrobiol J 2004, 21, 123–34.
[19] Ortiz M, Legatzki A, Neilson JW, et al. Making a living while starving in the dark: metagenomic insights into the energy dynamics of a carbonate cave. ISME J 2014, 8, 478–91.
[20] Tetu SG, Breakwell K, Elbourne LD, Holmes AJ, Gillings MR, Paulsen IT. Life in the dark: metagenomic evidence that a microbial slime community is driven by inorganic nitrogen metabolism. ISME J 2013, 7, 1227–36.
[21] Shabarova T, Pernthaler J. Karst pools in subsurface environments: collectors of microbial diversity or temporary residence between habitat types. Environ Microbiol 2010, 12, 1061–74.
[22] Cuezva S, Fernandez-Cortes A, Porca E, et al. The biogeochemical role of *Actinobacteria* in Altamira Cave, Spain. FEMS Microbiol Ecol 2012, 81, 281–90.
[23] Porca E, Jurado V, Žgur-Bertok D, Saiz-Jimenez C, Pašic L. Comparative analysis of yellow microbial communities growing on the walls of geographically distinct caves indicates a common core of microorganisms involved in their formation. FEMS Microbiol Ecol 2012, 81, 255–66.
[24] Portillo MC, Saiz-Jimenez C, Gonzalez JM. Molecular characterization of total and metabolically active bacterial communities of 'white colonizations' in the Altamira Cave, Spain. Res Microbiol 2009, 160, 41–17.
[25] Stomeo F, Portillo MC, Gonzalez JM, Laiz L, Saiz-Jimenez C. *Pseudonocardia* in white colonizations in two caves with Paleolithic paintings. Int Biodeterior Biodegr 2008, 62, 483–6.
[26] Barton HA, Northup DE. Geomicrobiology in cave environments: Past, current and future prospectives. J Cave Karst Stud 2007, 69, 163–78.
[27] Engel AS. Microbial diversity in cave ecosystems. In: Loy A, Mandl M, Barton LL, eds. Geomicrobiology: Molecular and Environmental Perspective, Springer, 2010, 219–38.
[28] Lee NM, Meisinger DB, Aubrecht R, et al. Caves and karst environments. In: Bell EM, ed. Life at Extremes: Environments, Organisms and Strategies for Survival. Egham, UK, CAB International, 2012, 320–44.
[29] Koch AL. Oligotrophs versus copiotrophs. BioEssays 2001, 23657–61.
[30] Charlier J-P, Bertrand C, Mudry J. Conceptual hydrogeological model flow and transport of dissolved organic carbon in a small Jura karst system. J Hydro 2012, 52–64.
[31] Mudarra M, Andreo B. Relative importance of the saturated and the unsaturated zones in the hydrogeological functioning of karst aquifers: The case of Alta Cadena (Southern Spain). J Hydro 2011, 397, 263–80.

[32] Northup DE, Carr DL, Crocker MT, et al. Biological investigations in Lechuguilla Cave. J Cave Karst Stud 1994, 56, 54–63.

[33] Laiz L, Groth I, Gonzalez I, Saiz-Jimenez C. Microbiological study of the dripping water in Altamira Cave (Santillana del Mar, Spain). J Microbiol Methods 1999, 36, 129–38.

[34] Rusterholtz KJ, Mallory LM. Density, activity, and diversity of bacteria indigenous to a karstic aquifer. Microbial Ecol 1994, 28, 79–99.

[35] Laiz L, Gonzalez JM, Saiz-Jimenez C. Microbial communities in caves: Ecology, physiology, and effects on Paleolithic paintings. In: Koestler RJ, Koestler VH, Charola AE, Nieto-Fernandez FE, eds. Art, Biology and Conservation: Biodeterioration of Works of Art. New York, Metropolitan Museum of Art, 2003, 211–25.

[36] Cray JA, Bell ANW, Bahaganna P, Mswaka AY, Timson DJ, Hallsworth JE. The biology of habitat dominance; can microbes behave as weeds? Microbial Biotech 2013, 6, 453–92.

[37] Eilers H, Pernthaler J, Glöckner FO, Amann RI. Culturability of *in situ* abundance of pelagic bacteria from the North Sea. Appl Environ Microbiol 2000, 66, 3044–51.

[38] Koch AL. Microbial physiology and ecology of slow growth. Microbiol Mol Biol Rev 1997, 61, 305–18.

[39] Amann RI, Snaidr J, Wagner M, Ludwig W, Schleifer KH. *In situ* visualization of high genetic diversity in a natural community. J Bacteriol 1996, 178, 3496–500.

[40] Pace NR. A molecular view of microbial diversity and the biosphere. Science 1997, 276, 734–40.

[41] Schabereiter-Gurtner C, Saiz-Jimenez C, Pinar G, Lubitz W, Rolleke S. Altamira Cave Paleolithic paintings harbor partly unknown bacterial communities. FEMS Microbiol Lett 2002, 211, 7–11.

[42] Vlasceanu L, Popa R, Kinkle BK. Characterization of *Thiobacillusthioparus* LV43 and its distribution in a chemoautotrophically based groundwater ecosystem. Appl Environ Microbiol 1997, 63, 3123–7.

[43] Groth I, Saiz-Jimenez C. Actinomycetes in hypogean environments. Geomicrobiol J 1999, 16, 1–8.

[44] Ivanova V, Tomova I, Kamburov A, Tomova A, Vasilev-Tonkova E, Kambourova M. High phylogenetic diversity of bacteria in the area of prehistoric paintings in Magura Cave, Bulgaria. J Cave Karst Stud 2013, 75, 218–28.

[45] Rusznyak A, Akob DM, Nietzsche S, et al. Calcite biomineralization by bacterial isolates from the recently discovered pristine karstic Herrenberg Cave. Appl Environ Microbiol 2012, 78, 1157–67.

[46] Barton HA, Taylor NM, Kreate M, Springer AJ, Oehrle SA, Bertog JL. The impact of host rock geochemistry on bacterial community structure in oligotrophic cave environments. Int J Speleol 2007, 36, 93–104.

[47] Fuerst JA, Sagulenko E. Beyond the bacterium: *Planctomycetes* challenge our concepts of microbial structure and function. Nature Rev Microbiol 2011, 9, 403–13.

[48] Jones RT, Robeson MS, Lauber CL, Hamady M, Knight R, Fierer N. A comprehensive survey of soil acidobacterialdiversity using pyrosequencing and clone library analysis. ISME J 2009, 3, 442–253.

[49] Moeseneder MM, Arrieta JM, Muyzer G, Winter C, Herndl GJ. Optimization of terminal-restriction fragment length polymorphism analysis for complex marine bacterioplankton communities and comparison with denaturing gradient gel electrophoresis. Appl Environ Microbiol 1999, 65, 3518–25.

[50] de Araujo JC, Schneider RP. DGGE with genomic DNA: suitable for detection of numerically important organisms but not for identification of the most abundant organisms. Water Res 2008, 42, 5002–10.

[51] La Duc MT, Dekas A, Osman S, Moissl C, Newcombe D, Venkateswaran K. Isolation and characterization of bacteria capable of tolerating the extreme conditions of clean room environments. Appl Environ Microbiol 2007, 73, 2600–11.
[52] La Duc MT, Osman S, Vaishampayan P, et al. Comprehensive census of bacteria in clean rooms by using DNA microarray and cloning methods. Appl Environ Microbiol 2009, 75, 6559–67.
[53] Campbell BJ, Engel AS, Porter ML, Takai K. The versatile Epsilon-proteobacteria: key players in sulphidic habitats. Nature Rev Microbiol 2006, 4, 458–68.
[54] Martinez-Romero E, ed. Dinitrogen-Fixing Prokaryotes. New York, Springer, 2006.
[55] Desai MS, Assig K, Dattagupta S. Nitrogen fixation in distinct microbial niches within a chemoautotrophy-driven cave ecosystem. ISME J 2013, 7, 2411–23.
[56] Rui Z, Wang H, Dong Q, Jingjing L, Jiang H, Barton HA. High abundance of ammonia-oxidizing prokaryotes and their role in nitrification in a dark and oligotrophic karst cave, central China. Appl Environ Microbiol 2015, Submitted.
[57] Hathaway JJ, Sinsabaugh RL, de Lurdes M, Dapkevicius NE, Northup DE. Diversity of ammonia oxidation (*amoA*) and nitrogen fixation (*nifH*) genes in lava caves of Terceira, Azores, Portugal. Geomicrobiol J 2014, 31, 221–35.
[58] Capone DG, Popa R, Flood B, Nealson KH. Geochemistry. Follow the nitrogen. Science 2006, 312, 708–9.
[59] Hodgson DA. Primary metabolism and its control in streptomycetes: a most unusual group of bacteria. Adv Microbial Physiol 2000, 42, 47–238.
[60] Bernardet J-F, Nakagawa Y. An introduction to the Family *Flavobacteriaceae*. In: The Prokaryotes. New York, Springer, 2006, 455–80.
[61] de Vos P, Garrity GM, Jones D, et al. Bergey's Manual of Systematic Bacteriology: The Firmicutes. 2nd ed. New York, Springer, 2009.
[62] Bayer EA, Shoham Y, Lamed R. Cellulose-decomposing bacteria and their enzyme systems. In: The Prokaryotes. New York, Springer, 2006, 578–617.
[63] Turnbaugh PJ, Hamady M, Yatsunenko T, et al. A core gut microbiome in obese and lean twins. Nature 2009, 457, 480–4.
[64] Scott KP, Gratz SW, Sheridan PO, Flint HJ, Duncan SH. The influence of diet on the gut microbiota. Pharmacological research, the official journal of the Italian Pharmacological Society 2013, 69, 52–60.
[65] Barton HA, Taylor NM, Lubbers BR, Pemberton AC. DNA extraction from low-biomass carbonate rock: an improved method with reduced contamination and the low-biomass contaminant database. J Microbiol Method 2006, 66, 21–31.
[66] Gottschal JC. Growth kinetics and competition – some contempory comments. Ant Leeuwenhoek 1993, 63, 299–313.
[67] DeBruyn JM, Nixon LT, Fawaz MN, Johnson AM, Radosevich M. Global biogeography and quantitative seasonal dynamics of Gemmatimonadetes in soil. Appl Environ Microbiol 2011, 77, 6295–300.
[68] Hanada S, Pierson BK. The Family *Chloroflexaceae*. In: The Prokaryotes. New York, Springer, 2006, 815–42.
[69] Fukunaga Y, Kurahashi M, Sakiyama Y, Ohuchi M, Yokota A, Harayama S. *Phycisphaera mikurensis* gen. nov., sp. nov., isolated from a marine alga, and proposal of Phycisphaeraceae fam. nov., Phycisphaerales ord. nov. and Phycisphaerae classis nov. in the phylum Planctomycetes. J Gen Appl Microbiol 2009, 55, 267–75.
[70] Griepenburg U, Ward-Rainey N, Mohamed S, et al. Phylogenetic diversity, polyamine pattern and DNA base composition of members of the order Planctomycetales. Int J Syst Bacteriol 1999, 49(Pt 2), 689–96.

[71] Quaiser A, Ochsenreiter T, Lanz C, et al. Acidobacteria form a coherent but highly diverse group within the bacterial domain: evidence from environmental genomics. Mol Microbiol 2003, 50, 563–75.

[72] Wagner M, Horn M. The *Planctomycetes, Verrucomicrobia, Chlamydiae* and sister phyla comprise a superphylum with biotechnological and medical relevance. Curr Op Biotechnol 2006, 17, 241–9.

[73] Jenkins C, Staley JT. History, classification and cultivation of the *Planctomycetes*. In: Fuerst JA, ed. *Planctomycetes:* Cell structures, origins and biology. New York, Springer, 2013, 1–38.

[74] Zekker I, Rikmann E, Tenno T, et al. Effect of HCO_3^- concentration on ANAMMOX nitrogen removal rate in a moving bed biofilm reactor. Environ Technol 2012, 33, 2263–71.

[75] Spear JR, Barton HA, Roberts JK, Francis CA, Pace NR. Microbial community biofabrics in a geothermal mine adit. Appl Environ Microbiol 2007, 73, 6172–80.

[76] Madrid VM, Taylor GT, Scranton MI, Chistoserdov AY. Phylogenetic diversity of bacterial and archaeal communities in the anoxic zone of the Cariaco Basin. Appl Environ Microbiol 2001, 67, 1663–74.

[77] Schmidt JM, Sharp WP, Starr MP. Metallic-oxide encrustations of the nonprosthecate stalks of naturally occuring populations of *Planctomyces bekefii*. Curr Microbiol 1982, 7, 389–94.

[78] Hug LA, Castelle CJ, Wrighton KC, et al. Community genomic analyses constrain the distribution of metabolic traits across the Chloroflexi phylum and indicate roles in sediment carbon cycling. Microbiome 2013, 1, 22.

[79] Tas N, Van Eekert MHA, De Vos WM, Smidt H. The little bacteria that can – diversity, genomics and ecophysiology of *'Dehalococcoides'* spp. in contaminated environments. Microb Biotechnol 2010, 3, 389–402.

[80] Kishimoto N, Kosako Y, Tano T. *Acidobacterium capsulatum* gen. nov., sp. nov.: an acidophilic chemoorganotrophic bacterium containing menaquinone from acidic mineral environments. Curr Microbiol 1991, 22, 1–7.

[81] Castro HF, Classen AT, Austin EE, Norby RJ, Schadt CW. Soil microbial community responses to multiple experimental climate change drivers. Appl Environ Microbiol 2010, 76, 999–1007.

[82] Barns SM, Cain ET, Sommervile L, Kuske CR. Acidobacteria phylum sequences in unarium-contaminated subsurface sediments greatly expand the known diversity within the phylum. Appl Environ Microbiol 2007, 73, 3113–6.

[83] Huber KJ, Wust PK, Rhode M, Overmann J, Foesel BU. *Aridibacter famidurans* and *Aridibacter kavangonensis*, 2 novel species of Acidobacteria subdivision 4 isolated from semiarid savanna soil. Intl J Sys Evol Microbiol 2014, 64, 1866–75.

[84] Janssen PH, Yates PS, Grinton BE, Taylor PM, Sait M. Improved culturability of soil bacteria and isolation in pure culture of novel members of the divisions *Acidobacteria, Actinobacteria, Proteobacteria* and *Verrucomicrobia*. Appl Environ Microbiol 2002, 68, 2391–6.

[85] Kielak AM, van Veen JA, Kowalchuk GA. Comparative analysis of acidobacterial genomic fragments from terrestrial and aquatic metagenomic libraries, with emphasis on acidobacteria subdivision 6. Appl Environ Microbiol 2010, 76, 6769–77.

[86] Eichorst SA, Kuske CR, Schmidt TM. Influence of plant polymers on the distribution and cultivation of bacteria in the phylum Acidobacteria. Appl Environ Microbiol 2011, 77, 586–96.

[87] Rawat SR, Männisto KM, Bromberg Y, Häggblom MM. Comparative genomic and physiological analysis provides insights into the role of *Acidobacteria* in organic carbon utilization in Arctic tundra soils. FEMS Microbiol Ecol 2012, 82, 341–55.

[88] Zimmermann J, Portillo MC, Serrano L, Ludwig W, Gonzalez JM. *Acidobacteria* in freshwater ponds at Doñana National Park, Spain. Microbial Ecol 2012, 63, 844–55.

[89] Zimmermann J, Gonzalez JM, Saiz-Jimenez C, Ludwig W. Detection and phylogenetic relationships of highly diverse uncultured Acidobacterial communities in Altamira Cave using 23S rRNA sequence analysis. Geomicrobiol J 2005, 22, 379–88.
[90] Meisinger DB, Zimmermann J, Ludwig W, et al. In situ detection of novel Acidobacteria in microbial mats from a chemolithoautotrophically based cave ecosystem (Lower Kane Cave, WY, USA). Environ Microbiol 2007, 9, 1523–34.
[91] Coates JD, Ellis DJ, Gaw CV, Lovley DR. *Geothrix fermentans* gen. nov., sp. nov., a novel Fe(III)-reducing bacterium from a hydrocarbon-contaminated aquifer. Int J Sys Bacteriol 1999, 49, 1615–22.
[92] Sait M, Davis KER, Janssen PH. Effect of pH on isolation and distribution of members of subdivision 1 of the phylum *Acidobacteria* occuring in soil. Appl Environ Microbiol 2006, 72, 1852–7.
[93] Ward NL, Challacombe JF, Janssen PH, et al. Three genomes from the phylum Acidobacteria provide insight into the lifestyles of these microorganisms in soils. Appl Environ Microbiol 2009, 75, 2046–56.
[94] Crowe MA, Power JF, Morgan XC, et al. *Pyrinomonas methylaliphatogenes* gen. nov., sp. nov., a novel group 4 thermophilic member of the phylum Acidobacteria from geothermal soils. Int J Sys Evol Microbiol 2014, 64, 220–7.
[95] Liesack W, Bak F, Kreft JU, Stackebrandt E. *Holophaga foetida* gen. nov., sp. nov., a new, homoacetogenic bacterium degrading methoxylated aromatic compounds. Arch Microbiol 1994, 162, 85–90.
[96] Jurkevitch E, Davidov Y. Phylogenetic diversity and evolution of predatory prokaryotes. Microbiol Monogr 2007, 4, 11–56.
[97] Zhou J, Huang Y, Mo M. Phylogenetic analysis on the soil bacteria distributed in karst forest. Brazil J Microbiol 2009, 40, 827–37.
[98] Liu X, Zhang J, Gu T, et al. Microbial community diversities and taxa abundances in soils along a seven-year gradient of potato monoculture using high throughput pyrosequencing approach. PloS ONE 2014, 9,e86610.
[99] Nemergut DR, Andersen SP, Cleveland CC, et al. Microbial community succession in an unvegetated, recently deglaciated soil. Microbial Ecol 2007, 53, 110–22.
[100] Schneider T, Keiblinger KM, Schmid E, et al. Who is who in litter decomposition? Metaproteomics reveals major microbial players and their biogeochemical functions. ISME J 2012, 6, 1749–62.
[101] Foesel BU, Rohde M, Overmann J. *Blastocatella fastidiosa* gen. nov., sp. nov., isolated from semiarid savanna soil – the first described species of Acidobacteria subdivision 4. Sys Appl Microbiol 2013, 36, 82–9.
[102] Simon KS, Pipan T, Culver DC. A conceptual model of the flow and distribution of organic carbon in caves. J Cave Karst Stud 2007, 69, 279–84.
[103] Ainsworth CC, Brockman FJ, Jardine PM. Biogeochemical considerations and complexities. In: Loomey BB, Falta RW, eds. Vadose Zone Science and Technology Solutions. Columbus, OH, Battelle Press, 2000, 589.
[104] Kieft TL, Brockman FJ. Vadose Zone Microbiology. In: Fredrickson JK, Fletcher M, eds. Subsurface Microbiology and Biogeochemistry. Wilmington, DE, Wiley-Liss, Inc., 2001, 352.
[105] Mudarra M, Andreo B, Baker A. Characterisation of dissolved organic matter in karst spring waters using intrinsic fluorescence: Relationship with infiltration processes. Sci Total Environ 2011, 409, 3448–62.
[106] Veni G. A framework for assessing the role of karst conduit morphology, hydrology, and evolution in the transport and storage of carbon and associated sediments. Acta Carsologica 2013, 42, 203–11.

[107] Long AJ, Valder JF. Multivariate analyses with end-member mixing to characterize groundwater flow: Wind Cave and associated aquifers. J Hydro 2011, 409, 315–27.

[108] Bullen HA, Oehrle SA, Bennett AF, Taylor NM, Barton HA. Use of attenuated total Reflectance Fourier Transform Infrared Spectroscopy to identify microbial metabolic products on carbonate mineral surfaces. App Environ Microbiol 2008, 74, 4553–9.

[109] Nelson CE, Carlson CA. Differential response of high-elevation planktonic bacterial community structure and metabolism to experimental nutrient enrichment. PloS ONE 2011, 6,e18320.

[110] Bhullar K, Waglechner N, Pawlowski A, et al. Antibiotic resistance is prevalent in an isolated cave microbiome. PloS ONE 2012, 7,e34953.

[111] Tranvik LJ. Bacterioplankton growth on fractions of dissolved organic carbon of different molecular weights from humic and clear waters. Appl Environ Microbiol 1990, 56, 1672–7.

[112] Lovley DR, Coates JD, Blunt-Harris EL, Phillips EJP, Woodward JC. Humic substances as electron acceptors for microbial respiration. Nature 1996, 382, 445–8.

[113] Rinke C, Schwientek P, Sczyrba A, et al. Insights into the phylogeny and coding potential of microbial dark matter. Nature 2013, 499, 431–7.

[114] Pitkanen E, Rousu J, Ukkonen E. Computational methods for metabolic reconstruction. Curr Op Biotechnol 2010, 21, 70–7.

[115] Zomorrodi AR, Suthers PF, Ranganathan S, Maranas CD. Mathematical optimization applications in metabolic networks. Metabolic Eng 2012, 14, 672–86.

[116] Barton HA, Pemberton A, Millette J. Comparative study of oligotrophic bacterial species cultivated from Jack Bradley Cave, Kentucky. In: Proceedings of the 14th International Congress of Speleology, Athens, Greece, 2005, O62, 1–6.

[117] Banks ED, Taylor NM, Gulley J, et al. Bacterial calcium carbonate precipitation in cave environments: A function of calcium homeostasis. Geomicrobiol J 2010, 27, 444–54.

[118] Hoehler TM, Jørgensen BB. Microbial life under extreme energy limitation. Nature Rev Microbiol 2013, 11, 83–94.

[119] Kayser A, Weber J, Hecht V, Rinas U. Metabolic flux analysis of *Escherichia coli* in glucose-limited continuous culture. I. Growth-rate-dependent metabolic efficiency at steady state. Microbiol 2005, 151, 693–706.

[120] Weber J, Kayser A, Rinas U. Metabolic flux analysis of *Escherichia coli* in glucose-limited continuous culture. II. Dynamic response to famine and feast, activation of the methylglyoxal pathway and oscillatory behaviour. Microbiol 2005, 151, 707–16.

[121] Sylvia DM, Fuhrmann JJ, Hartel PG, Zuberer DA, eds. Principles and Applications of Soil Microbiology. Upper Saddle River, NJ, Prentice Hall, 1999.

[122] Germida JJ, Casida LE. *Ensifer adhaerens* predatory activity against other bacteria in soil, as monitored by indirect phage analysis. Appl Environ Microbiol 1983, 45, 1380–8.

[123] Hibbing ME, Fuqua C, Parsek MR, Peterson SB. Bacterial competition: Surviving and thriving in the microbial jungle. Nature Rev Microbiol 2010, 8, 15–25.

[124] Foster KR, Kokko H. Cheating can stabilize cooperation in mutualisms. Proc Biol Sci Royal Soc 2006, 273, 2233–9.

[125] Bachmann BO, Van Lanen SG, Baltz RH. Microbial genome mining for accelerated natural products discovery: is a renaissance in the making? J Indust Microbiol Biotechnol 2014, 41, 175–84.

[126] Goodwin CR, Fenn LS, Derewacz DK, Bachmann BO, McLean JA. Structural mass spectrometry: rapid methods for separation and analysis of peptide natural products. J Nat Products 2012, 75, 48–53.

[127] Derewacz DK, Goodwin CR, McLean JA, Barton HA, Bachmann BO. Lechacyclines, new angucyclines and their mycothiol adducts from a deep hypogean streptomycete. 2013, unpublished.

[128] Begon M, Harper JL, Townsend CR. Ecology: Individuals, Populations and Communities. 3rd ed. Cambridge, Mass., Blackwell Scientific Publications, 1998.

[129] Hutchinson GE. The paradox of the plankton. Am Naturalist 1961, 882, 137–45.

[130] Czaran TL, Hoekstra RF, Pagie L. Chemical warfare between microbes promotes biodiversity. Proc Nat Acad Sci USA 2002, 99, 786–90.

[131] Kerr B, Riley MA, Feldman MW, Bohannan BJM. Local dispersal promotes biodiversity in a real-ilfe game of rock-paper-scissors. Nature 2002, 418, 171–4.

[132] Torsvik V, Ovreas L, Thingstad TF. Prokaryotic diversity – magnitude, dynamics, and controlling factors. Science 2002, 296, 1064–6.

[133] MacCarther RH, Wilson EO. The Theory of Island Biogeography. Princeton, NJ, Princeton University Press, 1967.

[134] Fierer N, Bradford MA, Jackson RB. Toward an ecological classification of soil bacteria. Ecology 2007, 88, 1354–64.

[135] Barton MD, Petronio M, Giarrizzo JG, Bowling BV, Barton HA. The genome of *Pseudomonas fluorescens* strain R124 demonstrates phenotypic adaptation to the mineral environment. J Bacteriol 2013, 195, 4793–803.

Marianyoly Ortiz, Julia W. Neilson, Antje Legatzki, and
Raina M. Maier

5 Bacterial and Archaeal Diversity on Cave Speleothem and Rock Surfaces: A Carbonate Cave Case Study from Kartchner Caverns

Abstract: Caves are stable, subterranean ecosystems ideal for evaluating microbial adaptations to darkness and nutrient limiting conditions. In this chapter, we review the bacterial and archaeal diversity and potential energy dynamics of microbial communities colonizing cave speleothem and rock wall surfaces. The primary focus is on recent studies in Kartchner Caverns, a carbonate cave near Benson, Arizona, USA. A combination of molecular- and culture-based techniques was used to provide a more complete picture of the microbial dynamics of cave ecosystems. Work in Kartchner demonstrated that each cave surface has a distinct taxonomic composition mostly dominated by the bacterial phyla Proteobacteria and Actinobacteria. Metagenomic analysis of stalactite samples identified genes for all six carbon fixation pathways and nitrification, which suggests that the cave microbial communities are, at least partially, supported by chemolithoautotrophic strategies.

5.1 Introduction

Caves provide an excellent window into the unique microbial dynamics of subterranean ecosystems. These ecosystems are characterized by the absence of light, low nutrient conditions, high humidity (95–100%), and constant temperature; conditions that potentially support novel assemblages of organisms that influence mineral transformations and ecosystem dynamics. This novelty is demonstrated by Fairy Cave in Colorado (USA) where 19% of the bacterial phylotypes identified from 16S rRNA gene sequences had less than 96% similarity to sequences deposited in the GenBank database [1]. This chapter will review the microbiology of speleothems and cave walls and conclude with a specific focus on current research in Kartchner Caverns in southern Arizona, USA.

The complete darkness and resulting lack of photosynthesis in most areas of caves results in limited nutrients to sustain life. Some shallow caves are supported by penetrating plant roots, but most caves depend on allochthonous sources of organic material. This allochthonous material is produced on the surface and reaches the cave as dissolved or particulate organic matter through the entrance, the epikarst (zone in the upper few meters of the bedrock above the cave and characterized by enlarged fissures and pores) and sinkholes [2]. Water infiltrating through the epikarst can also bring invertebrates that serve as food for other organisms. Another major source of organic matter is provided by the fecal material of crickets, birds, rats, raccoons, and bats. However, communities that live on fecal material are concentrated within a few hundred meters of the cave entrance or on the guano piles that collect under bat roosts [2]. Thus, the fecal material has a relatively small sphere of influence as a food source for cave life.

5.2 Bacterial and Archaeal Diversity on Caves Surfaces

Research using cultivation methods has found Actinobacteria to be the most abundant bacterial phylum in cave ecosystems [3–5] and has identified several isolates that are novel species [6–9]. Actinobacteria are generally described as common soil heterotrophs [3, 4] that are able to degrade humic acids and lignin-related compounds in the soil [10, 11]. In fact, humic acids have been successfully used to isolate Actinomycetes from environmental samples [12]. Humic acids and other recalcitrant soil compounds are the main type of organic material in cave drip water [13]. Thus, the ability of actinobacterial groups to live on recalcitrant soil compounds may offer a competitive advantage in some cave ecosystems.

Over the past few decades, the development of culture-independent methods, such as 16S rRNA gene clone libraries and denaturing gradient gel electrophoresis (DGGE) profiling, has revealed a broader bacterial diversity in caves [1, 14–17] by including members of the community that are difficult to culture. A phylogenetic tree constructed from publicly available 16S rRNA gene sequences obtained using molecular methods from 60 different caves from around the world, including Kartchner Caverns, found Proteobacteria to be the most abundant phylum [18]. The phylum Proteobacteria represents a metabolically diverse group of bacteria including chemolithoautotrophs such as the sulfur-oxidizing *Acidithiobacillus thiooxidans*, as well as model oligotrophic heterotrophs such as *Sphingopyxis alaskensis* [19]. The abundance of this phylum in caves may reflect the variable potential among its members to exploit a wide range of environmental conditions. In the phylogenetic analysis, Lee et al. [18] also found Actinobacteria, Chlorobi/Bacteroidetes, Chloroflexi, Acidobacteria, and Nitrospirae to be abundant and consistently present in caves.

The geographic or regional diversity of cave microbial communities was highlighted by a series of studies focused on visible spots of bacteria that colonize cave rock walls [5, 20, 21]. Three different types of communities have been found, characterized by gray, yellow, or white-colored colonies. Each spot is formed by a large number of bacteria and the community composition and distribution varies depending on the location in the cave and the organic inputs received [5, 20, 21]. Interestingly, analyses based on 16S rRNA gene clone libraries of yellow colonies isolated from caves in Spain, Slovenia, and the Czech Republic suggest that there are core groups of microorganisms involved in the formation of these colonies [22]. Three dominant groups represented 60% of the 16S rRNA gene sequences analyzed: the actinobacterial suborder

Pseudonocardinae and the γ-proteobacterial orders Chromatiales and Xanthomonadales. Additionally, phylotypes found in at least two caves were associated with Nitrospirae, δ-Proteobacteria, Actinobacteria, α-Proteobacteria, Bacteroidetes, and Acidobacteria. Each of these bacterial groups has been widely found in other cave studies around the world.

Studies focused on Archaea in caves are more limited but suggest that this group has a lower abundance and diversity than bacteria in most caves [15, 23–27]. The distribution and abundance of archaeal groups varies among geographically distinct caves; however, conclusions about factors affecting their distribution are difficult to make. Studies suggest that the archaeal communities on epigenic cave surfaces (epigenic caves are formed by carbonate dissolution due to dripping water and stream flow from the surface) are dominated by Crenarchaeota [15, 23, 24, 26], while sulfidic caves have been found to be dominated by Euryarchaeota [25, 27].

5.3 Microbial Energy Dynamics in Caves

Due to the absence of a photosynthesis-based primary production strategy in caves, early research presumed that cave microbial communities consisted of heterotrophic organisms living on limited energy inputs sourced from the surface. Caves are now known to sustain a greater diversity of microorganisms that exploit an unexpected variety of possible energy sources [28]. Identification of potential energy sources for microbial growth in cave environments has been done through the detection of functional or marker genes for specific metabolic pathways and through metagenomic analyses. Most of our current understanding of cave microbial functional dynamics comes from studies of sulfidic caves and water mat samples where microbial communities are supported by abundant reduced energy sources such as hydrogen sulfide and methane. For example, the presence of genes for the CO_2 fixation enzyme RubisCO and genes for bacterial sulfur oxidation were detected in water samples from Movile Cave [25] and in highly acidic biofilms from Frasassi Caves [29] using functional genes and metagenomics, respectively. The suggested presence of chemolithoautotrophic metabolism in epigenic caves and on cave surfaces is largely based on phylogenetic associations [5, 14, 18, 30]. Metagenomic strategies applied to these types of caves have been constrained by low biomass because large amounts of DNA template are required for high-throughput next generation sequencing technologies.

Nitrogen, like organic carbon, is a limiting nutrient in caves [28, 31, 32]; however, potential nitrogen cycling activities of microbial communities on cave surfaces have been observed. Barton et al. [31] identified potential denitrifying bacterial groups on rock walls in Carlsbad Caverns, New Mexico, although their role in caves is not yet understood. Archaeal *amoA* genes were found on speleothem surfaces from a mine adit in Colorado, [23] suggesting that ammonia oxidation may be a potential energy strategy for cave microorganisms. Oxidation of nitrite to nitrate, is hypothesized to occur in caves, as well, based on the repeated observance of two nitrite-oxidizing bacteria, *Nitrospira* spp. [5, 18, 33] and *Nitrobacter* spp. [32]. Likewise, archaeal and bacterial genes for nitrite and ammonia oxidation were detected on cave surface slimes from Weebubbie Cave, an underwater cave in Australia [34]. See Chapter 6 for more details about this cave. Evidence from sulfidic caves also supports the presence of nitrification in caves by the identification of bacterial *amoA* genes on floating microbial mats from Movile Cave [25].

5.4 Kartchner Caverns: An Epigenic Limestone Cave Case Study

Our understanding of speleothem and cave wall microbiology in highly oligotrophic epigenic limestone caves has been expanded by studies of the energy dynamics and microbial diversity in Kartchner Caverns, located in the Sonoran Desert of southern Arizona, USA (▸ Fig. 5.1). This system was designated a National Science Foundation Microbial Observatory in 2006. The cave developed in a block of Mississippian Escabrosa Limestone in the base-bounding fault zone between the Whetstone Moun-

Fig. 5.1. Sampling of formations in the Echo Passage (left) and Big Wall (right) regions of Kartchner Caverns. Sterile swabs were used by researchers to sample microbial populations on the surfaces of stalactites and cave wall surfaces.

tains and the San Pedro Valley [35, 36]. The cave is 3.9 km long and passages occur on one level at approximately 1408 m above mean sea level, with a total relief of 36.3 m [35]. Kartchner Caverns is ranked among the top 10 caves in the world for its mineralogical diversity and significance [37] based on the presence of nontronite and rectorite, both silicate minerals that have not been reported in any other cave previously, and a brushite moonmilk flowstone deposit not found in any other place in the world.

Environmental conditions in Kartchner Caverns include 99.4% average humidity, elevated CO_2 levels that range from 1000 to 5000 ppm, and an average mean annual temperature of 19.8 °C [38]. The cave's desert location results in particularly oligotrophic conditions characterized by drip water organic carbon concentrations of 0.79–2.7 mg organic C/L [45]. These levels are three orders of magnitude lower than in caves located in temperate regions [39, 40]. There is a nursery roost of *Myotis velifer* bats from late April to mid-September, and bat guano provides a major source of organic material for several species of invertebrates that inhabit localized areas of the cave.

5.4.1 Speleothem community diversity analysis

5.4.1.1 Microbial community structure survey

An initial 16S rRNA PCR-DGGE study characterized the microbial diversity of two cave speleothems from the Strawberry Alcove area of the cave to determine whether similar microbial communities colonized all cave formations, or whether different cave formations supported distinct microbial communities (SA, ▸ Fig. 5.2) [14]. The bacterial communities on two adjacent stalactites were more diverse than archaeal ones and the community profiles for each speleothem were distinct. Phylochip analysis revealed that bacterial populations belonged to 21 phyla and 16 candidate phyla. Phylogenetic analysis of excised DGGE bands showed that 42% of the bacterial and 50% of the archaeal bands were less than 96% similar to sequences found in microbial databases. This suggests the presence of previously uncharacterized microorganisms on the speleothem surfaces [14].

The observed divergence in community structure on adjacent speleothems fueled further questions concerning factors that influence microbial community distribution on speleothem surfaces [41]. Communities from 10 speleothems located within a second area of the cave (BW1, ▸ Fig. 5.2) were sampled and correlated to physical (i.e. color and dimension) and chemical (i.e. elemental profile and organic carbon concentration) speleothem parameters. The speleothem elemental profile varied among formations, and results confirmed that each speleothem bacterial community structure was distinct. Different patterns in bacterial community profiles from the 10 speleothems could not be statistically explained by elemental profile, the organic carbon concentrations, or speleothem color. However, distance between speleothems

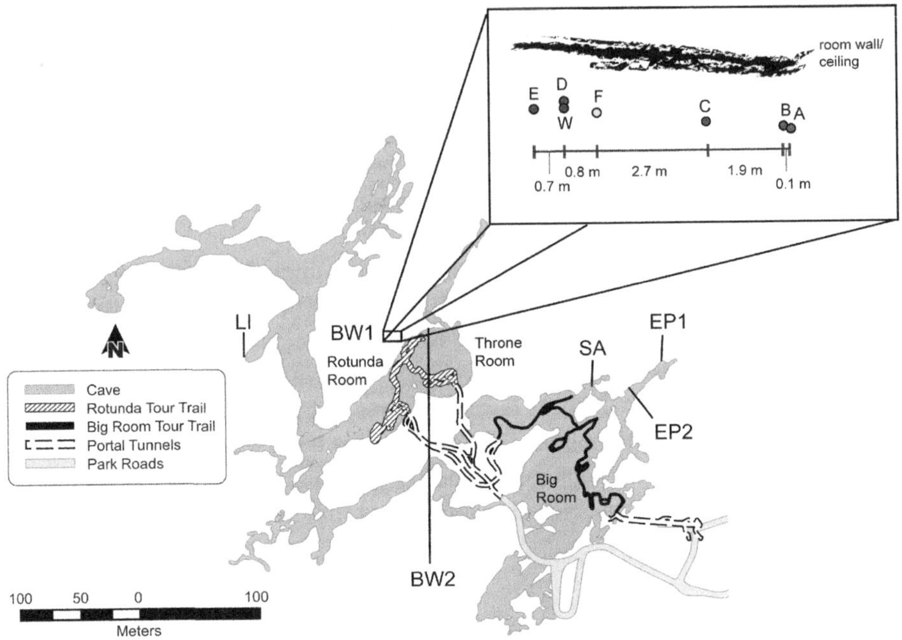

Fig. 5.2. Map of Kartchner Caverns. The sampling areas used in the different studies performed in Kartchner Caverns are shown. SA, Strawberry Alcove region; BW1, Big Wall room, first sampling area; BW2, Big Wall room, second sampling area; EP1, Echo Passage speleothem sampled for metagenomic analysis; EP2, Echo Passage shield sampled for culture studies. LI shows the sample site used in the previous culture study by Ikner et al. [42]. Inset: Map of the Big Wall room (BW1) formations showing the horizontal distance between the speleothem surfaces evaluated in the 454-pyrotag study. The formations are identified by their alphabetical letters and the different colored circles indicate the three community types discovered in this room: (●) Type 1 community, (●) Type 2 community, and ( ) Type 3 community.

significantly influenced bacterial community profiles; proximal formations had more similar profiles. An association was also observed between the drip lines feeding each speleothem and the bacterial profiles observed. Taken together, the data suggested that the drip water source was a critical factor shaping speleothem surface bacterial community structure [41].

5.4.1.2 Bacterial community taxonomic composition

Gene sequence data sets of 16S rRNA genes generated using 454 pyrosequencing (V6 region) and clone libraries (nearly full-length 16S rDNA Sanger sequences) were analyzed to characterize BW1 speleothem and cave wall bacterial diversity and to seek taxonomic explanations for the variations in microbial community structure detected from PCR-DGGE profiles [30]. Unexpectedly high bacterial richness was retrieved from

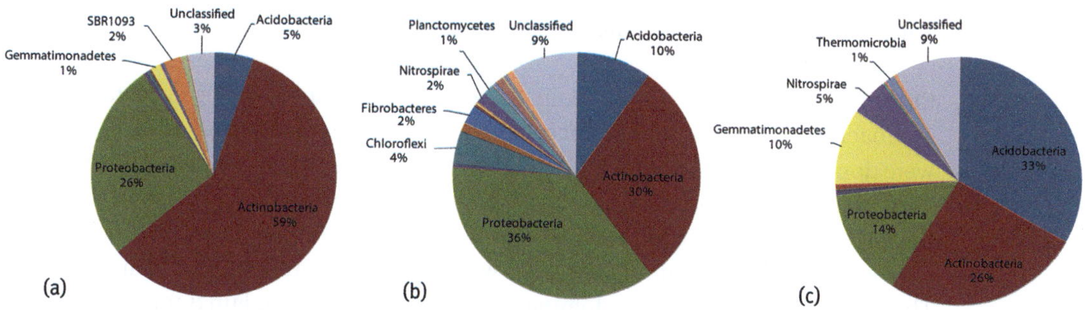

Fig. 5.3. Average phylum distribution of the three community types identified by 454-pyrotag analysis in the Big Wall region (BW1) of Kartchner Caverns. (a) Community Type 1 dominated by Actinobacteria, (b) community Type 2 dominated by Proteobacteria, and (c) community Type 3 dominated by Acidobacteria. Phyla representing 1% or more of the respective community are labeled.

seven speleothems and rock wall surfaces, with an average of 1994 operational taxonomic units (OTUs) per sample. Cave richness approached half the richness found in the soil sample taken above the cave, and there was minimal overlap between the cave and soil bacterial communities. Speleothem bacterial communities clustered into three distinct types based on the dominant taxonomic group (▸ Fig. 5.3). A Type 1 community was dominated by Actinobacteria belonging to Actinomycetales, Rubrobacteriales, and Acidimicrobiales orders. Type 2 communities were dominated by α-, δ-, and γ-Proteobacteria, with δ-Proteobacteria responsible for the major differences in proteobacterial abundance. Acidobacteria dominated the Type 3 community. The more diverse microbial communities were those in which Proteobacteria were most abundant, whereas the least diverse communities were those dominated by Actinobacteria. Communities in this area of the cave (BW1, ▸ Fig. 5.2) were taxonomically more similar to each other than to communities sampled from stalactites in a different room (SA, ▸ Fig. 5.2). The taxonomic profiles identified in this study are similar to those of microbial communities from other caves; however, this degree of taxonomic variability has not been demonstrated in such close proximity within a single cave.

A clone library examination of one of the Type 2 stalactites (stalactite D) revealed an abundance of heterotrophic bacteria closely associated with classic oligotrophs including *Polaromonas aquatica* (β-Proteobacteria), a genus that is commonly found in cold, nutrient-poor environments like glacial melt waters [43], and *Sphingopyxis alaskensis* (α-Proteobacteria) [19, 44], a marine taxon. A strain of *S. alaskensis* was also isolated from a different region of the cave [42], suggesting that oligotrophic bacteria are common in different regions of Kartchner Caverns.

5.4.2 Cave functional dynamics – a metagenomic approach

Chemolithoautotrophic potential in the cave was explored using a metagenomic approach [45]. Speleothem surfaces sampled in the Echo Passage (EP1, ▶ Fig. 5.2) were dominated by Bacteria (85%), followed by Archaea (10%), Eukaryota (5%), and viruses (0.13%). As found in the Big Wall and Strawberry Alcove regions of the cave, the bacterial community was dominated by Proteobacteria (52%) and Actinobacteria (13%); however, Planctomycetes (7.5%) were much more abundant in Echo Passage. The archaeal population was dominated by Thaumarchaeota. Fungi represented just 23% of the eukaryotes that could be classified. The EP1 taxonomic distribution was supported by a quantitative PCR (qPCR) analysis of two speleothem and two rock surfaces in the Big Wall area (BW2, ▶ Fig. 5.2) that showed Bacteria to be the most abundant group. Fungi were below detection, and Archaea were significantly more abundant in cave samples than in soil samples from above the cave [45].

The EP1 cave metagenome included genes for all six known CO_2 fixation pathways and an over-representation of RubisCO genes was observed in the cave relative to a selection of 12 bulk soil, rhizosphere soil, and open ocean metagenomes [45]. The CO_2 fixation genes identified were related to diverse groups of Archaea and Bacteria, including Thaumarchaeota and Nitrospirae, which are also known for their autotrophic ability to oxidize ammonia and nitrite, respectively. Cave drip water had NO_3-N concentrations that consistently exceeded dissolved organic carbon levels [45] over a 1-year sampling period, thus we hypothesize that inorganic nitrogen is an important energy driver in this oligotrophic cave. This theory was supported by the identification of both archaeal *amoA* genes, and bacterial nitrite oxidoreductase genes associated with "*Candidatus* Nitrospira defluvii" in the metagenome.

A comparative analysis conducted using whole metagenomes from four Kartchner Caverns metagenomes (EP1, two BW2 stalactites, and one BW2 calcite-coated rock wall) and the 12 cave samples, bulk soil, rhizosphere soil, and deep ocean samples [45] revealed that cave communities were more similar to soil and rhizosphere samples than to the ocean, despite the ocean being an oligotrophic environment. However, comparisons of clusters of orthologous group (COG) categories showed specific patterns that differentiated the nutrient poor environments (i.e. cave and ocean) from typically richer ones (i.e. bulk and rhizosphere soils). Signal transduction (T) and defense mechanism (V) COGs were under-represented in the ocean and cave samples. These two categories are trophic strategy indicators for uncultured marine oligotrophs [19]. Although the cave communities may have originated from soil ecosystems, the oligotrophic nature of the cave has selected for functional profiles that parallel those of similarly oligotrophic marine systems. Genes primarily associated with DNA repair mechanisms were significantly over-represented in the cave communities. Given the absence of environmental factors known to damage DNA in caves, such as UV light, we hypothesized that this over-representation is due to high calcium concentrations that have been previously linked to DNA damage [45].

5.4.3 The importance of culture-based characterizations

5.4.3.1 Evaluating impacts of cave development and tourism: a culture-based approach

When compared to molecular microbial analyses, culture-based studies are known to present an extremely limited profile of microbial community diversity because culturing recovers just 1% of the community. However, microbiological studies in Kartchner Caverns demonstrate the importance of developing strategies to culture these organisms. In May 2001, Kartchner Caverns State Park representatives contacted the University of Arizona (UA) about a slime problem in the cave. The slime was growing on the painted fiberglass material used to cover plumbing and electrical wires in the developed areas of the cave [42]. Park rangers regularly cleaned the surfaces with 10% bleach (Na hypochlorite) to remove the slime, but it consistently grew back. UA researchers investigated the origin of the slime by comparing the culturable bacterial communities on the fiberglass with those found on the adjacent natural rock surfaces [42]. The cultured fiberglass bacterial diversity was lower than the natural rock surface diversity, however the majority of the fiberglass isolates were closely related to ones on the adjacent rocks. In addition, seven of the nine bacterial isolates cultured from the fiberglass were able to use acrylic paint as a sole carbon source and three of these seven isolates produced copious amounts of slime during growth. These results indicated that the paint coating served as a food source thus selecting for the slime-producing bacterial populations that colonized the fiberglass surfaces. In the same study, the impact of tourism on culturable cave bacterial diversity was evaluated. Areas highly impacted by tourism were characterized by a lower and different bacterial diversity than the less impacted areas of the cave [42]. Tourists entering the cave introduce lint, hair, and skin cells that, like the paint, function as alternate food sources that have the potential to change the microbial community structure. The information gained from this culture-based study was used to change cave development and management strategies to limit human impacts on the cave microbial community.

5.4.3.2 Culturing the bacterial diversity of Kartchner Caverns

A major challenge in generating culture-based libraries is capturing a representative microbial diversity. Cave microorganisms have developed metabolic strategies to scavenge nutrients from nutrient-poor environments, so when grown on traditional nutrient-rich media, the microbes are unable to shutdown these metabolic pathways and frequently die from osmotic pressure [46, 47]. Heterotrophic bacteria from Kartchner Caverns were initially isolated on R2A medium (Becton, Dickinson and Company, Sparks, MD) from speleothems in two different regions of the cave, the Echo Passage and the Big Wall room (EP2 and BW1, ▸ Fig. 5.2). Isolates recovered included 15 α-Proteobacteria (68%), 4 Actinobacteria (18%), 1 each β- and γ-Proteobacteria, and 1 Firmicutes (▸ Tab. 5.1). These results can be compared to the earlier report by Ikner et

Table 5.1. Assignments of 16S rRNA gene sequences for bacteria recovered from Big Wall stalactites F, W, H, soda straw SS, and Echo Passage shield formation S on R2A media. Putative groups were based on classifications by RDP Classifier [51] (100% confidence level). Nearest neighbors were based on a Blast [52] search (NCBI database).

Sample site/ Isolate ID	Accession #	Putative group	Nearest neighbor (accession #)	Source of recovered nearest neighbor	Similarity (%)
Stalactite F					
KC-IT-F1	FJ711195	α-Proteobacteria	*Candidatus* Reyranella massiliensis strain URTM1 (EF394922)	River sample	99
KC-IT-F2	FJ711196	α-Proteobacteria	Uncultured alphaproteobacterium clone OS-C92 (EF612404)	Arizona desert soil	99
KC-IT-F4	FJ711197	α-Proteobacteria	*Afipia lausannensis* strain CRIB-05 (DQ123622)	Hospital water sample	100
KC-IT-F6	FJ711198	α-Proteobacteria	Alphaproteobacterium CRIB-02 (DQ123619)	Hospital water sample	96
			Uncultured alphaproteobacterium clone G08-1 (FM253623)	Rock biofilm	96
KC-IT-F7	FJ711199	α-Proteobacteria	Uncultured ferromanganous micronodule bacterium MNC12 (AF293001)	Green Bay sediment	96
KC-IT-F8	FJ711200	Actinobacteria	*Actinomadura* sp. TFS 455 (EF212022)	Fjord water sediment	97
Stalactite H					
KC-IT-H1	FJ711201	α-Proteobacteria	Uncultured bacterium clone RT_57 (EU644207)	Arctic permafrost	98
KC-IT-H2	FJ711202	α-Proteobacteria	*Bosea* sp. TPR12 (EU373419)	Root	94
			Bosea sp. BMA-4 (DQ855064)	Not determined	94
KC-IT-H4	FJ711203	α-Proteobacteria	Uncultured bacterium clone I-9 (AY625143)	Ground water	99
			Bacterium Ellin361 (AF498743)	Soil	99
			Bacterium Ellin333 (AF498715)	Soil	99
KC-IT-H5	FJ711204	Actinobacteria	*Actinomadura* sp. TFS 455 (EF212022)	Fjord water sediment	97
KC-IT-H6	FJ711205	Actinobacteria	*Pseudonocardia* sp. M1 (AY247276)	Wastewater plant	99
KC-IT-H8	FJ711208	α-Proteobacteria	*Sphingomonas* sp. JEM-14 (AB219361)	Activated sludge	96
KC-IT-H9	FJ711207	α-Proteobacteria	Uncultured Fe–Mn micronodule bacterium MNC12 (AF293001)	Green Bay sediment	96
Stalactite W					
KC-IT-W1	FJ711208	α-Proteobacteria	Uncultured alphaproteobacterium clone OS-C92 (EF612404)	Arizona desert soil	98
KC-IT-W2	FJ711209	α-Proteobacteria	Uncultured sludge bacterium A39 (AF234724)	Activated sludge	99
KC-IT-W4	FJ711210	α-Proteobacteria	Uncultured *Bradyrhizobium* sp. clone TM11_63 (DQ303337)	Ascomycota associated	99
			Uncultured *Bradyrhizobium* sp. clone TM11_36 (DQ303336)	Ascomycota associated	99
KC-IT-W5	FJ711211	α-Proteobacteria	Uncultured bacterium clone I-9 (AY625143)	Groundwater sample	99
Soda straw SS					
KC-IT-SS1	FJ711212	Actinobacteria	Uncultured soil bacterium clone TIID6 (DQ297955)	Soil	97
KC-IT-SS2	FJ711213	β-Proteobacteria	*Hydrogenophaga* sp. BAC306 (EU130968)	Water filter	99
Echo Passage Shield S					
KC-EP-S11	FJ711214	Firmicutes	*Brevibacillus levickii* strain R-12318 (AJ715382)	Geothermal Antarctic soil	97
KC-EP-S12	FJ711215	α-Proteobacteria	Uncultured Fe–Mn micronodule bacterium MNC12 (AF293001)	Green Bay sediment	96
KC-EP-S13	FJ711216	γ-Proteobacteria	*Pseudomonas* sp. IMT40 (AF302796)	Oil field soil	97

Table 5.2. Comparison of bacteria recovered from Echo Passage shield formation S on different media. Assignments of 16S rRNA gene sequences for putative groups of bacteria were based on classifications by the RDP Classifier [51] (100% confidence level). Nearest neighbors were based on a Blast [52] search (NCBI database).

Medium/Isolate ID	Accession #	Putative group	Nearest neighbor (accession #)	Source of recovered nearest neighbor	Similarity (%)
R2A media					
KC-EP-S11	FJ711214	Firmicutes	*Brevibacillus levickii* strain R-12318 (AJ715382)	Geothermal Antarctic soil	97
KC-EP-S12	FJ711215	α-Proteobacteria	Uncultured Green Bay ferromanganous micronodule bacterium MNC 12 (AF293001)	Green Bay sediment	96
KC-EP-S13	FJ711216	γ-Proteobacteria	*Pseudomonas sp.* IMT40 (AF302796)	Oil field soil	97
VL55/xylan[a]					
KC-EP-S1	FJ711217	Actinobacteria	*Marmoricola sp.* CNJ780 PL04 (DQ448720)	Marine sediment	98
KC-EP-S2	FJ711218	γ-Proteobacteria	*Pseudomonas sp.* HR 26 (AY032726)	Soil	99
KC-EP-S3	FJ711219	α-Proteobacteria	Uncultured *Bradyrhizobium* sp. clone TM11_63 (DQ303337)	Ascomycota associated	99
			Uncultured *Bradyrhizobium* sp. clone TM11_36 (DQ303336)	Ascomycota associated	99
KC-EP-S4	FJ711220	α-Proteobacteria	*Aminobacter aminovorans* strain DSM7048T (AJ011759)	Soil	99
KC-EP-S5	FJ711221	α-Proteobacteria	Uncultured Green Bay ferromanganous micronodule bacterium MNC12 (AF293001)	Green Bay sediment	96
KC-EP-S6	FJ711222	Actinobacteria	Uncultured *Catellatospora* sp. clone Plot21-G04 (EU193042)	Agricultural soil	96
KC-EP-S9	FJ711223	Actinobacteria	*Nocardioides oleivorans* (AB365060)	Coastal sea ice	99
KCM[b]					
KC-EP-S8	FJ711224	γ-Proteobacteria	*Lysobacter sp.* EM0470 (FJ392837)	Soil	98
KC-EP-S10	FJ711225	γ-Proteobacteria	*Lysobacter daejeonensis* strain GH1-9 (DQ191178)	Soil	98

[a] Non-agar based VL55 with gellan and 0.05% xylan [53]

[b] Kartchner Caverns medium with 0.01% yeast extract, 0.625% ground speleothem, and 1.5% noble agar

Fig. 5.4. Stalactite formations and cultured R2A plates used for plate-wash experiments. (Top) Stalactite formations A, E, D, F, H. The length for the formations was 26.7, 45.7, 62.2, 119.4, and 116.8 cm, respectively. (Bottom) Example of R2A plates from stalactites A, E, D, F, and H (A and E are 10^{-1} dilution plates and D, F, and H are 10^{-2} dilution plates).

al. [42] who recovered primarily Firmicutes (66%) with Actinobacteria (19%), and α-, β-, and γ-Proteobacteria (15%) on R2A medium from a rock surface in another remote area (LI, ▸ Fig. 5.2). Two additional media, including media targeting cave oligotrophs generated from ground cave stalactites supplemented with 0.01% yeast extract (KCM), were also tested to increase the diversity of heterotrophic bacteria recovered. Specifics of the alternate media are included in ▸ Tab. 5.2. Once again, all bacteria isolated belonged to the Proteobacteria, Actinobacteria, and Firmicutes phyla (▸ Tab. 5.2). However, the taxonomic distribution within these phyla was different for each of the three media tested. In fact, only one organism (belonging to α-Proteobacteria) was cultured on more than one medium. Results of this study confirm the importance of using a variety of culture media to increase the recovered diversity. Whereas the use

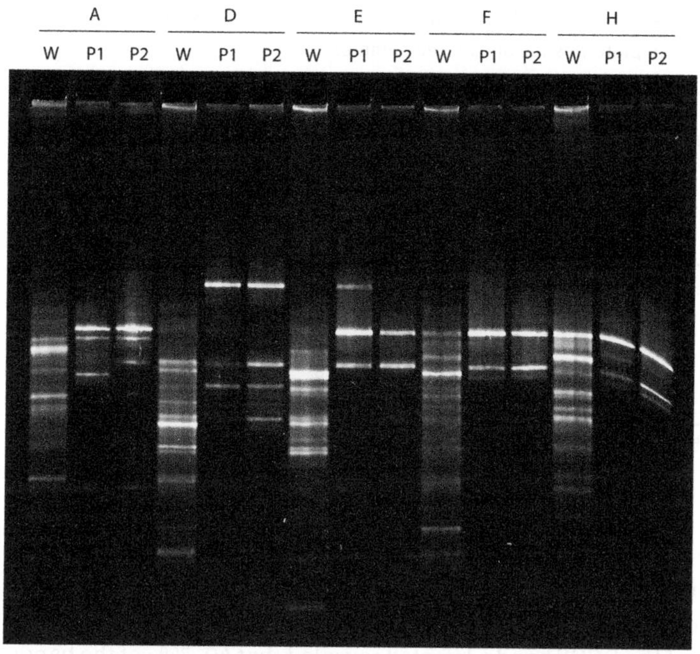

Fig. 5.5. PCR-DGGE comparison of bacterial community profiles from five separate stalactites (A, D, E, F, and H), as determined by plate-wash and whole community DNA extraction. For each stalactite, W represents the uncultured whole community profile, and P1 and P2 represent the profiles of cultured communities from plate wash of 10^{-1} and 10^{-2} dilutions, respectively.

of multiple media increased the diversity of the cultured heterotrophs, efforts have not been made to culture chemolithoautotrophs. Thus, a greater diversity of medium types would be required to capture the full metabolic diversity revealed by molecular analyses in the cave.

A comparison of the diversity recovered using R2A medium versus molecular techniques was evaluated by DGGE profile analysis as described by Legatzki et al. [14]. The study employed samples taken from five different stalactites (A, D, E, F, and H) in the Big Wall area (BW1, ▸ Fig. 5.2 and 5.4). A sample from each stalactite was plated onto R2A medium and incubated for 60 days (▸ Fig. 5.4), representing the cultured community (CC). A portion of each sample was simultaneously processed for whole community (W) DNA extraction. The CC bacteria were enumerated at regular intervals and a 50% increase in colony-counts was observed after the 60-day incubation period as compared to counts at 14 days. This result emphasized the importance of extended incubation times for the culturing of oligotrophs. Following the 60-day incubation period, the CC was washed from the R2A plates, the DNA extracted and 16S rRNA gene DGGE profiles were generated and compared to profiles generated from the parallel W DNA extracts. Four to eleven populations were detected in each CC profile, but the

Table 5.3. A comparison of DGGE profiles obtained from whole community DNA extracts from the original swab samples and from plate washes of cultured samples.[a]

Stalactite	Whole community DNA extract profile	Cultured community profile (10^{-1})		Cultured community profile (10^{-2})	
	Number of bands	Number of bands	Number of matching bands	Number of bands	Number of matching bands[a]
A	27	8	2	7	3
D	28	11	5	10	6
E	27	6	2	4	2
F	25	7	3	7	3
H	18	4	2	5	2

[a]The number of matching bands equals number of bands that match with bands in the whole community DNA extract profile.

DGGE band-number from the W extract profiles exceeded the cultured diversity by 2.5 to 6.8-fold (▸ Tab. 5.3, ▸ Fig. 5.5). Thus, a low percentage of the total bacterial communities could be recovered on R2A medium under the chosen growth conditions. Interestingly, a comparison of the DGGE bands detected revealed that 40–75% of the bands present in CC profiles were not present in the W profiles (▸ Tab. 5.3). One explanation for this could be that the actual population cell numbers represented by CC bands were very low in the original sample. However, despite their low number in the initial community, these populations responded rapidly to nutrients and out-competed other populations during growth on R2A medium.

5.4.3.3 Relevant metabolic activities of cultured isolates

A selection of the Kartchner Caverns cultured library was screened for activities related to mineral deposition in speleothems, including calcium carbonate precipitation and the production of siderophores [48] and biosurfactants [49]. Results indicated that 86% of the isolates (i.e. from at least eight different genera) that were able to grow on a traditional test medium for calcite precipitation (B4 medium with 16 mM calcium acetate and no glucose [50]) produced crystals (▸ Tab. 5.4). Crystal formation was observed within, and in close proximity to, the bacterial colonies (▸ Fig. 5.6). X-ray diffraction (XRD) analysis confirmed the composition of the crystals to be calcium carbonate (▸ Fig. 5.7). Just 43% and 23% of the isolates tested grew on siderophore and biosurfactant test media, respectively; six isolates were positive for siderophore production (e.g. KC-EP-S2, KC-EP-S4, KC-IT-F7, KC-IT-SS1, KC-IT-SS2, and KC-IT-W1; ▸ Tab. 5.1 and 5.2). No biosurfactant producers were identified.

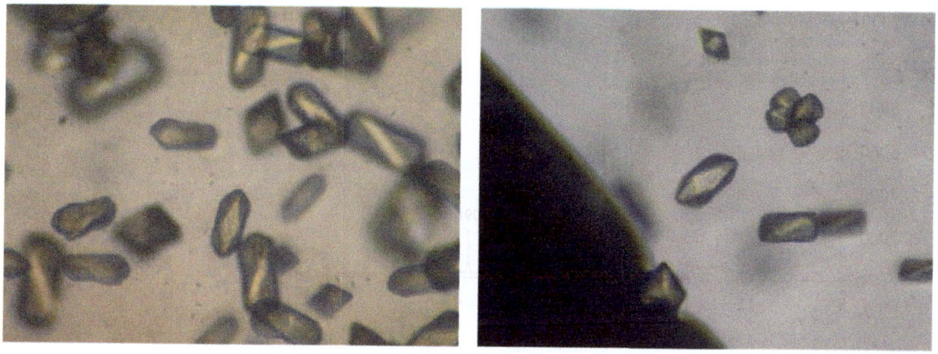

Fig. 5.6. Images of crystal formations in culture medium. (Left) Isolate KC-EP-S8 (shield S, Echo Passage) and (right) isolate KC-IT-H4 (stalactite H, Big Wall) grown on B4 media without glucose. Images taken with a 10× objective. The average crystal diameter for nine strains after 3 weeks of growth was 67 ± 28μm.

Table 5.4. Production of crystal structures by cave isolates on B4 media without glucose. Assignments of 16S rRNA gene sequences for the putative genera were based on classifications by the RDP Classifier [51] (100% confidence level).[a]

Isolate ID	Accession number	Putative genus or higher taxonomic group	Crystal production
KC-IT-F1	FJ711195	α-Proteobacteria	+
KC-IT-F4	FJ711197	*Afipia*	+
KC-IT-F6	FJ711198	Rhizobiales	–
KC-IT-F7	FJ711199	α-Proteobacteria	+
KC-IT-H4	FJ711203	α-Proteobacteria	+
KC-IT-H6	FJ711205	*Pseudonocardia*	+
KC-IT-SS1	FJ711212	*Arthrobacter*	+
KC-IT-SS2	FJ711213	*Hydrogenophaga*	+
KC-IT-W5	FJ711211	α-Proteobacteria	+
KC-EP-S2	FJ711218	*Pseudomonas*	+
KC-EP-S4	FJ711220	*Aminobacter*	+
KC-EP-S5	FJ711221	α-Proteobacteria	+
KC-EP-S8	FJ711224	*Lysobacter*	+
KC-EP-S11	FJ711214	*Brevibacillus*	–
KC-EP-S13	FJ711216	*Pseudomonas*	+

[a] When the confidence level was less than 100%, the next taxonomic group with 100% confidence level, whether order or class, is listed.

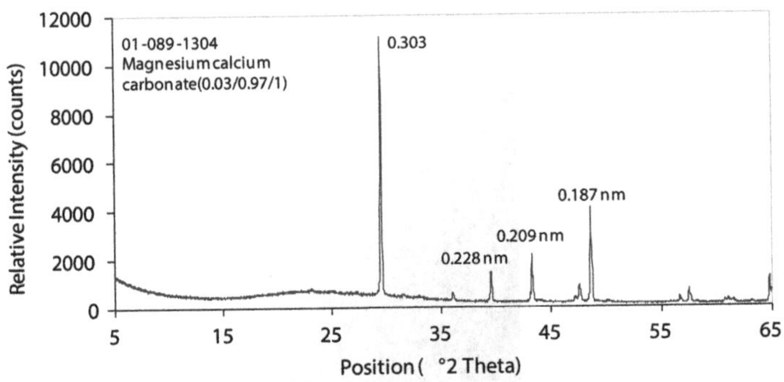

Fig. 5.7. X-ray diffraction pattern for crystals precipitated by isolate KC-EP-S2 on B4 media without glucose. The *d*-spacings of the major peaks (>10% relative intensity) are shown in the diagram. The reference number was obtained from the ICDD PDF-2 (powder diffraction) database.

5.5 Conclusions

The range of studies conducted in Kartchner Caverns demonstrates both the importance of integrating multiple culture-dependent and culture-independent techniques when characterizing the biological dynamics of a cave ecosystem, as well as the challenges of culturing the true diversity of an oligotrophic ecosystem. An unexpectedly high bacterial and archaeal diversity was found on speleothem and rock surfaces in Kartchner Caverns and minimal overlap was observed between the cave populations and those from the soil above the cave. This limited community overlap challenges the previous notion that cave communities are comprised primarily of translocated soil heterotrophs. The metagenomes reveal that the genetic potential of cave communities represents a complete ecosystem of chemolithoautotrophic primary producers and consumers. Primary productivity appears to be supported by inorganic nitrogen, with Thaumarchaeota and Nitrospirae as possible key players. Heterotrophic consumers were identified that are specifically adapted to oligotrophic conditions. However, the research to date has generated new questions and hypotheses that are dependent on the availability of cultured microorganisms for future analyses. Strategies to increase the number of cultured isolates recovered from cave habitats will come from combined molecular- and culture-based characterizations of microbial diversity, such as demonstrated by the Kartchner Caverns studies. A diverse culture-library will help to elucidate the role that microbes may have in calcium carbonate precipitation in caves, and facilitate the discovery of novel microorganisms, as well as new microbial byproducts with environmental and biotechnological applications.

Acknowledgments

We would like to recognize the contributions of Karis N. Nelson for characterizing calcium carbonate precipitation by cave isolates, and Codie E. Banczak for screening the culture library for biosurfactant and siderophore production. We also thank Rickard Toomey, Robert Casavant, Ginger Nolan and Steve Willsey of Arizona State Parks for their insights and assistance in Kartchner Caverns. Funding for this work was provided by the National Science Foundation (NSF) Microbial Observatory grant MCB0604300, a University of Arizona NSF IGERT Genomics Initiative fellowship awarded to Marianyoly Ortiz, grant no. DGE0654423, and NSF grant CHE1339597 co-funded with EPA for the study of Green Glycoplipids.

References

[1] Barton HA, Taylor MR, Pace NR. Molecular phylogenetic analysis of a bacterial community in an oligotrohic cave environment. Geomicrobiol J 2004, 21, 11–20.
[2] Hobbs HH, Culver DC. Cave Biology. In: Palmer AN, Palmer MV, eds. Caves and Karst of the USA. Huntsville, AL, National Speleological Society, 2009, 392–8.
[3] Groth I, Saiz-Jimenez C. Actinomycetes in hypogean environments. Geomicrobiol J 1999, 16, 1–8.
[4] Groth I, Vetterman R, Schuetze B, Schumann P, Saiz-Jimenez C. Actinomycetes in karstic caves of northern Spain (Altamira and Tito Bustillo). J Microbiol Meth 1999, 36, 115–22.
[5] Saiz-Jimenez C. Microbiological and environmental issues in show caves. World J Microb Biot 2012, 28, 2453–64.
[6] Jurado V, Groth I, Gonzalez JM, Laiz L, Saiz-Jimenez C. *Agromyces subbeticus* sp. nov., isolated from a cave in southern Spain. Int J Syst Evol Micr 2005, 55, 1897–901.
[7] Jurado V, Groth I, Gonzalez JM, Laiz L, Saiz-Jimenez C. *Agromyces salentinus* sp. nov. and *Agromyces neolithicus* sp. nov. Int J Syst Evol Micr 2005, 55, 153–7.
[8] Jurado V, Groth I, Gonzalez JM, Laiz L, Schuetze B, Saiz-Jimenez C. *Agromyces italicus* sp. nov., *Agromyces humatus* sp. nov. and *Agromyces lapidis* sp. nov., isolated from Roman catacombs. Int J Syst Evol Microbiol 2005, 55, 871–5.
[9] Jurado V, Kroppenstedt RM, Saiz-Jimenez C, et al. *Hoyosella altamirensis* gen. nov., sp. nov., a new member of the order *Actinomycetales* isolated from a cave biofilm. Int J Syst Evol Microbiol 2009, 59, 3105–10.
[10] Dari K, Bechet M, Blondeau R. Isolation of soil *Streptomyces* strains capable of degrading humic acids and analysis of their peroxidase activity. FEMS Microbiol Ecol 1995, 16, 115–21.
[11] Ball AS, Betts WB, Mccarthy AJ. Degradation of lignin-related compounds by actinomycetes. Appl Environ Microbol 1989, 55, 1642–4.
[12] Hayakawa M, Nonomura H. Humic acid vitamin agar, a new medium for the selective isolation of soil actinomycetes. J Ferment Bioeng 1987, 65, 501–9.
[13] van Beynen PE, Schwarcz HP, Ford DC, Timmins GT. Organic substances in cave drip waters: studies from Marengo Cave, Indiana. Can J Earth Sci 2002, 39, 279–84.
[14] Legatzki A, Ortiz M, Neilson J, et al. Bacterial and archaeal community structure of two adjacent calcite speleothems in Kartchner Caverns, Arizona, USA. Geomicrobiol J 2011, 28, 99–117.

[15] Gonzalez JM, Portillo MC, Saiz-Jimenez C. Metabolically active Crenarchaeota in Altamira Cave. Naturwissenschaften 2006, 93, 42–5.
[16] Engel AS, Meisinger DB, Porter ML, et al. Linking phylogenetic and functional diversity to nutrient spiraling in microbial mats from Lower Kane Cave (USA). ISME J 2010, 4, 98–110.
[17] Schabereiter-Gurtner C, Saiz-Jimenez C, Pinar G, Lubitz W, Rolleke S. Phylogenetic diversity of bacteria associated with paleolithic paintings and surrounding rock walls in two Spanish caves (Llonin and La Garma). FEMS Microbiol Ecol 2004, 47, 235–47.
[18] Lee NM, Meisinger DB, Aubrecht R, et al. Cave and karst environments. In: Bell EM, ed. Life at Extremes: Environments, Organisms and Strategies for Survival. Sterling, VA, CAB International, 2012, 320–44.
[19] Lauro FM, McDougald D, Thomas T, et al. The genomic basis of trophic strategy in marine bacteria. Proc Natl Acad Sci USA 2009, 106, 15527–33.
[20] Cuezva S, Sanchez-Moral S, Saiz-Jimenez C, Canaveras JC. Microbial communities and associated mineral fabrics in Altamira Cave, Spain. Int J Speleol 2009, 38, 83–92.
[21] Saiz-Jimenez C, Cuezva S, Jurado V, et al. Paleolithic art in peril: policy and science collide at Altamira Cave. Science 2011, 334, 42–3.
[22] Porca E, Jurado V, Zgur-Bertok D, Saiz-Jimenez C, Pasic L. Comparative analysis of yellow microbial communities growing on the walls of geographically distinct caves indicates a common core of microorganisms involved in their formation. FEMS Microbiol Ecol 2012, 81, 255–66.
[23] Spear JR, Barton HA, Robertson CE, Francis CA, Pace NR. Microbial community biofabrics in a geothermal mine adit. Appl Environ Microbiol 2007, 73, 6172–80.
[24] Northup DE, Barns SM, Yu LE, et al. Diverse microbial communities inhabiting ferromanganese deposits in Lechuguilla and Spider Caves. Environ Microbiol 2003, 5, 1071–86.
[25] Chen Y, Wu L, Boden R, et al. Life without light: microbial diversity and evidence of sulfur- and ammonium-based chemolithotrophy in Movile Cave. ISME J 2009, 3, 1093–104.
[26] Chelius M, Moore J. Molecular phylogenetic analysis of Archaea and Bacteria in Wind Cave, South Dakota. Geomicrobiol J 2004, 21, 123–34.
[27] Macalady JL, Jones DS, Lyon EH. Extremely acidic, pendulous cave wall biofilms from the Frasassi cave system, Italy. Environ Microbiol 2007, 9, 1402–14.
[28] Barton HA, Jurado, V. What's up down there? Microbial diversity in caves. Microbe 2007, 2, 132–8.
[29] Jones DS, Albrecht HL, Dawson KS, et al. Community genomic analysis of an extremely acidophilic sulfur-oxidizing biofilm. ISME J 2012, 6, 158–70.
[30] Ortiz M, Neilson JW, Nelson WM, et al. Profiling bacterial diversity and taxonomic composition on speleothem surfaces in Kartchner Caverns, AZ. Microb Ecol 2013, 65, 371–83.
[31] Barton HA, Taylor NM, Kreate MP, Springer AC, Oehrle SA, Bertog JL. The impact of host rock geochemistry on bacterial community structure in oligotrophic cave environments. Int J Speleol 2007, 36, 93–104.
[32] Barton HA, Northup DE. Geomicrobiology in cave environments: past, current and future perspectives. J Cave Karst Stud 2007, 69, 163–78.
[33] Holmes AJ, Tujula NA, Holley M, et al. Phylogenetic structure of unusual aquatic microbial formations in Nullarbor caves, Australia. Environ Microbiol 2001, 3, 256–64.
[34] Tetu SG, Breakwell K, Elbourne LD, Holmes AJ, Gillings MR, Paulsen IT. Life in the dark: metagenomic evidence that a microbial slime community is driven by inorganic nitrogen metabolism. ISME J 2013, 7, 1227–36.
[35] Jagnow DH. Geology of Kartchner Caverns State Park, Arizona. J Cave Karst Stud 1999, 61, 49–58.
[36] Hill C. Mineralogy of Kartchner Caverns, Arizona. J Cave Karst Stud 1999, 61, 73–8.

[37] Ford DC, Hill C. Dating of speleothems in Kartchner Caverns, Arizona. J Cave Karst Stud 1999, 61, 84–8.

[38] Buecher RH. Microclimate study of Kartchner Caverns, Arizona. J Cave Karst Stud 1999, 61, 108–20.

[39] Laiz L, Groth I, Gonzalez I, Saiz-Jimenez C. Microbiological study of the dripping waters in Altamira cave (Santillana del Mar, Spain). J Microbiol Meth 1999, 36, 129–38.

[40] Levy DB. Oxidation-reduction chemistry of Lechuguilla Cave seepage. J Cave Karst Stud 2007, 69, 351–8.

[41] Legatzki A, Ortiz M, Neilson JW, et al. Factors influencing observed variations in the structure of bacterial communities on calcite formations in Kartchner Caverns, AZ, USA. Geomicrobiol J 2012, 29, 422–34.

[42] Ikner LA, Toomey RS, Nolan G, Neilson JW, Pryor BM, Maier RM. Culturable microbial diversity and the impact of tourism in Kartchner Caverns, Arizona. Microb Ecol 2007, 53, 30–42.

[43] Irgens RL, Gosink JJ, Staley JT. *Polaromonas vacuolata* gen. nov., sp. nov., a psychrophilic, marine, gas vacuolate bacterium from Antarctica. Int J Syst Bacteriol 1996, 46, 822–6.

[44] Ostrowski M, Fegatella F, Wasinger V, Guilhaus M, Corthals GL, Cavicchioli R. Cross-species identification of proteins from proteome profiles of the marine oligotrophic ultramicrobacterium, *Sphingopyxis alaskensis*. Proteomics 2004, 4, 1779–88.

[45] Ortiz M, Legatzki A, Neilson JW, et al. Making a living while starving in the dark: metagenomic insights into the energy dynamics of a carbonate cave. ISME J 2014, 8, 478–91.

[46] Barton HA. Introduction to cave microbiology: a review for the non-specialist. J Cave Karst Stud 2006, 68, 43–54.

[47] Koch AL. Microbial physiology and ecology of slow growth. Microbiol Mol Biol R 1997, 61, 305–18.

[48] Schwyn B, Neilands JB. Universal chemical assay for the detection and determination of siderophores. Anal Biochem 1987, 160, 47–56.

[49] Bodour AA, Miller-Maier RM. Application of a modified drop-collapse technique for surfactant quantitation and screening of biosurfactant-producing microorganisms. J Microbiol Meth 1998, 32, 273–80.

[50] Boquet E, Boronat A, Ramoscor A. Production of calcite (calcium-carbonate) crystals by soil bacteria is a general phenomenon. Nature 1973, 246, 527–9.

[51] Wang Q, Garrity GM, Tiedje JM, Cole JR. Naive Bayesian classifier for rapid assignment of rRNA sequences into the new bacterial taxonomy. Appl Environ Microbiol 2007, 73, 5261–7.

[52] Altschul SF, Madden TL, Schaffer AA, et al. Gapped BLAST and PSI-BLAST: a new generation of protein database search programs. Nucleic Acids Res 1997, 25, 3389–402.

[53] Joseph SJ, Hugenholtz P, Sangwan P, Osborne CA, Janssen PH. Laboratory cultivation of widespread and previously uncultured soil bacteria. Appl Environ Microbiol 2003, 69, 7210–15.

Sasha G. Tetu, Liam D. H. Elbourne, Andrew Cronan,
Andrew J. Holmes, Michael R. Gillings, and Ian T. Paulsen

6 Microbial Slime Curtain Communities of the Nullarbor Caves

6.1 Introduction

6.1.1 The Nullarbor Cave environment

The Nullarbor Plain is a vast limestone karst system that covers a large part of Central Southern Australia. It spans more than 200,000 km^2, making it one of the world's great karst areas. The carbonate rocks underlying the Nullarbor Plain were laid down in cool to subtropical ocean environments during the Eocene (ca. 40 million years ago), as the soft Wilson Bluff Limestone encountered at the current water table, to Middle Miocene epochs (ca. 14 million years ago), as the harder Nullarbor Limestone [1]. Carbonate rock development ended with the rapid retreat of the sea coincident with a period of tilting and uplift in the Late Miocene to Early Pliocene [2]. The present surface of the Nullarbor Plain slopes slightly seaward, from 240 m above sea level in the northwest, and ends in a steep 40–90 m high ocean cliff line that extends for around 900 km [3]. The flat plain is sparsely vegetated and receives relatively little rainfall, with average precipitation away from the coastal fringe of between 150 and 250 mm and potential evaporation exceeding 2,000–3,000 mm [1]. Below the plain is a system of water-filled subterranean caves that house complex microbial communities in hanging slime curtains (▸ Fig. 6.1).

A series of large, deep caves between 50 and 150 m below the surface are found in a band along the southern edge of the plain, mostly within 60 km of the coast (▸ Fig. 6.2). There are around 100 caves in the Nullarbor Plain that have extensive passages and, in some cases, these penetrate deeply enough to intersect the water table (around 90 m at the coast). This results in large underground lakes and flooded passages [3]. Cave divers have explored a considerable distance into many of these passages, achieving world-record cave dives. For example, exploration of Cocklebiddy Cave recorded more than 6 km of flooded tunnels up to 20 m in diameter and intersected by large breakdown chambers. The striking blue-green color and clarity of the water in these cave systems have made these sites increasingly popular with cave divers, despite the difficulties in accessing entrance points. Entry generally requires rappelling into a collapsed doline, followed by an extensive underground traverse and portage of diving gear (▸ Fig. 6.1).

Fig. 6.1. Photographs of the Nullarbor Plain surface environment, large underground caves and microbial slime communities within submerged cave sections from the Nullarbor Plain. Pictures from top left to bottom right are: (1) The view along the south coast of Australia, showing the flat Nullarbor Plain ending in coastal cliffs which show stratigraphy of limestone; (2) A typical Nullarbor karst system collapsed doline and entrance point for cavers exploring submerged cave systems; (3) A large chamber within one of the cave systems showing a diver entrance point; (4) Cave divers exiting from a narrow tunnel in a submerged section of Warbla Cave into a large, completely water filled chamber; (5) Microbial slime curtains in a side alcove within Warbla Cave; (6) Collection of microbial material in a sterile tube during a diving expedition in Warbla Cave. Photographs courtesy of Steve Trewavas and David Rhea.

6.2 Microbial Slime Curtains

During the exploration of these submerged cave environments, divers reported the widespread occurrence of unusual microbial biofilms. These were referred to as dense mantles or curtains of biological material, known colloquially as Nullarbor cave slimes. The slime communities consist of long, mucoid, tendril-like structures or amorphous biofilms attached to cave surfaces. Microbial formations have been observed by divers in the majority of explored caves, but only within the fully submerged, completely dark passages. Early reports indicated that the curtain formations were up to 1 m in length [4]; however, more recent expeditions record formations mostly confined to side passages and, although still extensive, have shorter tendrils of ~20–40 cm. Despite there being no published accounts of aquatic macrofauna in these parts of the caves [5], one cave diving group has recently observed copepods in the vicinity of the microbial curtains in one cave system (S. Eberhard, personal communication).

Examination of cave systems to survey and collect microbial mantle material for scientific study was first conducted in 1999 and 2000 in the Western Australian caves Cocklebiddy (31°57′S, 125°54′E), Murra-el-Elevyn (32°02′S, 126°02′E), and Tommy Grahams (32°05′S, 126°11′E), and in South Australia caves Weebubbie (31°39′S, 128°46′E)

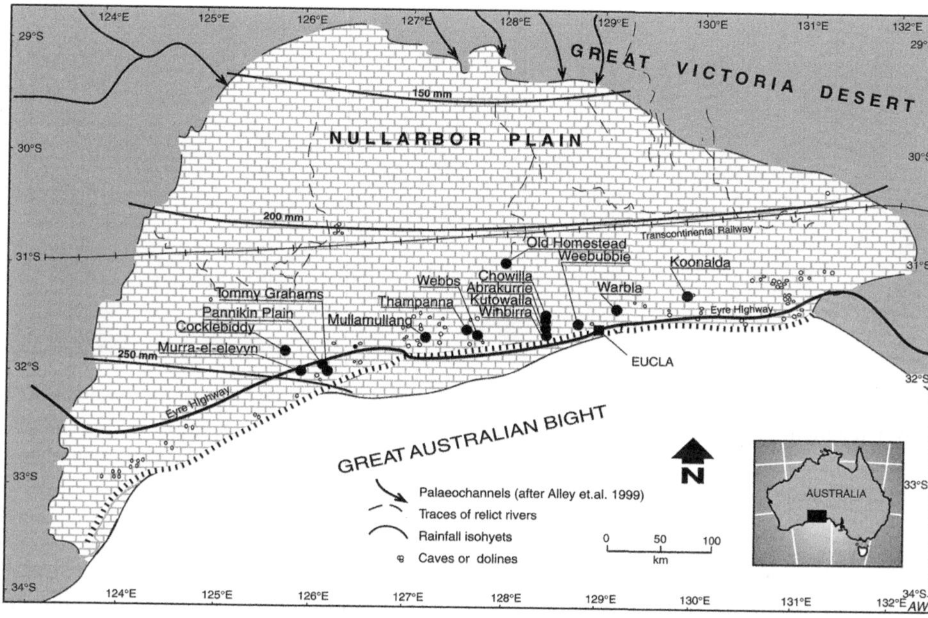

Fig. 6.2. Map of the Nullarbor Plain region of southern Australia, showing rainfall isohyets and locations of major caves, dolines, and paleochannels (reproduced from [3]).

and Warbla (31°31′S, 129°07′E) [4, 6]. All but Tommy Grahams Cave contain distinctive microbial slime communities. Microbial material was morphologically similar in all caves, being mainly opaque white in appearance. In a small number of instances, slime communities observed in the near-entrance passages had a slight red-brown color. Specimens were sampled using sterile syringes and water samples were collected and filtered (to 0.22 μm) to conduct chemical analyses for each cave. In all caves, water column measurements showed a pH close to 7, with moderate temperatures (between ~19 and 24 °C) and salinity [4]. Water in these caves contained levels of sulfate between 520 and 1,830 ppm, nitrite between 5 and 10 ppm, and nitrate levels between 25 and 900 ppm (detailed in Table 6 of reference [4]). The water was saturated with respect to calcite [4]. Earlier testing of filtered water collected from portions of the caves around microbial mantles showed no detectable organic carbon [7]. The observations of low organic carbon, high levels of inorganic nutrients, and relatively high microbial biomass led to the hypothesis that these microbial slimes represented unusual chemolithoautotrophic communities [4].

6.2.1 Microscopy and association of calcite crystals

Material collected from Cocklebiddy, Murra-el-Elevyn, Warbla, and Weebubbie cave systems during the 1999–2000 surveys was subjected to microscopy [4]. In all cases,

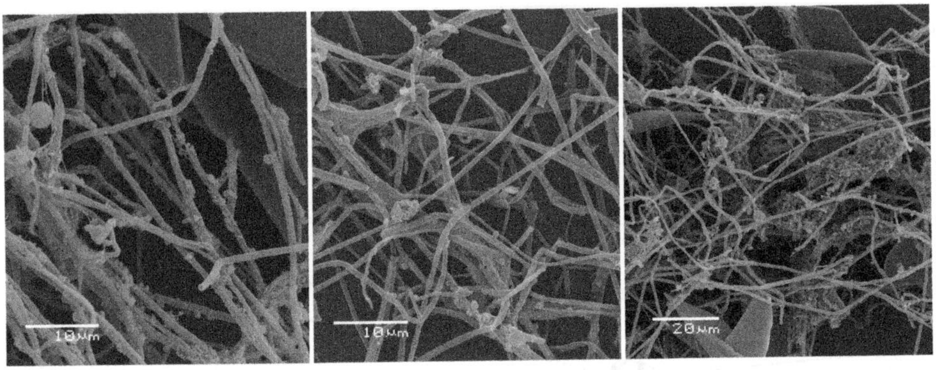

Fig. 6.3. Scanning electron photomicrograph of microbial curtain material from submerged Warbla Cave (left panel) and Weebubbie Cave (central and right panel). These representative images show a dense tangle of filaments interspersed with numerous cell-like shapes and large crystals (hypothesized to be calcite).

the material was dominated by thin filaments (1–2 µm diameter) that were best observed under differential interference contrast (DIC) microscopy. Within the filament network, other characteristic morphologies were observed and presumed to represent single-celled organisms, including cocco-bacillary, spiral, and donut shapes. Also within this network were a large number of inorganic crystals, hypothesized to be calcite, which were extracellular but closely associated with the filaments. Scanning electron microscopy (SEM) revealed these same morphologies, as well as additional, more regularly shaped crystals, which were thought to be formed as a result of sample preparation.

Consistent with microscopy-based observations of inorganic crystals associated with the cave slimes, cave divers reported the presence of dense accumulations of crystals within the microbial mantles, as well as snowfields of microcrystals on the rocks below the communities in the caves [7]. Examination of this material using X-ray diffraction and Fourier transform infrared spectroscopy (FTIR) confirmed the crystals to be calcite [6]. Research was conducted to determine if the microbial communities were linked to production of these crystals including those previously observed [4]. By using SEM, Contos et al. [6] showed that the crystals recovered from the cave had a distinctive morphology, not observed in crystals grown under the sole influence of inorganic ions. They posited that crystal formation in the caves was likely a consequence of biological activity. These researchers hypothesized that the presence of calcite crystals could not be explained purely by precipitation from a supersaturated solution because of activation barriers to calcite nucleation and that the close association between the crystals and microbial colonies suggests microorganisms play a role in crystal nucleation [6].

Additional microbial slime material was collected in 2010 from two of the caves visited in the earlier survey (Warbla and Weebubbie caves). Similar microscopic examination was done by using both DIC and SEM (Tetu et al., unpublished data). As before, the main morphotype observed was a complex network of long tangled filaments, interspersed with crystals and the characteristically common bacterial and archaeal cell morphologies (▸ Fig. 6.3). This morphology was relatively consistent across examined samples.

6.3 Community Membership

Samples of microbial community material from Murra-el-Elevyn, Warbla, and Weebubbie caves were used by Holmes et al. [4] for DNA extraction, with the aim to provide the first phylogenetic assessment of the microbial communities associated with the Nullarbor cave slimes. Bacterial 16S ribosomal RNA (rRNA) clone libraries were generated (using primers f27 and r1492 [8]). Libraries were analyzed using restriction fragment length polymorphism (RFLP) profile comparison of rRNA genes to investigate the variability and complexity of each community. Comparison of the RFLP band pattern for each cave indicated simple communities, conservation within a cave sample, and considerable overlap between the caves [4]. As all RFLP bands observed for Murra-el-Elevyn and Warbla caves were also observed for Weebubbie Cave, 16 representative clones were selected for sequencing from the Weebubbie Cave sample. Analysis of sequencing data indicated that the majority of phylotypes belonged to the Proteobacteria, with sequences related to *Pseudomonas* and *Pseudoalteromonas* spp. being particularly abundant. A relatively high proportion (~12%) of clones was classified to the phylum Nitrospirae, of which all characterized members are able to oxidize nitrite to nitrate [4]. This finding suggested that nitrite oxidation could play a significant role in the trophic structure of the slime communities. There were also a large number of clones that represented novel phylotypes and highlighted the unusual nature of this microbial ecosystem [4].

To extend the findings of Holmes and colleagues, a new analysis of the slime microbial community from Weebubbie Cave was carried out in 2012, using shotgun metagenomics together with a more extensive 16S rRNA gene survey [9]. This study used the same sample of DNA extracted from Weebubbie Cave microbial material in 2001 and utilized next generation sequencing technologies (Roche 454 GS-FLX) to sequence genes to a much greater depth than was previously feasible. A key goal in carrying out this new analysis was to gain a more comprehensive view of the taxonomic diversity present in this unusual microbial assemblage. The repeated observation of abundant filamentous cells within these communities, whose morphology did not match with the known cell shape of any abundant clone type in the study of Holmes et al. [4], indicated that additional key organism(s) could be present that had not been detected using the PCR-based approach.

Shotgun metagenomics has the advantage of not requiring PCR with primer sets that may amplify only a subset of the true population. However, to generate sufficient material for sequencing from the limited remaining sample, a multiple strand displacement (Phi29 DNA polymerase) amplification step was necessary prior to sequencing. A large library of 16S rRNA gene amplicons was also constructed using the original DNA, with the primer set F515 and R816, which amplify a region of the 16S rRNA gene from both Archaea and Bacteria [10]. Taxonomic binning of the roughly half a million, high quality, processed read sequences generated by shotgun metagenomic sequencing revealed that the Weebubbie Cave slime communities were comprised of a mix of archaeal and bacterial members [9] (▸ Fig. 6.4).

A sizable proportion of archaeal members was also observed in the 16S rRNA amplicon analysis, with 45% of all sequence reads being assigned to the archaeal phylum Thaumarchaeota. This phylum, which was proposed as recently as 2008 [11] and was previously referred to a "mesophilic Crenarchaeota," includes all known ammonia-oxidizing archaea, as well as clusters of environmental sequences from organisms with unknown energy metabolism [12]. This phylum has been the focus of considerable research effort in the past few years and there is now genome sequence information available for a number of isolates, including the tropical aquarium isolate *Nitrosopumilus maritimus* SCM1 [13], the low salinity isolate *Nitrosoarchaeum limnia* SFB1 [14], the soil isolate *Nitrososphaera viennensis* [15], and more recently representatives from deep marine sediment [16]. All of these species derive energy from the oxidation of ammonia to nitrite.

The abundance of Thaumarchaeota, which were not observed in the taxonomic survey of Holmes et al. [4], opens up the possibility that this group could comprise the "missing" filamentous morphotype observed in microscopy-based analyses of these samples. Examples of filamentous Thaumarchaeotes have been reported [17], thereby indicating the potential for the microbial slime filaments in the Nullarbor caves to similarly represent examples of novel thaumarchaeotal cell types.

Taxonomic breakdown of the bacterial component of the metagenomic and 16S rRNA amplicon data indicated that this community also includes a diverse array of bacterial phyla, with the Proteobacteria consistently having the highest relative abundance [9] (▸ Fig. 6.4). Among the sequences assigned to the Proteobacteria, more than half belonged to the class Gammaproteobacteria. The remaining proportion was relatively evenly distributed among the alpha-, beta-, and deltaproteobacterial classes. Further examination of the Gammaproteobacteria at the genus level showed *Pseudoalteromonas* to be the most abundant group. The abundance of Proteobacteria, particularly *Pseudoalteromonas*, corresponded well to the results observed in the earlier bacterial 16S rRNA clone library analysis of this community, where this group was observed to make up ~16% of clone library [4]. However, interestingly, this previous work indicated a high proportion of clones belonged to the genus *Pseudomonas*, comprising ~23% of clone library. Both the more recent analyses assigned a relatively small proportion of reads to this genus.

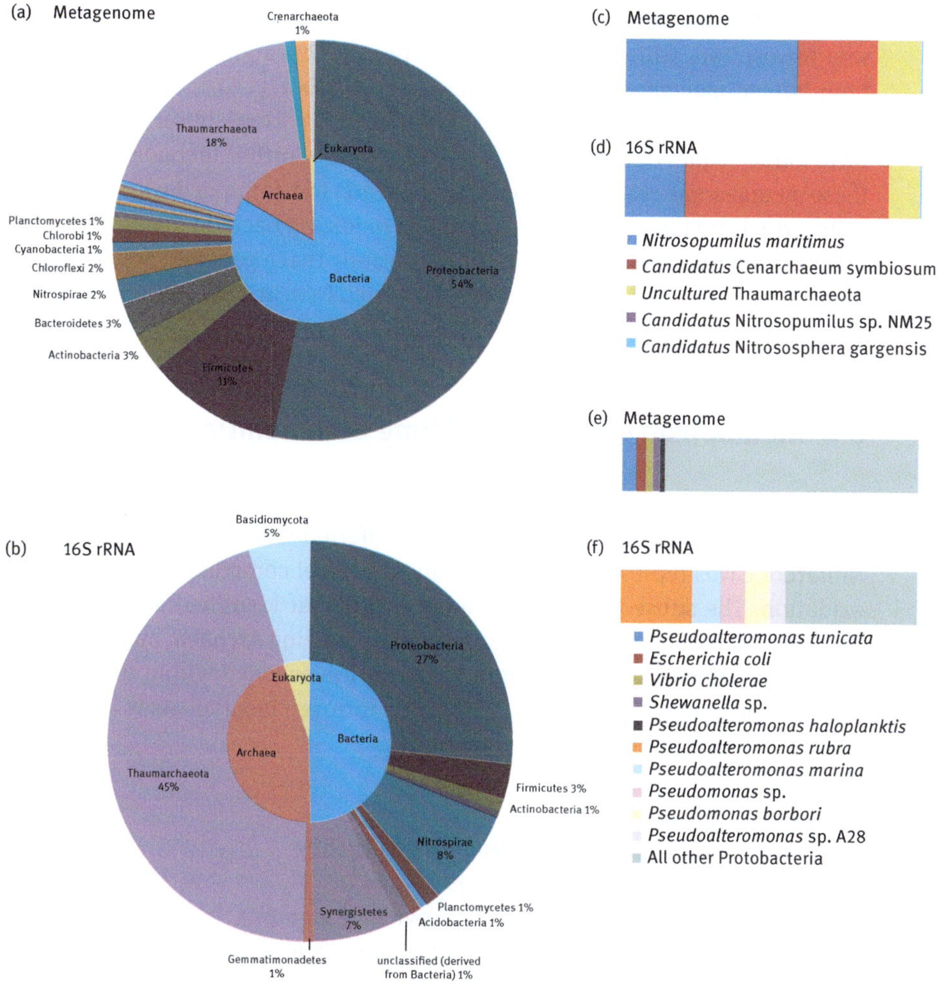

Fig. 6.4. The inferred taxonomic composition of the Weebubbie Cave community from Tetu et al. [9]. Taxonomic designations were assigned at phylum level based on the (a) complete metagenomic data set, and (b) 16S rRNA amplicon data set. Species level designation for the phylum Thaumarchaeota was based on the (c) complete metagenomic data set, and (d) 16S rRNA amplicon data set. Species level assignments for the phylum Proteobacteria were based on (e) the complete metagenomic data set and (f) the 16S rRNA amplicon data set. Taxonomic analyses based on the metagenome data were done with MG-RAST, and sequences were compared with the M5 nonredundant protein database with a maximum e-value cutoff of 10^{-15}. Unclassified sequences accounted for 7.31% of metagenome reads and are not included in this figure. The taxonomic analyses based on the 16S rRNA data according to MG-RAST, and sequences were compared with the M5 rRNA nonredundant rRNA database with a maximum e-value cutoff of 10^{-15}. Unclassified sequences accounted for 26.35% of 16S rRNA amplicon reads. Full color breakdown of the organisms represented in the pie charts in (a) and (b) can be found in the Supplementary material of Tetu et al. [9].

The other bacterial phylum that was consistently well represented in these data was Nitrospirae, which accounted for between ~2% and 8% of sequence reads in both binned metagenome and 16S rRNA samples (▸ Fig. 6.4). This phylum was also inferred to be abundant from the Holmes et al. [4] clone library results, in which Nitrospirae sequences comprised ~12% of clones. Further taxonomic breakdown of these metagenome reads indicated that the reads were affiliated with the genus *Nitrospira*, which are diverse, widespread nitrite-oxidizing bacteria in natural ecosystems, including marine environments. It is only recently that a genome sequence for "*Candidatus* Nitrospira defluvii" has been reconstructed from an activated sludge enrichment culture [18].

6.4 Metabolism of Microbial Slime Communities

Based on water chemistry and their data regarding the phylotypic structure of the Weebubbie Cave community, Holmes et al. [4] inferred that the microbial mantles from Nullarbor caves represent a biogeochemically novel community supported by nitrite oxidation. The subsequent metagenomic study that identified additional abundant known chemolithoautotrophs, the ammonia-oxidizing Archaea, has considerably extended our knowledge of the metabolic pathways encoded within this unusual community. The findings indicate that inorganic nitrogen metabolism is key to life in this dark, low organic nutrient environment [9].

6.4.1 Weebubbie Cave nitrogen and carbon cycling

Analysis of metagenome data can provide fascinating insights into the ecology and metabolic potential of unculturable organisms and communities. Weebubbie Cave slime community metagenomic reads were analyzed to determine the abundance and likely taxonomic affiliation of key genes for nitrogen metabolism enzymes (▸ Fig. 6.5) [9]. Genes encoding enzymes for each step of the nitrogen cycle, except nitrogen fixation, were present in this community. Of the reads matching ammonia monooxygenase (AMO) subunit-encoding genes, 82% were most likely of thaumarchaeotal origin, while the remainder were most likely encoded by representatives of the Nitrosomonadales. A relatively small number of reads matching hydroxylamine oxidase, responsible for the next step in nitrification in bacteria were observed in the metagenome, which also indicates that bacteria were responsible for only a small proportion of ammonia oxidation in this environment. There is also evidence for the second stage of nitrification, oxidation of nitrite to nitrate, in a number of sequence hits to nitrate oxidoreductase, with both bacterial and archaeal forms observed in the metagenome. A relatively large number of putative nitrate reductase encoding genes

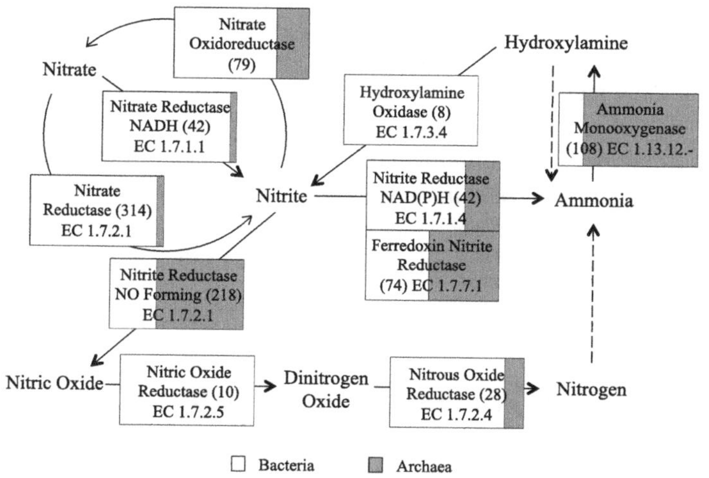

Fig. 6.5. Bacterial and archaeal abundances observed for key nitrogen metabolism enzymes in the metagenome from Weebubbie Cave [9]. The total abundance of gene fragments for each enzyme is shown in parentheses next to the enzyme name. The gray and white shaded boxes indicate the relative proportions of bacterial and archaeal reads, respectively, that match each enzyme. Solid arrows represent reactions performed by a known enzyme that was found in the metagenome. Dotted lines represent known nitrogen transformations for which no evidence was obtained from the metagenome.

were also observed in the Weebubbie metagenome sequence, the majority of which were likely bacterial in origin.

N. maritimus SCM1 has been hypothesized to fix carbon by using the 3-hydroxy-propionate/4-hydrobutyrate pathway, with homologs of all the genes implicated in this pathway observed within this organism's genome [13]. Reads recruiting to all of these genes (biotin dependent acetyl-CoA/propionyl-CoA carboxylase, methylmalonyl-CoA epimerase, and mutase and 4-hydroxybutyrate dehydratase) were observed in the Weebubbie Cave metagenome. Evidence for ribulose-1,5-bisphosphate carboxylase-oxygenase (RuBisCO) was also recovered from the bacterial-assigned metagenome reads.

From the metagenome, a high proportion of reads were assigned to the phylum Thaumarchaeota. Alignment of metagenomic contigs with the *N. maritimus* SCM1 genome was possible and revealed numerous syntenous regions in the Weebubbie Cave thaumarchaeotal contigs with the *N. maritimus* SCM1 genome. However, in other regions, nucleic acid sequence similarities were relatively low [9]. There were also a number of instances where genes with sequence similarity to those of *N. maritimus* SCM1 were present in the Weebubbie Cave metagenomes, but showed a different genomic arrangement. A large number of indels were also evident from the *N. maritimus* SCM1 genome but absent in the Weebubbie Cave metagenome, as well as from Weebubbie thaumarchaeotal contigs but not in the *N. maritimus* SCM1 genome.

One of the gene regions of most interest was the ammonia monooxygenase operon, which indicates the potential for organisms to play a role in nitrogen cycling. In the Weebubbie Cave metagenome, contigs were recovered that span large regions of this operon, including a single large contig that covers both the *amoB* and *amoC* genes, and a second contig that contains part of the *amoA* gene. Alignments of each of these translated genes from *N. maritimus* SCM1 with the relevant region of Weebubbie Cave contigs showed protein sequence identities of 85% for AmoB and 95% for AmoC, while the fragment of AmoA showed protein identities of 92% (over 131 of 216 amino acids).

6.5 Comparison of Metabolic Profiles from Other Habitats

The Weebubbie Cave environment represents an unusual aquatic habitat, with brackish salinity levels, high levels of sulfate, nitrate and nitrite, no detectable organic carbon, and a complete absence of sunlight [4]. The metabolic profile of this community was compared to that of other aquatic metagenomes, including representatives from a freshwater aquifer located in South Australia [19], the deep-sea (1,000 m Marmara Sea [20]), and other marine environments (Chesapeake Bay Estuary, Isabella Island Mangroves, Coastal Caribbean Sea [21]). The metabolic profile of the Weebubbie Cave metagenome overlapped most closely with the deep Marmara Sea metagenome [9]. These two metabolic profiles grouped together with the South Australia Groundwater Aquifer metagenome to form a separate branch from the other examined marine metagenomes, which all represented photic zone habitats.

The grouping of the Weebubbie metagenome metabolic profile with that of the Marmara deep-sea environment reflects similar abundances for many metabolic categories. The "iron acquisition and metabolism" group in particular stands out as relatively abundant in these two environments, with the Weebubbie Cave metagenome showing the highest relative abundance of iron acquisition gene fragments and suggesting an environment low in biologically available iron. Examination of the predicted gene products within this subsystem indicated that the cave metagenome includes numerous iron transporter components, as well as receptors and siderophores. Further investigation of the water chemistry of these caves specifically examining ferrous iron levels would be particularly interesting in light of these findings.

6.6 Conclusions

The Nullarbor cave slime communities are comprised of unusual chemolithoautotrophs that utilize multiple nitrogen species in energy metabolism. The communities provide a fascinating example of the adaptability of microbial life, which has successfully occupied a niche that appears to be completely isolated from sunlight or photosynthetically derived carbon. In 2001, Holmes and colleagues applied up-to-date molecular methods to examine the phylotypic structure of this community. The findings at that time provided the first glimpse into the ecology of this rare system [4]. A little more than 10 years later, biologists can make use of a range of new technologies to take an increasingly in-depth look into the ecology, taxonomy, and metabolic composition of these unusual and complex communities. Next generation sequencing technologies used to carry out shotgun metagenomic and large scale 16S rRNA amplicon sequencing have revealed fascinating insights into the taxonomically diverse community within Weebubbie Cave. Metagenomic sequencing allowed us to look directly for genes relating to nitrogen cycling in this community and indicated that both stages of nitrification are likely key to energy generation in this community. As such, the Weebubbie Caves slime curtains represent a diverse assemblage of chemolithoautotrophic bacterial and archaeal communities driven by nitrification. However, a number of questions remain to be explored:

- How similar are microbial slime communities, both within and between Nullarbor cave systems?
- How do changes in the microbial community reflect the variations in water chemistry observed between the caves?
- Can a link be made between the morphotypes observed via microscopy and the phylotypes detected via DNA sequencing?
- Can we experimentally confirm the carbon and nitrogen cycling models predicted from the metagenomic data?
- Are these extensive microbial communities able to support animal communities?

Additional sampling expeditions in these cave systems should provide further insight into these interesting chemolithoautotrophic assemblages.

Acknowledgments

We would like to acknowledge the cave divers who have provided samples, photography, and useful discussions: Peter Rogers, Cheryl Bass, Bob Davis, Phil Prust, Steve Trewavas, David Rhea, Paul Hosie, and Liz Rogers.

References

[1] Miller CR, James NP, Bone Y. Prolonged carbonate diagenesis under an evolving late cenozoic climate; Nullarbor Plain, southern Australia. Sediment Geol, 2012, 261, 33–49.

[2] Burnett S, Webb JA, White S. Shallow caves and blowholes on the Nullarbor Plain, Australia—Flank margin caves on a low gradient limestone platform. Geomorphology, 2013, 201, 246–253.

[3] Webb JA, James JM. Karst evolution of the Nullarbor Plain, Australia. In: Harmon RS, Wicks CM, eds. Karst geomorphology, hydrology and geochemistry. Geol Soc Am Special Paper 2006, 404, 65–78.
[4] Holmes AJ, Tujula NA, Holley M, et al. Phylogenetic structure of unusual aquatic microbial formations in Nullarbor caves, Australia. Environ Microbiol, 2001, 3, 256–264.
[5] Richards AM. An ecological study of the cavernicolous fauna of the Nullarbor Plain Southern Australia. J Zool Lond, 1971, 164, 1–60.
[6] Contos AK, James JM, Heywood B, Pitt K, Rogers P. Morphoanalysis of bacterially precipitated subaqueous calcium carbonate from Weebubbie Cave, Australia. Geomicrobiol J, 2001, 18, 331–343.
[7] James J, Rogers P. The "mysterious" calcite precipitating organism of the Nullarbor caves, Australia. In: Sasowsky ID, Palmer M, eds. Breakthroughs in Karst Geomicrobiology and Redox Geochemistry, Special Publication 1, Leesburg, Virginia, Karst Waters Institute, 1994, 34–35.
[8] Lane DJ. 16S/23S rRNA sequencing. In: Stackebrandt E, Goodfellow M, eds. Nucleic Acid Techniques in Bacterial Systematics. Chichester: John Wiley and Sons, 1991, 115–175.
[9] Tetu SG, Breakwell K, Elbourne LD, Holmes AJ, Gillings MR, Paulsen IT. Life in the dark: metagenomic evidence that a microbial slime community is driven by inorganic nitrogen metabolism. ISME J, 2013, 7, 1227–1236.
[10] Bates ST, Berg-Lyons D, Caporaso JG, Walters WA, Knight R, Fierer N. Examining the global distribution of dominant archaeal populations in soil. ISME J, 2011, 5, 908–917.
[11] Brochier-Armanet C, Boussau B, Gribaldo S, Forterre P. Mesophilic Crenarchaeota: proposal for a third archaeal phylum, the Thaumarchaeota. Nature Rev, 2008, 6, 245–252.
[12] Pester M, Schleper C, Wagner M. The Thaumarchaeota: an emerging view of their phylogeny and ecophysiology. Curr Opin Microbiol, 2011, 14, 300–306.
[13] Walker CB, de la Torre JR, Klotz MG, et al. *Nitrosopumilus maritimus* genome reveals unique mechanisms for nitrification and autotrophy in globally distributed marine crenarchaea. Proc Nat Acad Sci USA 2010, 107, 8818–8823.
[14] Blainey PC, Mosier AC, Potanina A, Francis CA, Quake SR. Genome of a low-salinity ammonia-oxidizing archaeon determined by single-cell and metagenomic analysis. PLoS ONE, 2011, 6, e16626.
[15] Tourna M, Stieglmeier M, Spang A, et al. *Nitrososphaera viennensis*, an ammonia oxidizing archaeon from soil. Proc Nat Acad Sci USA, 2011, 108, 8420–8425.
[16] Park SJ, Ghai R, Martin-Cuadrado AB, et al. Genomes of two new ammonia-oxidizing archaea enriched from deep marine sediments. PLoS ONE, 2014, 9, e96449.
[17] Muller F, Brissac T, Le Bris N, Felbeck H, Gros O. First description of giant Archaea (Thaumarchaeota) associated with putative bacterial ectosymbionts in a sulfidic marine habitat. Environ Microbiol, 2010, 12, 2371–2383.
[18] Lucker S, Wagner M, Maixner F, et al. A *Nitrospira* metagenome illuminates the physiology and evolution of globally important nitrite-oxidizing bacteria. Proc Nat Acad Sci USA 2010, 107, 13479–13484.
[19] Smith RJ, Jeffries TC, Roudnew B, et al. Metagenomic comparison of microbial communities inhabiting confined and unconfined aquifer ecosystems. Environ Microbiol, 2012, 14, 240–253.
[20] Quaiser A, Zivanovic Y, Moreira D, Lopez-Garcia P. Comparative metagenomics of bathypelagic plankton and bottom sediment from the Sea of Marmara. ISME J, 2011, 5, 285–304.
[21] Rusch DB, Halpern AL, Sutton G, et al. The Sorcerer II Global Ocean Sampling expedition: northwest Atlantic through eastern tropical Pacific. PLoS Biol, 2007, 5, e77.

Sarah K. Carmichael and Suzanna L. Bräuer

7 Microbial Diversity and Manganese Cycling: A Review of Manganese-oxidizing Microbial Cave Communities

Abstract: Microorganisms have long been known to mediate manganese (Mn) oxidation in a variety of environments, including caves. Microbial Mn oxide minerals are typically dark brown to black in color, nm-scale, and poorly crystalline, with birnessite (layer) or todorokite (tunnel) crystal structures. Both bacteria and fungi produce Mn oxide minerals, although the exact mechanism for Mn oxidation remains elusive. Chemolithoautotrophic Mn oxidation is highly unlikely to be carried out with the enzymes currently known, although indirect oxidation of Mn during heterotrophic growth or reproduction has been observed in both bacteria and fungi. Differences in nutrient availability in caves influence not only the microbial community structure associated with ferromanganese deposits, but also Mn cycling and microbial function. This chapter serves as a comprehensive review of all known Mn-oxidizing bacteria and fungi, as well as all known mechanisms of microbial Mn oxidation, with a special emphasis on the microbial populations found in cave environments.

7.1 Introduction

Ferromanganese crusts are common in both terrestrial and marine environments on the Earth's surface and in the subsurface. Although iron (Fe) oxide minerals form through both biological and abiotic processes ([1] and references therein), the presence of manganese (Mn) oxides, hydroxides, and oxyhydroxides (generally referred to as Mn oxides) on Earth's surface is usually associated with the presence of Mn-oxidizing microbes [2–5], as abiotic Mn oxidation is kinetically inhibited even in the presence of free oxygen [6, 7]. Microorganisms increase the rate of Mn oxidation by up to five orders of magnitude [4, 8]. Microbial Mn oxidation on early Earth has been reported to be a causal mechanism for the evolution of modern photosynthesis and the rise of atmospheric O_2 [7, 9], although this finding has generated some controversy [10, 11].

The first evidence for biological oxidation of Mn was found in soils in 1946 [12]. In marine environments, biologically mediated Mn oxide deposits are commonly associated with hydrothermal vents and found in seafloor nodules and concretions [see review by 13]. Direct and indirect evidence for the biological formation of Mn coatings, crusts, and nodules can be found in diverse terrestrial environments, such as deserts [14–19], rivers and streams [20–24], springs [25, 26], soils [27–30], ore deposits [31, 32], and caves (discussed further in detail). Mn oxide mixtures have also been used as media in prehistoric cave art [33–36].

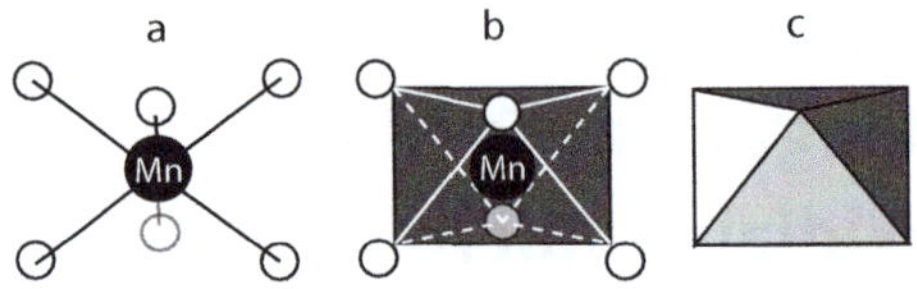

Fig. 7.1. Mn oxide octahedral structure.

Mn oxide minerals are built from Mn oxide octahedra (▸ Fig. 7.1) and form more than 30 distinct minerals [37]. Currently, all known biological Mn oxides are nm-scale and poorly crystalline [2], formed from sheets of Mn oxide octahedra (birnessite [37]) (▸ Fig. 7.2 (a)) or 3 × 3 tunnels of octahedra (todorokite [38]) (▸ Fig. 7.2 (b)). Even with speculation that Mn oxide crystal structures are dependent on an individual microbial species [39], experimental evidence suggests that substrate geochemistry plays a significant role in determining crystal structures [40]. Biologically deposited Mn oxides are highly reactive due to their many vacancies, negatively charged surfaces, and adsorptive properties [41–43]. This makes them environmentally important tools for remediation of contaminated surface and groundwaters [42, 44, 45].

Fe- and Mn-oxidizing microbial strains have been successfully isolated from the environment for many years [13, 46–50]. Results from molecular phylogenetic analyses [51], as well as culturing studies [15, 45, 52–54], suggest that Mn-oxidizing microbial diversity is far greater than previously recognized. Despite these recent advances, we have only begun to scratch the surface of microbial diversity, including fungi (which have been grossly understudied), especially in caves where microbial diversity among Mn-oxidizers is astounding [54–60].

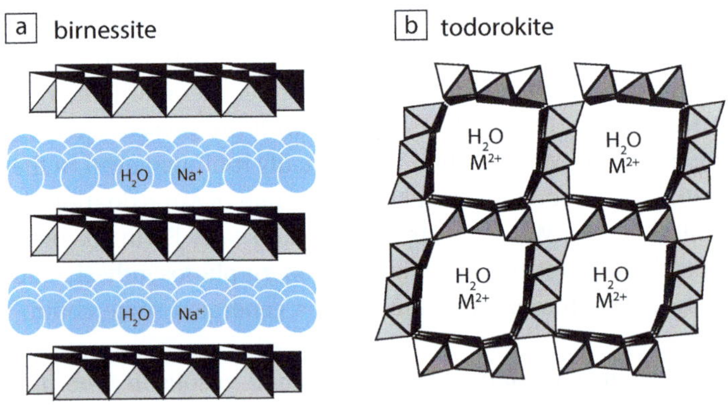

Fig. 7.2. (a) Birnessite structure (layers of octahedra), and (b) todorokite (3 × 3 tunnels of octahedra) with water molecules or cations in the spaces between octahedral sheets and tunnels (modified from [37, 38]).

7.2 Manganese Oxides in Caves

Manganese oxides in caves are typically found as fine-grained, poorly crystalline black or dark brown coatings and crusts, or as biofilms on cave walls and rock surfaces (▸ Fig. 7.3). They can also fill voids or vugs in karst and paleokarst. They generally belong to the birnessite or todorokite crystal groups, although other minerals, such as ranciete and pyrolusite, have also been observed (▸ Tab. 7.1). Microbes associated with Mn oxide production have long been described from cave systems [see review by 61], and microbial association with Mn oxidation has often been observed via electron microscopy and light microscopy, or described using culture-based techniques [62–70]. Recent studies provide detailed molecular identification of the particular organisms responsible for Mn oxide deposition in specific cave systems [54, 56, 58, 59, 71–74] (▸ Tab. 7.1). Many studies have inferred a biological origin for cave ferromanganese deposits from the existing literature [75–83], while others document the presence of ferromanganese deposits but do not discuss potential biological origins for them [84–87].

The influence of microbes on speleogenesis and cave geochemistry is apparent [see review by 60]. Different microbial communities are present in caves with diverse formation histories. Epigenic cave systems form via carbonic acid during interaction with descending meteoric water in the shallow subsurface. In contrast, hypogene cave systems form when ascending groundwater interacts with reduced gases, such as hydrogen sulfide (typically derived from kerogen-rich deposits), and the gas oxidizes to sulfuric acid. Hypogene caves are typically deeper and contain significant chemolithoautotrophic microbial communities. Although based on relatively few studies, initial molecular (16S RNA) gene sequencing indicates that hypogene com-

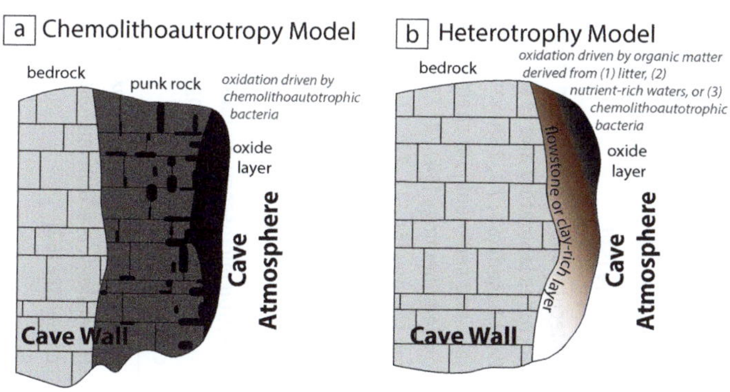

Fig. 7.3. Cave Mn oxides are typically found as fine-grained, poorly crystalline black or dark brown coatings, crusts, and biofilms on rock surfaces. Schematic diagram of (a) oligotrophic cave systems showing chemolithoautotrophic coatings on cave walls (modified from [56]), and (b) heterotrophic systems showing surface Mn oxidation driven by nutrient input from exogenous carbon (either via nutrient-rich water, litter, feces, or other detrital material), or through interactions with chemolithoautotrophic bacteria within the community.

Table 7.1. Published studies of cave Mn oxide deposits with molecular, cultured, or microscopic evidence for a microbial origin.

Cave location	Methodologies and References	Observations	Mn-oxidizing microbes present or inferred
Lechuguilla Cave and Spider Cave, New Mexico (USA)	Molecular- and culture-based studies [56], electron microscopy [56, 63], Mn oxide mineralogy, rock chemistry [56, 71]	Hypogene caves with black and brown crusts common on cave walls; SEM imaging shows sheet-like, crumpled tissue paper, boxwork and star morphology, and associated microbial filaments. Mineralogy consistent with birnessite.	*Hyphomicrobium*, *Pedomicrobium*, and *Pantoea*, as well as *Bacillus* and *Alcaligenes*
Wind Cave, South Dakota (USA)	Molecular and culture-based studies [59], electron microscopy [80], FT-IR [80], water and sediment chemistry [59]	Hypogene cave, susceptible to agricultural runoff and anthropogenic impact. Brown, earthy masses. Mineralogy inconclusive due to sample impurities.	*Pseudomonas* sp.
Carter Salt Peter Cave, Rockhouse Cave, and Worley's Cave, Tennessee, and Daniel Boone Caverns, Virginia (USA)	Molecular and culture-based studies (includes laboratory and *in situ* field-based) [54, 72, 73]; water [72], rock and sediment [54] chemistry, Mn oxide mineralogy [54, 72, 73], electron microscopy [54, 72, 73]	Epigene caves with black crusts common on cave walls and flowstone, biofilms common in nutrient-rich groundwater seeps. Carter Salt Peter Cave is susceptible to agricultural runoff and anthropogenic impact. SEM and light microscopy show crumpled tissue paper morphology, and intercellular Mn oxidation in and along fungal cells and hyphae. Mineralogy consistent with birnessite, buserite, and todorokite.	*Leptothrix*, *Janthinobacterium*, *Flavobacterium*, *Pseudomonas* spp., Plectosphaerellaceae, *Acremonium*, *Alternaria*, *Pleosporales*, *Leptosphaerulina*, two genus *incertae sedis* tentatively identified as a member of the genus *Mortierella* within the Zygomycota/deep branching fungal lineages
Dubuque Caves, Iowa (USA)	Electron microscopy, culture-based studies, Mn oxide mineralogy, rock chemistry [65]	Hypogene cave with black crusts, flowstones, and stalactites. Mineralogy consistent with birnessite and todorokite.	*Leptothrix* spp.
Matts Black Cave, West Virginia (USA)	culture-based studies, Mn oxide mineralogy, electron microscopy, water chemistry [62]	Dull, thick (up to 10 mm) soft black coatings on cave ceilings, with hard, glossy black coatings (up to 2 mm) on cave walls. Mineralogy consistent with birnessite.	*Leptothrix* spp.

Table 7.1. (cont.) Published studies of cave Mn oxide deposits with molecular, cultured, or microscopic evidence for a microbial origin.

Cave location	Methodologies and References	Observations	Mn-oxidizing microbes present or inferred
Paulter Cave, Iowa (USA)	Mn oxide mineralogy, electron microscopy, and water chemistry [64]	Epigene cave with shiny black and dull brown coatings on stream cobbles and in seeps. Mineralogy consistent with birnessite and buserite.	presence inferred (based on *Gallionella* spp. associated with Fe oxides)
Lascaux Cave (France)	molecular and culture-based studies, Mn oxide mineralogy, electron microscopy [74]	Irregular black stains associated with clay-rich sediments	*Acremonium nepalense*
Sima de la Higuera Cave (Spain)	Electron microscopy, Mn oxide mineralogy, rock chemistry and mineralogy [68]	Thermal cave formed from CO_2-rich fluids originating from magmatic activity. Botryoidal aggregates coating boxwork calcite, composed of todorokite and pyrolusite	Bacterial influence inferred from electron microscopy
El Soplao Cave, Cantabria (Spain)	Electron microscopy, Mn oxide mineralogy, rock and sediment chemistry and mineralogy [69]	Hypogene cave with black rimstone-dam deposits and oncoids with truncated laminae. Mineralogy consistent with birnessite ± ranceite.	Bacterial influence inferred from fossil microstructures
Grand Cayman Karst Terrain (British West Indies)	Electron microscopy, Mn oxide mineralogy, and rock chemistry and mineralogy [82]	Epigene cave and karst with various morphologies of Mn oxide deposits. Mineralogy consistent with birnessite or todorokite.	Bacterial influence inferred from electron microscopy
Jaskinia Czarna Cave (Poland)	Electron microscopy, Mn oxide mineralogy, and rock chemistry and mineralogy [70]	Black flowstones of poorly crystalline Mn oxide layers.	Fungal and bacterial influence inferred from electron and light microscopy
Azé Cave, Saône-Et-Loire (France)	Electron and light microscopy, Mn oxide mineralogy, molecular identification attempted [67]	Black coatings up to several cm in thickness on cave walls, flowstone, animal bones. Poorly crystalline material tentatively identified as birnessite.	Fungal and bacterial influence inferred from electron and light microscopy
Trémies Cave, 3PP Cave, Bagaud Cave (France)	Electron microscopy, Mn oxide mineralogy [66]	Submarine caves (limestone, sandstone and conglomerate) formed by differential erosion with black coatings on cave walls.	Bacterial and fungal influence inferred from microscopy

munities [56–58, 60 and references therein, 63, 88] may have little taxonomic overlap with communities found in caves that formed from interactions with descending surface waters and groundwaters [54, 72]. It is possible that these observed differences are primarily due to the sequencing methods used (cloning and Sanger sequencing), which generally yield low numbers of sequences. Thus, additional studies of microbial populations across both hypogene and epigenic cave systems incorporating next generation sequencing technologies to obtain greater sequencing depth are needed to confirm the role of speleogenesis in microbial community structure.

Molecular investigations of 16S RNA gene sequences from ferromanganese deposits in different types of cave systems have only been done from a handful of locations, but demonstrate that the microbial community structures of ferromanganese deposits are distinct. Epigenic caves in the southern Appalachian Mountains, USA, have abundant Mn-oxidizing bacteria, such as *Leptothrix* spp., *Janthinobacterium* spp., *Flavobacterium* spp., and *Pseudomonas* spp. [54]. Hypogenic caves that are now susceptible to agricultural runoff or human foot traffic, such as Wind Cave in South Dakota, USA, also contain *Pseudomonas* spp., as determined by cloning and sequencing 16S rRNA genes [59]. However, ferromanganese deposits in Lechuguilla and Spider Caves in New Mexico, USA (that formed from hypogenic processes) contain primarily chemolithoautotrophic bacteria as well as organisms related to the Mn-oxidizing bacterial taxa *Hyphomicrobium*, *Pedomicrobium*, and *Pantoea* [56, 58]. When cave waters contain elevated nutrient concentrations (e.g. nitrate, nitrite, and phosphate), the microbial community structure of known Mn- and Fe-oxidizing microbes closely resembles that of above-ground wetland communities [54, 64, 72], typically dominated by *Leptothrix* and *Gallionella* spp. [89, 90].

Molecular tools are powerful mechanisms for determining the associations of microbes with Mn oxidation, but culturing is necessary to provide specific information about which strains produce visible Mn oxide deposits. Bacterial cultures of *Leptothrix* spp. associated with ferromanganese deposits were first isolated in the 1980s from caves in Iowa and West Virginia [65, 91]. Mn- and Fe-oxidizing bacteria, including *Leptothrix*, *Janthinobacterium*, *Flavobacterium*, and *Pseudomonas* spp., have been isolated from caves in the southern Appalachian Mountains [54] and the southwestern USA [56, 58], although the cultures from the southwest are mixed rather than isolates. Culturing techniques for Mn-oxidizing bacteria are listed in [43].

In addition to bacteria, Mn-oxidizing fungi are also found in caves [72–74]. *Acremonium nepalense* has been implicated in spreading black stains in parts of Lascaux Cave, France [74], based on cultured isolates and fungal internal transcribed spacer (ITS) sequencing for ITS1, 5.8S rDNA, and ITS2 gene regions. *In situ* and *in vitro* enrichment and culturing studies in southern Appalachian cave systems retrieved Mn-oxidizing members of the orders Glomerellales and Pleosporales within the Ascomycota, as well as at least two Mn-oxidizing enrichments that were dominated almost entirely by a member of the *Mortierella* genus, within the Zygomycota/early diverging fungal lineages [73].

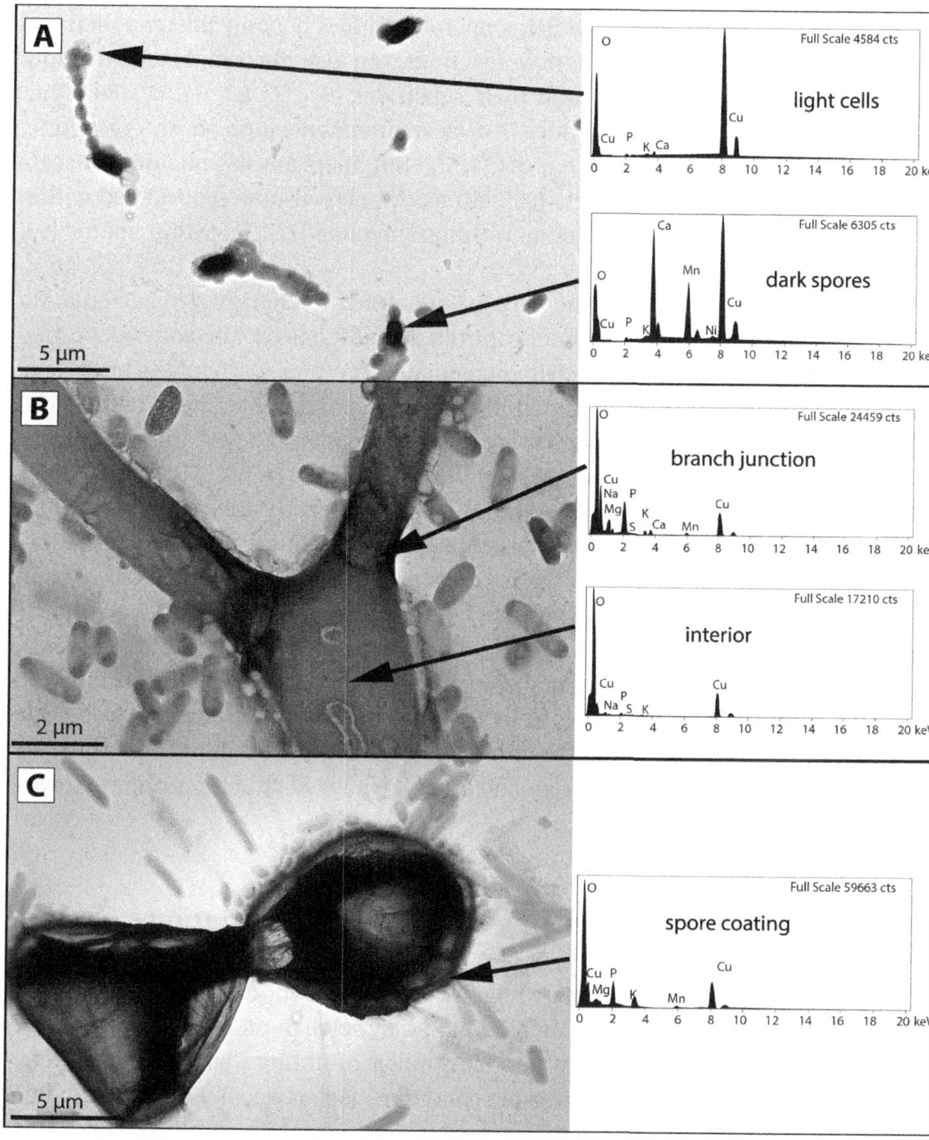

Fig. 7.4. Transmission electron microscopy (TEM) image of Mn oxidation (a) within cells and/or spores of an unknown organism cultured from a drip network [92], (b) within branches of fungal hyphae (genus *incertae sedis*, tentatively identified as a member of the genus *Mortierella* within the Zygomycota/deep branching fungal lineages) cultured from a piece of electrical tape (image modified from [73]), and (c) within cells or spores from an unknown organism cultured from a discarded cotton sock [92], all from Carter Saltpeter Cave in Tennessee, USA.

Electron microscopy of Mn-oxidizing microbes shows not only the microbial and mineral morphologies involved in these reactions, but also the complex relationship between microbial communities and their substrates [54, 56, 63, 64, 67, 72]. Typically, biologically deposited Mn oxides in cave environments form poorly crystalline, nanometer-scale crumped tissue shapes [54, 56]. But, there are also micrometer-scale star and boxwork shapes, or sheets [56]. Mn oxides may also be sequestered within cell membranes at the tips and joints of fungal hyphae [72], or within spores [92] (▸ Fig. 7.4).

Numerous laboratory experiments have attempted to characterize bacterial Mn oxide deposition and structures under controlled conditions [2]. Mn oxides deposited by bacteria initially form layered phyllomanganate structures, particularly birnessite, which may then be converted to tunnel-like todorokite and hollandite structures [93, 94]. Mn oxides produced by cave microorganisms are consistent with this experimental work, as they typically form poorly crystalline birnessites (▸ Tab. 7.1). Interestingly, tiny clumps of poorly crystalline biogenic Mn oxides can recrystallize and reorganize to form ordered Mn oxide crystals (particularly buserite, birnessite, and vernadite) in as few as 8 months, as long as the bacteria producing these minerals have sufficient nutrition [56]. Differences in nutrient content between caves could explain why some caves exhibit highly crystalline Mn oxide minerals besides birnessite (▸ Tab. 7.1), and others have thick deposits of poorly crystalline birnessite even after > 1 million years [69]. Although experiments to determine the structures of fungal Mn oxides are few [95–98], they do suggest that the todorokite and birnessite mineral structures deposited by fungi may be distinct from bacterially deposited structures and may even be species specific [39, 98].

7.3 Functions and Mechanisms of Manganese Oxidation

The function of Mn(II) oxidation has remained enigmatic for decades. For example, researchers have speculated that Mn oxidation may serve to protect cells against UV radiation or oxidative stress, or may be a mechanism of storing a powerful oxidant for use as an electron acceptor in anaerobic conditions [see review by 4]. Mn oxidation has long been considered a potential energy-yielding reaction, although conclusive evidence for chemolithoautotrophic Mn oxidation has remained elusive [99]. Indeed, Mn oxidation by one two-electron transfer is generally considered favorable above pH 1–3 (▸ Fig. 7.5 (a)) [6]. More recently however, it has become apparent that most bacterial mechanisms of Mn oxidation include two one-electron transfer reactions, from Mn(II) to Mn(III) and from Mn(III) to Mn(IV) [5, 100]. In the presence of Mn(III) chelating ligands or siderophores, such as pyroverdine [101, 102], Mn(III) can be stabilized and is an important intermediate in the environment [103]. Mn oxidation via two one-electron transfer steps is unfavorable with oxygen as a substrate, but is highly favorable with superoxide (O_2^-) or hydroxyl (OH^-), and is also favorable with peroxide

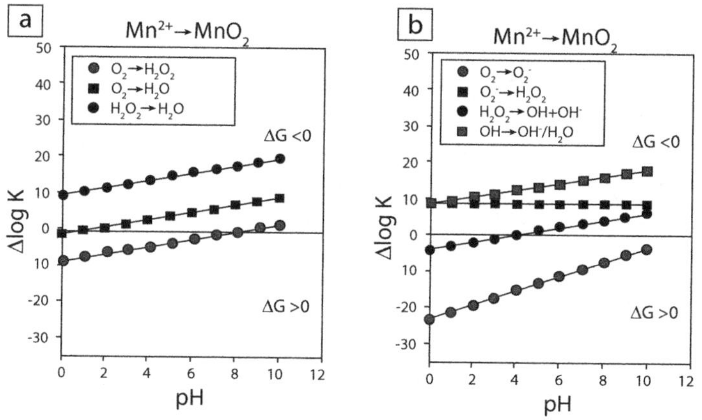

Fig. 7.5. Electron transfer reactions for Mn(II–IV) oxidation with different oxygen species; (a) represents reactions for one two-electron transfer reaction, while (b) represents two one-electron transfer reactions. Modified from [6].

(H_2O_2) above pH 4 (▸ Fig. 7.5 (b)) [6]. Thus, it is clear that chemolithoautotrophic Mn oxidation is highly unlikely to be carried out with the enzymes currently known. As an alternate mechanism, bacterial species are thought to oxidize Mn more indirectly via the production of superoxide during heterotrophic growth or reproduction, including *Roseobacter* sp. AzwK-3b [104] and several species of fungi in the Ascomycota, including *Stilbella aciculosa* [105], *Pyrenochaeta* spp. DS3sAY3a, and *Stagonospora* spp. SRC1lsM3a [106]. However, the connection between the proteins associated with antioxidant expression and Mn(II)-oxidizing activity is still elusive [107]. Regardless of oxidation mechanism, the relatively simple Mn oxidation process, outlined in [3], must now be reevaluated to account for this newer experimental data (▸ Fig. 7.6).

7.3.1 Enzymes associated with manganese oxidation

Several different putative Mn oxidase enzymes have been identified and studied in bacteria, including MopA in *Aurantimonas manganoxydans* SI85-9A1 and *Erythrobacter* spp. strain SD-21 [108], MofA in *Leptothrix discophora* SS-1 [109, 110], and MoxA in *Pedomicrobium* spp. strain ACM 3067 [111]. CotA has recently been suggested in *Bacillus pumilus* WH4 [112], and CueO in recombinant strain *Escherichia coli* strain ECueO [113], although results in these studies need to be verified with a knockout mutant as the high concentrations (5 mM) of Mn(II) utilized for both experiments make the results unclear. CumA was originally thought to be involved in Mn oxidation in *Pseudomonas putida*, GB-1 [114], although it was later determined not to be involved [115]. More definitive evidence for Mn-oxidizing activity has been collected for the enzymes McoA and MnxG in *P. putida* GB-1 [116] and for MnxG in *Bacillus* spp. SG1 [117]. In fact, the MnxDEFG holoenzyme has recently been shown to catalyze both

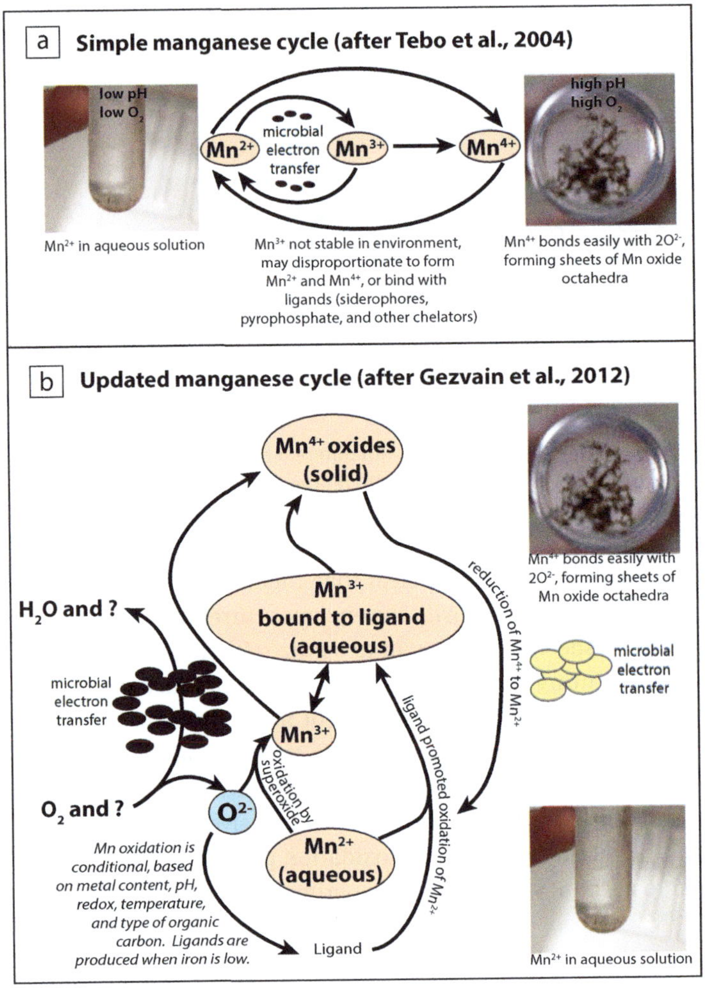

Fig. 7.6. Microbial Mn oxidation cycle. (a) Represents the original cycle as developed by [3], and (b) represents an updated cycle that incorporates the role of superoxide in Mn oxidation, as developed by [5].

Table 7.2. Enzymes associated with biological Mn oxidation [120].

Copies of putative manganese oxidases found in sequenced genomes	Copies in genome			
	mnx G	*mco A*	*mof A*	*mop A*
Pseudomonas putida GB-1	1	1	0	1
Aurantimonas manganoxydans SI85-9A1	0	1	0	1
Erythrobacter sp. SD-21	0	0	0	1
Leptothrix cholodnii SP-6	2	2	1	1
Roseobacter sp. AzwK-3b	0	0	0	2
Bacillus sp. SG1	1	0	0	0

oxidation steps, from Mn(II) to Mn(III) and from Mn(III) to MnO_2 [118]. Interestingly, many organisms carry redundant copies of these putative Mn oxidases (► Tab. 7.2), as a search in the IMG database [119] for oxidase homologues among sequenced Mn-oxidizers has revealed [120]. This redundancy may allow the organisms to oxidize Mn(II) over broader range of conditions, although the physiological purpose of Mn oxidation is still speculative [5], as discussed in Section 7.3.

7.3.2 Mangenese-oxidizing bacteria and fungi

Many different bacterial species from the phyla Proteobacteria, Firmicutes, Actinobacteria, and Bacteroidetes oxidize Mn(II) in axenic culture (► Tab. 7.3). *Crenothrix*, *Clonothrix* [121], and Planctomycetes [122] have also been reported to oxidize Mn and Fe, but have never been obtained in pure culture. Ascomycota and Basidomycota are known Mn(II)-oxidizers (► Tab. 7.4). Basidomycota oxidize Mn(II) to Mn(III), which is used to further oxidize recalcitrant phenolics, lignins, and humics via Mn peroxidases, or a versatile peroxidase that can also function independent of Mn [123]. Many different species of Basidiomycetes are capable of Mn(II) oxidation, including *Cryptococcus albidus* [124], as well as at least 56 other species, as reviewed by [125].

Table 7.3. Reported Mn(II)-oxidizing bacterial isolates. Cave isolates shown in bold.

Phylum (Class)	Closest representative genus/genera	Selected references (cave isolates in bold)
Actinobacteria	*Agromyces* *Cellulomonas* *Microbacterium*	[126]
	Arthrobacter[a]	[127–130]
	Geodermatophilus *Actinomycete*	[130]
	Micrococcus	[16, 130]
	Pseudonocardia *Lapillicoccus* *Leifsonia* *Terrabacter*	[28]
	Nocardia *Streptomyces*	[131]
	Geodermatoiphilus	[130]
	Corynebacterium	[132]
	Propionibacterium	[133]
	Rhodococcus	[134]

Table 7.3. (cont.) Reported Mn(II)-oxidizing bacterial isolates. Cave isolates shown in bold.

Phylum (Class)	Closest representative genus/genera	Selected references (cave isolates in bold)
Bacteroidetes	*Cytophaga*	[46, 135]
	Flavobacterium	[45, 46] **[54]**
	Riemerella	[136]
	Sphingobacterium *Terrimonas*	[15]
Firmicutes	*Bacillus*[a]	[45, 51, 130, 135, 137–139]
	Exiguobacterium[a]	[137]
	Lysinibacillus	[90]
	Planococcus	[16]
	Staphylococcus	[133, 140]
Alphaproteobacteria	*Afipia* *Rhizobium* *Shinella* *Sphingopyxis*	[15]
	Agrobacterium SRC1K2fb	[45]
	Aurantimonas *Fulvimarina*	[141]
	Caulobacter	[135]
	Erythrobacter	[142]
	Hyphomicrobium	[130, 143, 144]
	Methylarcula *Sulfitobacter*	[53]
	Methylobacterium *Mycobacterium*	[28]
	Pedomicrobium	[144, 145]
	Paracoccus *Sinorhizobium*	[126]
	Roseobacter	[146]
	Rhodobacter	[52]

Table 7.3. (cont.) Reported Mn(II)-oxidizing bacterial isolates. Cave isolates shown in bold.

Phylum (Class)	Closest representative genus/genera	Selected references (cave isolates in bold)
Betaproteobacteria	*Achromobacter*	[15]
	Duganella[b]	[15, 147]
	Albidiferax[b]	[147]
	Burkholderia	[28]
	Caldimonas	[148]
	Chromobacterium	**[130, 135]**
	Cupriavidus/Ralstonia *Variovorax*	[126]
	Janthinobacterium	[15]**[54]**
	Leptothrix	**[54, 65]**[149, 150]
	Oxalicibacterium strain MY14, AB008503	[136]
Gammaproteobacteria	*Aeromonas*	[46]
	Alteromonas *Halomona* *Pseudoalteromonas* *Marinobacter* *Microbulbifer*	[53]
	Citrobacter	[27, 151]
	Enterobacter *Proteus*	[132]
	Escherichia	[126]
	Klebsiella *Luteibactor*	[15]
	Oceanospirillum *Vibrio*	[48]
	Pantoea	[15, 28]
	Pseudomonas[a]	[45]**[54]**[130, 132, 151–153]
	Shewanella[a]	[52, 154]

[a] Can also reduce Mn(III/IV).

[b] Experiments in [147] were conducted with 17 mM Mn(II); additional experiments are needed at lower Mn(II) concentrations to confirm whether or not the Mn(II) oxidation was biologically rather than abiotically mediated.

Table 7.4. Reported Mn(II)-oxidizing Ascomycota fungi (for a review of Basidiomycota, see [125]).

Phylum (Class)	Closest representative genus/genera	Selected references
Ascomycota	*Acremonium strictum* strain DS1bioAY4a *Alternaria alternate* *Microdochium bolleyi* *Phoma* sp. *Pithomyces chartarum* *Plectosphaerella cucumerina* *Pyrenochaeta* sp. *Stagonospora* sp.	[45]
	Acremonium strictum strain KR21–2	[28, 96, 155]
	Bipolaris-related *Cladosporium*-related *Cyphellophora laciniata* *Ophiobolus*-related	[28]
	Coniothyrium	[124, 156]
	Nectria	[156]
	Oidium mangiferae	[157]
	Stilbella aciculosa	[45, 105]
Basidiomycota	*Various*	for a review of Basidiomycota, see [125]

7.3.3 Mechanisms of microbial manganese oxidation in caves

All of the Mn oxidizers discovered to date are heterotrophic. From epigenic caves in the southern Appalachian Mountains, field evidence indicates that Mn oxidation is dramatically stimulated by exogenous carbon input [72, 73] into what are otherwise carbon limited environments [55]. In one case, dilute sewage input into Carter Saltpeter Cave in Tennessee, USA, led to a massive bloom of a bacterially driven, Mn-oxidizing biofilm [72]. Most likely, this was due to input of organic acids common in sewage, such as acetate, butyrate, and propionate [158]. Although many Mn-oxidizing bacteria can grow on sugars, such as glucose, they often do not oxidize in the presence of glucose. Oxidation occurs during growth on acetate, succinate, and peptides that enter through the TCA cycle. These growth substrates are commonly found in casamino acids, yeast extract, and peptone [43]. In contrast, fungal oxidation is stimulated by sugars, such as glucose, or more complex molecules (i.e. cellulose or other polymers) found in cave litter, such as socks, electrical tape, etc. Consequently, where bacterial Mn oxidation is stimulated by contaminated runoff or non-point source sewage water (▸ Fig. 7.3 (b) and 7.7) [72], solid litter input often stimulates fungal oxidation (▸ Fig. 7.7) [73]. Moreover, because many Mn-oxidizing fungi produce an excess of organic acids, such as oxalate and pyrophosphate, Mn(III) chelation can facilitate Mn oxidation [159]. Under

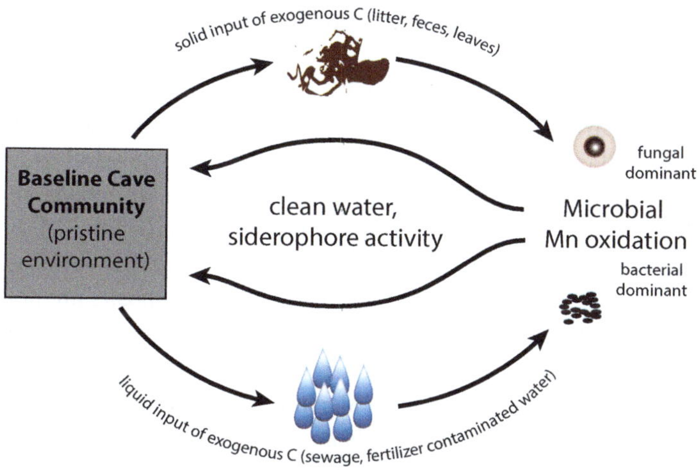

Fig. 7.7. Working model for stimulation of Mn oxidation by exogenous carbon inputs in shallow, epigene cave environments and in cave environments susceptible to anthropogenic impact (from [92]).

these circumstances, not only can rates of Mn oxidation increase, but total bacterial loads increase, as well.

Several field studies provide evidence to support the hypothesis that nutrient loading in caves via tourism [160], septic effluent [161], and input of rich carbon sources, such as guano, feces, and human traffic [162], affect total bacterial loads. In all three of the studies [160–162], cultivable counts of bacteria, either total colony forming units (CFUs) of total aerobic bacteria or total fecal coliform bacteria, were at least two orders of magnitude higher than similar counts in low impact zones within the caves. As observed at the Mn Falls site in Carter Saltpeter Cave, if exogenous carbon input subsides, then the black/brown Mn crust can gradually disappear [72]. It is unclear if this is due to Mn(IV)/Mn(III) reduction or another mechanism. A wide variety of organisms can reduce Mn oxides, including *Pseudomonas* spp. [153], which were abundant in the Mn Falls site [54]. In fact, several strains of Mn(II)-oxidizing bacteria have been isolated that can also reduce Mn(IV) oxides, such as *Arthrobacter* spp. [127], *Bacillus* spp. SG1 [139], *Shewanella* spp. [52], and *Exiguobacterium* spp. [137].

In deeper caves receiving less organic input, such as the hypogene caves of the southwestern USA, Mn oxidation may occur autotrophically [4, 58]. Alternatively, the Mn-oxidizers may be reliant upon autotrophic organisms, such as Fe-oxidizers, to provide organic substrates for oxidation (▸ Fig. 7.3 (b)). This may explain why it is not uncommon to find Mn oxides forming black crusts associated with orange iron oxide layers in many environments, including caves [56, 64, 65] and deep sea environments [53, 163].

7.4 Remaining Questions about Manganese Oxidation in Caves

The role of biological Mn oxidation has been well established in several cave systems, but there is still much to be learned. Caves with active biological Mn oxidation show poorly crystalline birnessite (layer) or todorokite (tunnel) structures. However, it is not yet known why some caves contain deposits of highly crystalline Mn oxides, or contain Mn oxides with structures other than birnessite or todorokite (▸ Tab. 7.1). Is it possible that abiotic crystallization of Mn oxides, which nucleated on the surfaces of biological Mn oxides, is occurring? Or, because the ages of these deposits are not well constrained, is it possible that aging promotes increased crystallinity? It is not yet known how Fe- and Mn-oxidizing organisms in a community interact, or if any of these organisms are truly chemolithoautotrophic. Mechanisms of Mn oxidation are still hotly debated and more experimental research is necessary, particularly regarding enzymatic pathways (or combinations thereof) that result in Mn oxidation, and particularly in relatively nutrient-limited environments, such as caves.

Acknowledgments

The authors appreciate the reviews provided by A. Engel, an anonymous reviewer, as well as input from R. E. Davis. Students M. J. Carmichael, B. T. Zorn, and L. A. Roble provided electron microscopy photos, laboratory analyses, and field support. Support for this chapter was provided by NC Space Grant New Investigator awards to both S. K. Carmichael and S. L. Bräuer. Additional support to both authors was provided through Appalachian State University.

References

[1] Fortin D, Langley S. Formation and occurrence of biogenic iron-rich minerals. Earth Sci Rev, 2005, 72, 1–19.

[2] Spiro TG, Bargar JR, Sposito G, Tebo BM. Bacteriogenic manganese oxides. Accounts Chem Res, 2010, 43, 2–9.

[3] Tebo BM, Bargar JR, Clement BG, et al. Biogenic manganese oxides: Properties and mechanisms of formation. Annu Rev Earth Planet Sci, 2004, 32, 287–328.

[4] Tebo BM, Johnson HA, McCarthy JK, Templeton AS. Geomicrobiology of manganese(II) oxidation. Trends Microbiol, 2005, 13, 421–428.

[5] Geszvain K, Butterfield C, Davis R, et al. The molecular biogeochemistry of manganese(II) oxidation. Biochem Soc Trans, 2012, 40, 1244.

[6] Luther GW. The role of one-and two-electron transfer reactions in forming thermodynamically unstable intermediates as barriers in multi-electron redox reactions. Aquat Geochem, 2010, 16, 395–420.

[7] Johnson JE, Webb SM, Thomas K, Ono S, Kirschvink JL, Fischer WW. Manganese-oxidizing photosynthesis before the rise of cyanobacteria. Proc Nat Acad Sci USA, 2013, 110, 11238–11243.

[8] Emerson D. Microbial oxidation of Fe(II) and Mn(II) at circumneutral pH. In: Lovley DR, ed. Environmental Microbe-Metal Interactions. Washington, DC: American Society for Microbiology, 2000, 31–52.

[9] Allen JF, Martin W. Out of thin air. Nature, 2007, 445, 610–612.

[10] Johnson JE, Webb SM, Thomas K, Ono S, Kirschvink JL, Fischer WW. Reply to Jones and Crowe: Correcting mistaken views of sedimentary geology, Mn oxidation rates, and molecular clocks. Proc Nat Acad Sci USA, 2013, 110, E4119-E20.

[11] Jones C, Crowe SA. No evidence for manganese-oxidizing photosynthesis. Proc Nat Acad Sci USA, 2013, 110, E4118.

[12] Mann P, Quastel J. Manganese metabolism in soils. Nature, 1946, 158, 154–156.

[13] Edwards KJ, Bach W, McCollom TM. Geomicrobiology in oceanography: microbe-mineral interactions at and below the seafloor. Trends Microbiol, 2005, 13, 449–456.

[14] Dorn RI, Oberlander TM. Microbial origin of desert varnish. Science, 1981, 213, 1245–1247.

[15] Northup DE, Snider JR, Spilde MN, et al. Diversity of rock varnish bacterial communities from Black Canyon, New Mexico. J Geophys Res [Biogeosci], 2010, 115, 19.

[16] Palmer FE, Staley JT, Murray RGE, Counsell T, Adams JB. Identification of manganese-oxidizing bacteria from desert varnish. Geomicrobiol J, 1986, 4, 343–360.

[17] Taylor-George S, Palmer F, Staley JT, Borns DJ, Curtiss B, Adams JB. Fungi and bacteria involved in desert varnish formation. Microb Ecol, 1983, 9, 227–245.

[18] Raymond JR, Guthrie JGD, Bish DL, Reneau SL, Chipera SJ. Biomineralization of manganese in rock varnish. In: Skinner HCW, Fitzpatrick RW, eds. Biomineralization Processes of Iron and Manganese. Cremlingen, Germany, Catena Verlag, 1992, 321–336.

[19] Parchert KJ, Spilde MN, Porras-Alfaro A, Nyberg AM, Northup DE. Fungal communities associated with rock varnish in Black Canyon, New Mexico: casual inhabitants or essential partners? Geomicrobiol J, 2012, 29, 752–766.

[20] Saunders JA, Pritchett MA, Cook RB. Geochemistry of biogenic pyrite and ferromanganese coatings from a small watershed: a bacterial connection? Geomicrobiol J, 1997, 14, 203–217.

[21] Bargar JR, Fuller CC, Marcus MA, et al. Structural characterization of terrestrial microbial Mn oxides from Pinal Creek, AZ. Geochim Cosmochim Acta, 2009, 73, 889–910.

[22] Tani Y, Miyata N, Iwahori K, et al. Biogeochemistry of manganese oxide coatings on pebble surfaces in the Kikukawa River System, Shizuoka, Japan. Appl Geochem, 2003, 18, 1541–1554.

[23] Vojak PWL, Edwards C, Jones MV. Evidence for microbiological manganese oxidation in the River Tamar Estuary, South West England. Estuar Coast Shelf Sci, 1985, 20, 661–671.

[24] Shiller AM, Stephens TH. Microbial manganese oxidation in the lower Mississippi River: methods and evidence. Geomicrobiol J, 2005, 22, 117–125.

[25] Miller AZ, Dionisio A, Sequeira Braga MA, et al. Biogenic Mn oxide minerals coating in a subsurface granite environment. Chem Geol, 2012, 322–323, 181–191.

[26] Mita N, Maruyama A, Usui A, Higashihara T, Hariya Y. A growing deposit of hydrous manganese oxide produced by microbial mediation at a hot spring, Japan. Geochem J, 1994, 28, 71–80.

[27] Douka C. Study of bacteria from manganese concretions. Soil Bio Biochem, 1977, 9, 89–97.

[28] Cahyani VR, Murase J, Ishibashi E, Asakawa S, Kimura M. Phylogenetic positions of Mn^{2+}-oxidizing bacteria and fungi isolated from Mn nodules in rice field subsoils. Biol Fert Soils 2009, 45, 337–346.

[29] He J, Zhang L, Jin S, Zhu Y, Liu F. Bacterial communities inside and surrounding soil iron-manganese nodules. Geomicrobiol J, 2008, 25, 14–24.
[30] Robbins EE, D'Agostino JP, Ostwald J, Fanning DS, Carter V, Van Hoven RL. Manganese nodules and microbial oxidation of manganese in the Huntley Meadows wetland, Virginia, USA. In: Skinner HCW, Fitzpatrick RW, eds. Biomineralization Processes of Iron and Manganese. Cremlingen, Germany, Catena Verlag, 1992, 179–202.
[31] Polgári M, Hein JR, Vigh T, et al. Microbial processes and the origin of the Úrkút manganese deposit, Hungary. Ore Geol Rev, 2012, 47, 87–109.
[32] Kilias SP. Microbial mat-related structures in the Quaternary Cape Vani manganese-oxide (-barite) deposit, NW Milos Island, Greece. Spec Publ Soc Sed Geol, 2012, 101, 97–110.
[33] Chalmin E, Menu M, Vignaud C. Analysis of rock art painting and technology of Palaeolithic painters. Meas Sci Tech, 2003, 14, 1590.
[34] Lahlil S, Lebon M, Beck L, et al. The first in situ micro-Raman spectroscopic analysis of prehistoric cave art of Rouffignac St-Cernin, France. J Raman Spectrosc, 2012, 43, 1637–1643.
[35] Nuevo M, Martín Sánchez A, Oliveira C, Oliveira J. In situ energy dispersive X-ray fluorescence analysis of rock art pigments from the 'Abrigo dos Gaivões' and 'Igreja dos Mouros' caves (Portugal). X-Ray Spectrom, 2012, 41, 1–5.
[36] Smith D, Bouchard M, Lorblanchet M. An initial Raman microscopic investigation of prehistoric rock art in caves of the Quercy District, SW France. J Raman Spectrosc, 1999, 30, 347–354.
[37] Post JE. Manganese oxide minerals: crystal structures and economic and environmental significance. Proc Nat Acad Sci USA, 1999, 96, 3447–3454.
[38] Post JE, Heaney PJ, Hanson J. Synchrotron X-ray diffraction study of the structure and dehydration behavior of todorokite. Am Mineral, 2003, 88, 142–150.
[39] Petkov V, Ren Y, Saratovsky I, et al. Atomic-scale structure of biogenic materials by total X-ray diffraction: a study of bacterial and fungal MnOx. ACS Nano, 2009, 3, 441–445.
[40] Zhu M, Ginder-Vogel M, Parikh SJ, Feng X-H, Sparks DL. Cation effects on the layer structure of biogenic Mn oxides. Environ Sci Technol, 2010, 44, 4465–4471.
[41] Villalobos M, Bargar J, Sposito G. Trace metal retention on biogenic manganese oxide nanoparticles. Elements, 2005, 1, 223–226.
[42] Hennebel T, De Gusseme B, Boon N, Verstraete W. Biogenic metals in advanced water treatment. Trends Biotechnol, 2009, 27, 90–98.
[43] Tebo BM, Clement BG, Dick GJ. Biotransformations of manganese. In: Hurst CJ, Crawford RL, Garland JL, Lipson DA, Mills AL, Stetzenbach LD, eds. Manual of Environmental Microbiology. 3rd ed. Washington, DC, ASM Press, 2007, 1223–1238.
[44] Luan F, Santelli CM, Hansel CM, Burgos WD. Defining manganese(II) removal processes in passive coal mine drainage treatment systems through laboratory incubation experiments. Appl Geochem, 2012, 27, 1567–1578.
[45] Santelli C, Pfister D, Lazarus D, Sun L, Burgos W, Hansel C. Promotion of Mn(II) oxidation and remediation of coal mine drainage in passive treatment systems by diverse fungal and bacterial communities. Appl Environ Microbiol, 2010, 76, 4871–4875.
[46] Nealson KH. The isolation and characterization of marine bacteria which catalyze manganese oxidation. In: Krumbein WE, ed. Environmental biogeochemistry and geomicrobiology Volume 3: Methods, Metals and Assessment. Ann Arbor, MI, Ann Arbor Science Publishers Inc., 1978, 847–858.
[47] Ehrlich H. Bacteriology of manganese nodules II. Manganese oxidation by cell-free extract from a manganese nodule bacterium. Appl Microbiol, 1968, 16, 197–202.

[48] Ehrlich HL. Manganese as an energy source for bacteria. In: Nriagu O, ed. Environmental Biogeochemistry, Volume 2. Ann Arbor, Michigan, Ann Arbor Science Publishers, 1976, 633–644.

[49] Emerson D, Moyer C. Isolation and characterization of novel iron-oxidizing bacteria that grow at circumneutral pH. Appl Environ Microbiol, 1997, 63, 4784–4792.

[50] Ghiorse WC, Hirsch P. Isolation and properties of ferromanganese-depositing budding bacteria from Baltic Sea ferromanganese concretions. Appl Environ Microbiol, 1982, 43, 1464–1472.

[51] Dick GJ, Lee YE, Tebo BM. Manganese(II)-oxidizing *Bacillus* spores in Guaymas Basin hydrothermal sediments and plumes. Appl Environ Microbiol, 2006, 72, 3184–3190.

[52] Bräuer SL, Adams C, Kranzler K, et al. Culturable *Rhodobacter* and *Shewanella* species are abundant in estuarine turbidity maxima of the Columbia River. Environ Microbiol, 2011, 13, 589–603.

[53] Templeton AS, Staudigel H, Tebo BM. Diverse Mn(II)-oxidizing bacteria isolated from submarine basalts at Loihi Seamount. Geomicrobiol J, 2005, 22, 127–139.

[54] Carmichael MJ, Carmichael SK, Santelli CM, Strom A, Bräuer SL. Mn(II)-oxidizing bacteria are environmentally relevant members of ferromanganese deposits in caves of the upper Tennessee River Basin. Geomicrobiol J, 2013, 30, 1–22.

[55] Barton HA, Jurado V. What's up down there? Microbial diversity in caves. Microbe, 2007, 2, 132–138.

[56] Spilde MN, Northup DE, Boston PJ, et al. Geomicrobiology of cave ferromanganese deposits: A field and laboratory investigation. Geomicrobiol J, 2005, 22, 99–116.

[57] Jones DS, Lyon EH, Macalady JL. Geomicrobiology of biovermiculations from the Frasassi Cave System, Italy. J Cave Karst Stud, 2008, 70, 78–93.

[58] Northup DE, Barns SM, Yu LE, et al. Diverse microbial communities inhabiting ferromanganese deposits in Lechuguilla and Spider Caves. Environ Microbiol, 2003, 5, 1071–1086.

[59] Chelius MK, Moore JC. Molecular phylogenetic analysis of archaea and bacteria in Wind Cave, South Dakota. Geomicrobiol J, 2004, 21, 123–134.

[60] Engel A. Microbial diversity of cave ecosystems. In: Barton LL, Mandl M, Loy A, eds. Geomicrobiology: Molecular and Environmental Perspective. Netherlands, Springer, 2010, 219–238.

[61] Northup DE, Lavoie KH. Geomicrobiology of caves: A review. Geomicrobiol J, 2001, 18, 199–222.

[62] Moore GW. Manganese deposition in limestone caves. In: Beck BF, editor. Proceedings of the International Congress of Speleology, 1981, 642–645.

[63] Cunningham KI, Northup DE, Pollastro RM, Wright WG, LaRock EJ. Bacteria, fungi and biokarst in Lechuguilla Cave, Carlsbad Caverns National Park, New Mexico. Environ Geol, 1995, 25, 2–8.

[64] Frierdich AJ, Hasenmueller EA, Catalano JG. Composition and structure of nanocrystalline Fe and Mn oxide cave deposits: Implications for trace element mobility in karst systems. Chem Geol, 2011, 284, 82–96.

[65] Peck SB. Bacterial deposition of iron and manganese oxides in North American caves. Nat Speleol Soc Bull, 1986, 48, 26–30.

[66] Allouc J, Harmelin JG. Mn-Fe deposits in shallow cryptic marine environment: examples in northwestern Mediterranean submarine caves. Bull Soc Geol Fr, 2001, 172, 765–778.

[67] Papier S, Baele J-M, Gillan D, Barriquand L, Barriquand J. Manganese geomicrobiology of the black deposits from the Azé Cave, Saône-et-Loire, France. Quaternaire, 2011, 4, 297–305.

[68] Gázquez F, Calaforra J-M, Rull F. Boxwork and ferromanganese coatings in hypogenic caves: An example from Sima de la Higuera Cave (Murcia, SE Spain). Geomorphology, 177–2012, 178, 158–166.

[69] Rossi C, Lozano RP, Isanta N, Hellstrom J. Manganese stromatolites in caves: El Soplao (Cantabria, Spain). Geology, 2010, 38, 1119–1122.
[70] Gradzinski M, Banas M, Uchman A. Biogenic origin of manganese flowstones from Jaskinia Czarna Cave, Tatra Mountains, Western Carpathians. Acta Soc Geol Pol, 1995, 65, 19–27.
[71] Northup DE, Dahm CN, Melim LA, et al. Evidence for geomicrobiological interactions in Guadalupe Caves. J Cave Karst Stud, 2000, 62, 80–90.
[72] Carmichael SK, Carmichael MJ, Strom A, Johnson KW, Roble LA, Bräuer SL. Sustained anthropogenic impact in Carter Salt Peter Cave, Carter County, Tennessee and the potential effects on Mn cycling. J Cave Karst Stud, 2013, 75, 189–204.
[73] Carmichael SK, Zorn BT, Santelli CM, Roble LA, Carmichael MJ, Bräuer SL. Nutrient input influences fungal community composition and size and can stimulate Mn(II) oxidation in caves. Environ Microbiol Reports, 2015, 10.1111/1758-2229.12291.
[74] Saiz-Jimenez C, Miller AZ, Martin-Sanchez PM, Hernandez-Marine M. Uncovering the origin of the black stains in Lascaux Cave in France. Environ Microbiol, 2012, 14, 3220–3231.
[75] Broughton P. Origin and distribution of mineral species in limestone caves. Earth Sci J, 1971, 5, 36–43.
[76] Crabtree P. Bog ore from Black Reef Cave. Cave Science, 1962, 4, 360–361.
[77] Laverty M, Crabtree S. Ranciéite and mirabilite: some preliminary results on cave mineralogy. Trans Brit Cave Res Assoc, 1978, 5, 135–142.
[78] Onac BP, Pedersen RB, Tysseland M. Presence of rare-earth elements in black ferromanganese coatings from Vantului Cave (Romania). J Cave Karst Stud, 1997, 59, 128–131.
[79] Onac BP, Tysseland M, Bengeanu M, Hofenpradli A. Deposition of black manganese and iron-rich sediments in Vântului Cave (Romania). Proceedings 12th International Congress of Speleology, 1997, 1, 235–238.
[80] White WB, Vito C, Scheetz BE. The mineralogy and trace element chemistry of black manganese oxide deposits from caves. J Cave Karst Stud, 2009, 71, 136–143.
[81] Schubert BW, Kaufmann JE. A partial short-faced bear skeleton from an Ozark cave with comments on the paleobiology of the species. J Cave Karst Stud, 2003, 65, 101–110.
[82] Jones B. Manganese precipitates in the karst terrain of Grand Cayman, British West Indies. Can J Earth Sci, 1992, 29, 1125–1139.
[83] López-González F, Grandal-d'Anglade A, Vidal-Romaní JR. Deciphering bone depositional sequences in caves through the study of manganese coatings. J Archaeol Sci, 2006, 33, 707–717.
[84] Gascoine W. The formation of black deposits in some caves of south east Wales. Cave Science, 1982, 9, 167–175.
[85] Kashima N. On the wad-minerals from the cavern environment. Int J Speleol, 1983, 13, 67–72.
[86] Zupančič N, Šebela S, Miler M. Mineralogical and chemical characteristics of black coatings in Postojna cave system. Acta Carsologica, 2011, 1, 40.
[87] Onac BP, Hess JW, White WB. The relationship between the mineral composition of speleothems and mineralization of breccia pipes: evidence from Corkscrew Cave, Arizona, USA. Can Mineral, 2007, 45, 1177–1188.
[88] Jones DS, Schaperdoth I, Macalady JL. Metagenomic evidence for sulfide oxidation in extremely acidic cave biofilms. Geomicrobiol J, 2014, 31, 194–204.
[89] Fleming EJ, Langdon AE, Martinez-Garcia M, et al. What's new is old: resolving the identity of *Leptothrix ochracea* using single cell genomics, pyrosequencing and FISH. PLoS ONE, 2011, 6, e17769.
[90] Johnson K, Carmichael M, Mcdonald W, et al. Increased abundance of *Gallionella* spp., *Leptothrix* spp. and total bacteria in response to enhanced Mn and Fe concentrations in a disturbed southern Appalachian high elevation wetland. Geomicrobiol J, 2012, 29, 124–138.
[91] Hill CA. Origin of black deposits in caves. Nat Speleol Soc Bull, 1982, 44, 15–19.

[92] Carmichael SK. unpublished data.

[93] Cui H, Liu F, Feng X, Tan W, Wang MK. Aging promotes todorokite formation from layered manganese oxide at near-surface conditions. J Soils Sediments 2010, 10, 1540–7.

[94] Feng XH, Zhu M, Ginder-Vogel M, Ni C, Parikh SJ, Sparks DL. Formation of nano-crystalline todorokite from biogenic Mn oxides. Geochim Cosmochim Acta, 2010, 74, 3232–3245.

[95] Grangeon S, Lanson B, Miyata N, Tani Y, Manceau A. Structure of nanocrystalline phyllomanganates produced by freshwater fungi. Am Mineral, 2010, 95, 1608–1616.

[96] Miyata N, Tani Y, Maruo K, Tsuno H, Sakata M, Iwahori K. Manganese(IV) oxide production by *Acremonium* sp. strain KR21–22 and extracellular Mn(II) oxidase activity. Appl Environ Microbiol, 2006, 72, 6467–6473.

[97] Saratovsky I, Gurr SJ, Hayward MA. The structure of manganese oxide formed by the fungus *Acremonium* sp. strain KR21-2. Geochim Cosmochim Acta, 2009, 73, 3291–3300.

[98] Santelli CM, Webb SM, Dohnalkova AC, Hansel CM. Diversity of Mn oxides produced by Mn(II)-oxidizing fungi. Geochim Cosmochim Acta, 2011, 75, 2762–2776.

[99] Dick GJ, Podell S, Johnson HA, et al. Genomic insights into Mn(II) oxidation by the marine alphaproteobacterium *Aurantimonas* sp strain SI85-9A1. Appl Environ Microbiol, 2008, 74, 2646–2658.

[100] Tebo B, Geszvain K, Lee S-W. The molecular geomicrobiology of bacterial manganese(II) oxidation. In: Barton LL, Mandl M, Loy A, eds. Geomicrobiology: Molecular and Environmental Perspective. Netherlands, Springer, 2010, 285–308.

[101] Parker DL, Sposito G, Tebo BM. Manganese(III) binding to a pyoverdine siderophore produced by a manganese(II)-oxidizing bacterium. Geochim Cosmochim Acta, 2004, 68, 4809–4820.

[102] Parker DL, Lee S-W, Geszvain K, et al. Pyoverdine synthesis by the Mn (II)-oxidizing bacterium *Pseudomonas putida* GB-1. Front Microbiol, 2014, 5, 202.

[103] Webb SM, Dick GJ, Bargar JR, Tebo BM. Evidence for the presence of Mn(III) intermediates in the bacterial oxidation of Mn(II). Proc Nat Acad Sci USA, 2005, 102, 5558–5563.

[104] Learman DR, Wankel SD, Webb SM, Martinez N, Madden AS, Hansel CM. Coupled biotic-abiotic Mn(II) oxidation pathway mediates the formation and structural evolution of biogenic Mn oxides. Geochim Cosmochim Acta, 2011, 75, 6048–6063.

[105] Hansel CM, Zeiner CA, Santelli CM, Webb SM. Mn(II) oxidation by an ascomycete fungus is linked to superoxide production during asexual reproduction. Proc Nat Acad Sci USA, 2012, 109, 12621–12625.

[106] Tang Y, Zeiner CA, Santelli CM, Hansel CM. Fungal oxidative dissolution of the Mn(II)-bearing mineral rhodochrosite and the role of metabolites in manganese oxide formation. Environ Microbiol, 2013, 15, 1063–1077.

[107] Learman D, Hansel C. Comparative proteomics of Mn (II)-oxidizing and non-oxidizing Roseobacter clade bacteria reveal an operative manganese transport system but minimal Mn (II)-induced expression of manganese oxidation and antioxidant enzymes. Environ Microbiol Rep, 2014, 6, 12164.

[108] Anderson CR, Johnson HA, Caputo N, Davis RE, Torpey JW, Tebo BM. Mn(II) oxidation is catalyzed by Heme Peroxidases in "*Aurantimonas manganoxydans*" Strain SI85-9A1 and *Erythrobacter* sp. Strain SD-21. Appl Environ Microbiol, 2009, 75, 4130–4138.

[109] Corstjens PLAM, de Vrind JPM, Goosen T, de Vrind-de Jong EW. Identification and molecular analysis of the *Leptothrix discophora* SS-1 *mofA* gene, a gene putatively encoding a manganese-oxidizing protein with copper domains. Geomicrobiol J, 1997, 14, 91–108.

[110] El Gheriany IA, Bocioaga D, Hay AG, Ghiorse WC, Shuler ML, Lion LW. An uncertain role for Cu(II) in stimulating Mn(II) oxidation by *Leptothrix discophora* SS-1. Arch Microbiol, 2011, 193, 89–93.

[111] Ridge JP, Lin M, Larsen EI, Fegan M, McEwan AG, Sly LI. A multicopper oxidase is essential for manganese oxidation and laccase-like activity in *Pedomicrobium* sp. ACM 3067. Environ Microbiol, 2007, 9, 944–953.

[112] Su J, Bao P, Bai T, et al. CotA, a multicopper oxidase from *Bacillus pumilus* WH4, exhibits manganese-oxidase activity. PLoS ONE, 2013, 8, e60573.

[113] Su J, Deng L, Huang L, Guo S, Liu F, He J. Catalytic oxidation of manganese (II) by multicopper oxidase CueO and characterization of the biogenic Mn oxide. Water Res, 2014, 56, 304–313.

[114] Brouwers G-J, de Vrind JPM, Corstjens PLAM, Cornelis P, Baysse C, de Vrind-de Jong EW. *CumA*, a gene encoding a multicopper oxidase, is involved in Mn^{2+}-oxidation in *Pseudomonas putida* GB-1. Appl Environ Microbiol, 1999, 65, 1762–1768.

[115] Geszvain K, Tebo BM. Identification of a two-component regulatory pathway essential for Mn(II) oxidation in *Pseudomonas putida* GB-1. Appl Environ Microbiol, 2010, 76, 1224–1231.

[116] Geszvain K, McCarthy JK, Tebo BM. Elimination of manganese(II,III) oxidation in *Pseudomonas putida* GB-1 by a double knockout of two putative multicopper oxidase genes. Appl Environ Microbiol, 2013, 79, 357–366.

[117] Soldatova AV, Butterfield C, Oyerinde OF, Tebo BM, Spiro TG. Multicopper oxidase involvement in both Mn(II) and Mn(III) oxidation during bacterial formation of MnO_2. J Biol Inorg Chem, 2012, 17, 1151–1158.

[118] Butterfield CN, Soldatova AV, Lee S-W, Spiro TG, Tebo BM. Mn(II, III) oxidation and MnO_2 mineralization by an expressed bacterial multicopper oxidase. Proc Nat Acad Sci USA, 2013, 110, 11731–11735.

[119] Markowitz VM, Korzeniewski F, Palaniappan K, et al. The integrated microbial genomes (IMG) system. Nucleic Acids Res, 2006, 34, D344-D8.

[120] Davis RE, Tebo BM. Culture-independent identification of manganese-oxidizing genes from deep-sea hydrothermal vent chemoautotrophic ferromanganese microbial communities using a metagenomic approach. American Geophysical Union Fall Meeting. San Francisco, CA, 2013, B14B-02.

[121] Ghiorse WC. Biology of iron- and manganese-depositing bacteria. Annu Rev Microbiol, 1984, 38, 515–550.

[122] Schmidt JM, Sharp WP, Starr MP. Metallic-oxide encrustations of the nonprosthecate stalks of naturally occurring populations of *Planctomyces bekefii*. Curr Microbiol 1982, 7, 389–394.

[123] Hofrichter M, Ullrich R, Pecyna MJ, Liers C, Lundell T. New and classic families of secreted fungal heme peroxidases. Appl Microbiol Biotechnol, 2010, 87, 871–897.

[124] van Veen WL. Biological oxidation of manganese in soils. A Van Leeuw, 1973, 39, 657–662.

[125] Hofrichter M. Review: lignin conversion by manganese peroxidase (MnP). Enzyme Microb Technol, 2002, 30, 454–466.

[126] Yang W, Zhang Z, Zhang Z, et al. Population structure of manganese-oxidizing bacteria in stratified soils and properties of manganese oxide aggregates under manganese-complex medium enrichment. PLoS ONE, 2013, 8, e73778.

[127] Bromfield SM, David DJ. Sorption and oxidation of manganous ions and reduction of manganese oxide by cell suspensions of a manganese oxidizing bacterium. Soil Bio Biochem, 1976, 8, 37–43.

[128] Ehrlich HL. Bacteriology of manganese nodules I. Bacterial action on manganese in nodule enrichments. Appl Microbiol, 1963, 11, 15–19.

[129] Dubinina G, Zhdanov A. Recognition of the iron bacteria "Siderocapsa" as Arthrobacters and description of *Arthrobacter siderocapsulatus* sp. nov. Int J Syst Bacteriol, 1975, 25, 340–350.

[130] Hungate B, Danin A, Pellerin NB, et al. Characterization of manganese-oxidizing (MnII→MnIV) bacteria from Negev Desert rock varnish: Implications in desert varnish formation. Can J Microbiol, 1987, 33, 939–943.

[131] Timonin M. Soil microflora in relation to manganese deficiency. Sci Ag, 1950, 30, 324–325.
[132] Bromfield S, Skerman V. Biological oxidation of manganese in soils. Soil Sci, 1950, 69, 337–347.
[133] Palanichamy S, Maruthamuthu S, Manickam S, Rajendran A. Microfouling of manganese-oxidizing bacteria in Tuticorin harbour waters. Curr Sci 2002, 2, 865–869.
[134] Ahmad M, Roberts JN, Hardiman EM, Singh R, Eltis LD, Bugg TD. Identification of DypB from *Rhodococcus jostii* RHA1 as a lignin peroxidase. Biochemistry (Mosc), 2011, 50, 5096–5107.
[135] Gregory E, Staley JT. Widespread distribution of ability to oxidize manganese among freshwater bacteria. Appl Environ Microbiol, 1982, 44, 509–511.
[136] Stein LY, La Duc MT, Grundi TJ, Nealson KH. Bacterial and archaeal populations associated with freshwater ferromanganous micronodules and sediments. Environ Microbiol, 2001, 3, 10–18.
[137] Sujith P, Mourya B, Krishnamurthi S, Meena R, Loka Bharathi P. Mobilization of manganese by basalt associated Mn(II)-oxidizing bacteria from the Indian Ridge System. Chemosphere, 2014, 95, 486–495.
[138] de Vrind JPM, de Vrind-de Jong EW, de Voogt J-WH, Westbroek P, Boogerd FC, Rosson RA. Manganese oxidation by spores and spore coats of a marine *Bacillus* species. Appl Environ Microbiol, 1986, 52, 1096–1100.
[139] de Vrind J, Boogerd FC, de Vrind-de Jong EW. Manganese reduction by a marine *Bacillus* species. J Bacteriol, 1986, 167, 30–34.
[140] Chandramohan D, Lokabharathi P, Nair S, Matondkar S. Bacteriology of ferromanganese nodules from the Indian Ocean. Geomicrobiol J, 1987, 5, 17–31.
[141] Anderson CR, Dick GJ, Chu ML, et al. *Aurantimonas manganoxydans*, sp. nov. and *Aurantimonas litoralis* sp. nov.: Mn(II)-oxidizing representatives of a globally distributed clade of *Alphaproteobacteria* from the order *Rhizobiales*. Geomicrobiol J, 2009, 26, 189–198.
[142] Francis CA, Co EM, Tebo BM. Enzymatic manganese(II) oxidation by a marine α-proteobacterium. Appl Environ Microbiol, 2001, 67, 4024–4029.
[143] Tyler Pa. *Hyphomicrobia* and the oxidation of manganese in aquatic systems. A Van Leeuw J Microb, 1970, 36, 567–578.
[144] Ghiorse WC, Hirsch P. An ultrastructural study of iron and manganese deposition associated with extracellular polymers of *Pedomicrobium*-like bacteria. Arch Microbiol, 1979, 123, 213–226.
[145] Gebers R. Enrichment, isolation, and emended description of *Pedomicrobium ferrugineum Aristovskaya* and *Pedomicrobium manganicum Aristovskaya*. Int J Syst Bacteriol, 1981, 31, 302–316.
[146] Hansel CM, Francis CA. Coupled photochemical and enzymatic Mn(II) oxidation pathways of a planktonic *Roseobacter*-like bacterium. Appl Environ Microbiol, 2006, 72, 3543–3549.
[147] Akob DM, Bohu T, Beyer A, et al. Identification of Mn(II)-oxidizing bacteria from a low-pH contaminated former uranium mine. Appl Environ Microbiol, 2014, 80, 5086–5097.
[148] Takeda M, Kamagata Y, Ghiorse WC, Hanada S, Koizumi J. *Caldimonas manganoxidans* gen. nov., sp. nov., a poly(3-hydroxybutyrate)-degrading, manganese-oxidizing thermophile. Int J Syst Evol Microbiol, 2002, 52, 895–900.
[149] Mulder EG, van Veen WL. Investigations on the *Sphaerotilus-Leptothrix* group. A Van Leeuw, 1963, 29, 121–153.
[150] Johnson A, Stokes J. Manganese oxidation by *Sphaerotilus discophorus*. J Bacteriol, 1966, 91, 1543–1547.
[151] Douka C. Kinetics of manganese oxidation by cell-free extracts of bacteria isolated from manganese concretions from soil. Appl Environ Microbiol, 1980, 39, 74–80.

[152] Jung WK, Schweisfurth R. Manganoxydierende bakterien III. Wachstum und manganoxydation bei *Pseudomonas manganoxidans* Schw. Z Allg Mikrobiol, 1976, 16, 587–597.
[153] Nealson KH, Myers CR. Microbial reduction of manganese and iron: new approaches to carbon cycling. Appl Environ Microbiol, 1992, 58, 439–443.
[154] Staudigel H, Hart SR, Pile A, et al. Vailulu'u seamount, Samoa: Life and death on an active submarine volcano. Proc Nat Acad Sci USA, 2006, 103, 6448–6453.
[155] Chang J, Tani Y, Naitou H, Miyata N, Seyama H. Fungal Mn oxides supporting Mn (II) oxidase activity as effective Mn (II) sequestering materials. Environ Technol, 2013, 34, 1–7.
[156] Schweisfurth R. Manganoxydierende Pilze II. Untersuchungen an Laboratoriumskulturen. Z Allg Mikrobiol, 1972, 12, 667–671.
[157] Thiel GA. Manganese precipitated by micro-organisms. Econ Geol Bull Soc Econ Geol, 1925, 20, 301–310.
[158] Cummings JH. Short chain fatty-acids in the human colon. Gut, 1981, 22, 763–779.
[159] Miyata N, Tani Y, Iwahori K, Soma M. Enzymatic formation of manganese oxides by an *Acremonium*-like hyphomycete fungus, strain KR21-2. FEMS Microbiol Ecol, 2004, 47, 101–109.
[160] Ikner LA, Toomey RS, Nolan G, Neilson JW, Pryor BM, Maier RM. Culturable microbial diversity and the impact of tourism in Kartchner Caverns, Arizona. Microb Ecol, 2007, 53, 30–42.
[161] Simon KS, Buikema AL. Effects of organic pollution on an Appalachian cave: Changes in macroinvertebrate populations and food supplies. Am Midl Nat, 1997, 138, 387–401.
[162] Mulec J, Kristufek V, Chronakova A. Comparative microbial sampling from eutrophic caves in Slovenia and Slovakia using RIDA ®COUNT test kits. Int J Speleol, 2011, 41, 1–8.
[163] Edwards KJ, Glazer B, Rouxel OJ, et al. Ultra-diffuse hydrothermal venting supports Fe-oxidizing bacteria and massive umber deposition at 5000 m off Hawaii. ISME J, 2011, 5, 1748–1758.

Diana E. Northup and Kathleen H. Lavoie

8 Microbial Diversity and Ecology of Lava Caves

Abstract: Lava caves are found around the Earth and on other planets. Lava caves are well suited for examining questions of what controls species diversity and for examining patterns of biogeography of microorganisms. Colonization of lava caves is a primary succession event since the cave was formed from molten lava, which simplifies species diversity studies. Analysis of bacterial 16S rRNA genetic sequences suggests a common core of phyla found worldwide, with high levels of endemism. At the operational taxonomic unit level, the data suggest biogeographical patterns may exist and factors such as concentration of nitrogen and other nutrients play an important role in shaping species diversity. Significant additional work needs to be done on nutrient cycling between the surface and the lava cave environment, especially nitrogen cycling and contributions of plant roots. Scanning electron microscopy (SEM) of microbial mats and secondary minerals that contain abundant microbial communities shows several common microbial morphologies, including cocci, rods, beads-on-a-string, and reticulated, smooth, and segmented filaments. Many cells show extensive pilus-like appendages. The study of lava cave microbial communities is especially relevant in improving life detection strategies with biomarkers for searching for extant or extinct life on other planets.

8.1 Introduction

Caves make excellent laboratories for examining questions of what controls species diversity and for examining patterns of biogeography of microorganisms. Compared to the surface, caves are simplified natural ecosystems with less variation in terms of temperature, humidity, nutrients, weathering, and light, which allows for the analysis of fewer variables. Caves lack sunlight beyond the twilight zone, removing photosynthesis as an energy source for the food base, so most caves are nutrient-limited (oligotrophic) environments based on detrital inputs of organic carbon with a limited number of trophic levels [1]. The exceptions are lava caves with surface plant roots growing through the ceilings [2] into the caves for a source of water, as seen with roots of the Ohi'a tree in ▸ Fig. 8.1 from a cave in the southern part of the Big Island, Hawai'i, USA. Despite these constraints, lava caves support a wealth of life, especially microbial.

Lava caves are widespread around the world and provide entry to the shallow subsurface. Oceanic islands including Hawai'i, the Azores, Canary Islands, Iceland, and Republic of Mauritius harbor many lava caves. Continental lava caves are abundant in Australia, Italy, Spain, South Korea, eastern and southern Africa, Argentina, western United States and Canada, and New Zealand. The most extensive lava cave in the world is Kazumura on the Kilauea volcano of the Big Island of Hawai'i [3]. It has a vertical extent of 1,102 m (3,614 ft) and a length of 65.5 km (40.7 mi), with a straight-line extent of 32.2 km (20 mi). The Kazumura lava cave formed only 350–500 years ago [4].

Fig. 8.1. Ohi'a tree roots growing into a lava cave in the southern part of the island of Hawai'i, USA. © Kenneth Ingham. Reproduced by permission of Kenneth Ingham. Permission to reuse must be obtained from the rights holder.

Why should we care about microbial diversity in caves? Microbes are of central importance to the biosphere and to biogeochemical cycling. Cave microbial diversity allows us to explore frontiers of strategies and limits of life. Microbes have been found in some of the most extreme environments including deep-sea thermal vents, within rock cores, and in caves. These microorganisms are important participants in the precipitation and dissolution of minerals in caves [5–7] and on the surface [8]. However, we have barely begun to characterize the microbial diversity of caves and the roles of microorganisms in the subsurface.

Caves can provide accessible model systems for the study of possible life on other planets [9]. Life is probably abundant in the universe, mostly microbial, based on sulfur, and located below the surface [10]. We can use microbes to monitor environmental change, the quality of an environment, and the recovery of a system to a particular stress. Thus, microbes can play a major role in conservation and restoration biology. Microbes and microbial communities also provide important models for understanding principles of ecology and evolution.

8.2 Geology and Ecology of Lava Caves

The study of lava caves is included in the science of vulcanospeleology. The speleogenesis of lava caves [11] is an interesting process that is still being studied. Molten rock called magma rises to the surface from depths of 50–700 km; on the surface, the molten and solidified form is called lava. A lava cave is a lava tube that is large enough to accommodate humans [11]. Many smaller lava tubes exist, forming mesocaverns in the lava that can serve as refuges for arthropods and microorganisms [2, 12]. Establishing the age or elemental composition of a lava cave should be done from rock inside the cave, not on the surface since lava flows can downcut into older lava below.

Lava forms a range of igneous rocks that vary in chemical composition and size of crystals. Most lava caves form in basalt, a dark, finely crystalline lava that is about 50% silicon dioxide (SiO_2), with varying amounts (in decreasing order) of aluminum (Al_2O_3), iron (FeO), calcium (CaO), magnesium (MgO), and small amounts of sodium (Na_2O), titanium (TiO_2), potassium (K_2O), with trace amounts of manganese (MnO) and phosphate (P_2O_5) [13].

Lava flows downhill in streams from calderas and vents. Lava exits the ground at a temperature of 700–1,200 °C and begins to cool at temperatures of 1,000 °C [11]. The opposite of karst caves, the oldest lava cave passages are deep and younger passages are closer to the surface. Most textbooks describe formation of lava caves as the cooling and crusting over of the lava flow channel, followed by the emptying out molten lava that leaves behind an empty conduit tube [11].

An alternate theory of how lava caves form was proposed by Kempe [14]. Most lava caves form in ropy pāhoehoe lava that is less viscous and only grows at the tip of the flow. As the flow advances, thin sheets of lava cover the ground. When a new sheet of molten lava comes along, it "lifts" the previous sheet, "inflating" it, and flows under it. Below the newly forming roof, hot lava continues to flow in a tunnel called a pyroduct and erodes downward. Because cooled lava contains bubbles from dissolved gases, it is less dense than the flowing lava, so solid lava rocks will float on liquid lava. Eventually, the lava tube empties of molten lava and the walls and ceiling cool to become the lava caves we see today.

The color of most lava cave walls ranges from dark gray to bright red due to iron oxides. We also find orange, brown, greenish, and black, depending on the chemical

Fig. 8.2. White and orange stripes of minerals deposited in Braided Cave, El Malpais National Monument, New Mexico, USA. © Kenneth Ingham. Reproduced by permission of Kenneth Ingham. Permission to reuse must be obtained from the rights holder.

composition of the lava and the environmental conditions when the lava solidified. The banding of white and orange stripes that can be seen on the ceilings and walls in ▸ Fig. 8.2 represents minerals that were deposited after the lava cave formed. The white areas are calcium carbonate or amorphous silica, while the orange areas are iron oxides or fulvic and humic acids leaching into the cave from overlying soils [15]. Secondary mineral deposits are most commonly of sulfur, oxidized and hydroxide minerals, and salts that form by evaporation of infiltrating surface water rich in sulfates and chlorites. Sodium minerals are commonly seen, while calcite and aragonite deposits are rare [11, 16]. Speleothems in lava caves tend to be rare [11, 17].

8.2.1 Physical conditions

Physical conditions in lava caves depend on the length of the cave, geometry of the passages, and the number of entrances. Some lava caves maintain the mean annual surface temperature (MAST) of the area as a function of latitude and elevation, while others are cold sinks with year-round ice. Ice caves have entrances above passages, allowing for inward cold air flow [18]. Variability in lava cave temperature is seen if we compare five caves from El Malpais in New Mexico that are all at the same elevation (2,200 m), receive the same amount of annual precipitation (281 mm), and are in native vegetation with a MAST of 10 °C. However, the temperatures in these five caves range from a low of −2 °C to a high of 11 °C due to passage geometry and entrance number and location.

Caves are impacted seasonally around entrances, but the deep cave zone is defined by nearly constant year-round temperatures, and biologically interesting caves have relative humidity (RH) of greater than 98%. As an example, Valentine Cave, a lava cave in Lava Beds National Monument, USA, is about 270 m long with one entrance.

(a) (b)

Fig. 8.3. (a) Braided cave passage in Golden Dome Cave in Lava Beds National Monument, California, USA. (b) Aerial view of Big Skylight Cave from El Malpais National Monument, New Mexico, USA. Both photographs © Kenneth Ingham. Reproduced by permission of Kenneth Ingham. Permission to reuse must be obtained from the rights holder.

Fig. 8.4. Characteristic benches in a Hawaiian lava cave. © Kenneth Ingham. Reproduced by permission of Kenneth Ingham. Permission to reuse must be obtained from the rights holder.

The surface temperature was 29.8 °C and 9% RH on the September day when measurements were taken. Valentine Cave was in total darkness within 12 m and reached the cave temperature of 11.3 °C in 125 m, but humidity had climbed to saturation in only 77 m.

Lava cave passages are oriented parallel to the direction of lava flow, which is controlled by surface topography. Steep gradients produce a single, straight conduit, while less steep gradients form sinuous passages that may partly braid (▸ Fig. 8.3 (a)) with other tubes. Lava caves are largest at the upper ends of the flow. ▸ Fig. 8.3 (b) shows an aerial photograph of Big Skylight Cave from El Malpais National Monument that was formed from a pāhoehoe lava flow that originated from the Bandera crater approximately 11,000 years ago. Part of the surface collapsed into the underlying tube which formed a trench and skylight.

As the name suggests, lava tube caves have a tube shape, often with curbs, gutters, benches (▸ Fig. 8.4), or step marks that denote later depths of flow. Passage widths can be up to 14–15 m (45–50 ft), though most are much narrower. Lava tubes are entered mostly through areas of ceiling collapse called a puka skylight (▸ Fig. 8.5) or puka. Lava covering most lava caves is only 1–15 m thick (3–50 ft) [11]. Lava caves are often found in segments due to roof collapse (▸ Fig. 8.3 (b)).

Fig. 8.5. Ceiling collapse forms a skylight or puka, in this case an entrance to the cave. © Kenneth Ingham. Reproduced by permission of Kenneth Ingham. Permission to reuse must be obtained from the rights holder.

8.3 Microbiological Studies in Lava Caves

Because lava caves form *de novo*, succession in lava caves is primary so there are no organisms of any type until there are inputs of microbes and organics from the surface. The maximum age of the lava flow is the maximum time for colonization and evolution of cave-adapted life. For example, the oldest island in the Azores is about 28 million years old (Ma) while Hawai'i formed about 0.4 Ma (400,000 years bp). This range of ages also allows for testing of the effect of cave age on microbial species diversity. However, across all groups of microorganisms, eukaryotic (Fungi, Algae, and Protozoa) and Archaea and Bacteria, studies of microbes in lava caves lag behind studies in karst caves.

Cave-adapted animals were not expected in tropical caves based on prevailing models of isolation in caves as refuges from glaciers. It was not until the pioneering work of Howarth [12, 19] showing speciation in tropical caves by adaptive shift that lava caves became of interest to biospeleologists. Like karst caves, animal species in lava caves show high levels of endemism. We are just beginning to explore the question of endemism in microbial species in lava caves, but Vickie Peck and Diana Northup (unpublished results) and Snider et al. [20] demonstrated that bacteria cultured from El Malpais lava caves lacked resistance to ultraviolet radiation in comparison to surface bacteria. The results suggest some degree of adaptation has occurred in these bacteria despite the shallow nature of the lava caves.

Fig. 8.6a. Collage of macroscopic images of microbial mats (a) and close-ups of white, tan, brown and yellow microbial colonies (b – *see following page*) in Azorean and Hawaiian lava caves. Note the variety of colony textures. © Kenneth Ingham. Reproduced by permission of Kenneth Ingham. Permission to reuse must be obtained from the rights holder.

Fig. 8.6b (cont.). *(see previous page)*

8.3.1 Lava cave microorganisms

Lava caves contain a diverse and often colorful array of microbes that range in size from small, scattered colonies to extensive mats covering walls and ceilings (▶ Fig. 8.6a). Coverage is more extensive in moist lava caves in Hawai'i and the Azores than in more arid caves in Hawai'i and New Mexico. Colors include white, yellow, tan, gold, orange, and pink, with shades in between [21–25]. Lava caves become colonized as soon as they cool down, and caves in lava flows in Hawai'i show diverse mat structure in tens of years. Donachie et al. [26] describes the Hawaiian archipelago as a microbial diversity hotspot. The aquatic sites they compared are globally and locally remote with great physical and chemical diversity. Each site hosts diverse microbial communities dominated by bacteria. Only seven operational taxonomic units (OTUs) at >97% sequence identity of 16S rRNA genes were found in more than one site, and only 52 cultures and 100 OTUs shared identity with published species. Habitat chemistry controls community structure such that Hawaiian aquatic habitats have novel microbes and microbial communities with unique evolutionary niches. This diversity is also found in the lava caves of Hawai'i. Hathaway et al. [27] identify Actinobacteria in Hawaiian lava caves using Sanger sequencing with 92% of the recovered bacterial clones less than 97% similar to any known organism based on 16S rRNA gene sequences. Microbial diversity is shaped by many factors associated with the lava cave habitat.

8.4 Eukaryotic Microorganisms

8.4.1 Fungi

In all ecosystems, fungi function as decomposers by breaking down and recycling organic nutrients and serving as food for animals. A few fungi are pathogenic. Most fungi in caves are associated with deposits of organic matter from flooding; guano, animal and insect droppings; roots; wood and leaves; and insects. Fungal distribution within caves is not uniform. Factors that influence fungal type, distribution, and activity are the type and location of organic material available to colonize; micro-environmental conditions, particularly water availability and temperature; and mineral composition of the rock. Most fungi are on the cave floor and walls where organic matter deposits are found and most are surface species brought into caves with animals or organic matter [28]. Caumartin [29] predicted that there would be no troglobitic fungi due to their high nutrient demands, and caves unimpacted by humans, animals, floods, or winds should have no native fungi.

Lava caves studied at El Malpais National Monument range in age from 11,000 to 110,000 years old. There are small amounts of soil and some piles of bat guano ranging from a few pellets to large piles. Fungal inhabitants of these caves have recently been studied because of the sudden and catastrophic bat fungal infection known as white-nose syndrome (WNS) caused by *Pseudogymnoascus destructans* (formerly *Geomyces destructans*) [30]. Vanderwolf et al. [28] did a global review of fungi, yeast, and slime molds in solution caves. Nearly 60% of the papers reviewed studied European caves, and *P. destructans* was not documented until it was targeted by researchers. It is now believed that *P. destructans* is widespread throughout Europe, but without the lethality seen in North American bats, probably due to a long history of co-evolution [30, 31]. In a study of bats from the Czech Republic, Zukal et al. [32] found WNS in a wide range of ecologically diverse hibernating bats, and suggested that all hibernating bats in the range of the fungus may be at risk, but a modeling study by Alves [33] predicts that WNS will not spread throughout North America, but could cause population declines in 32% of hibernating bat species.

Vespertilionid bats use seasonal hibernation to conserve energy during the winter, which allows them to survive periods of low food availability and high thermoregulatory demands [34]. In response to concerns about WNS spreading to New Mexico, Northup and Buecher [35] began a program to test soil and guano in El Malpais lava caves for the presence of *Geomyces* or *Pseudogymnoascus* species (in the class Leotiomycetes) using a combination of molecular genetic techniques and cultures (▸ Fig. 8.7). An analysis of cave microclimate where bats hibernate was done to see if conditions favored by the psychrophilic fungus (3–15 °C and >90% RH) existed in 10 lava caves in El Malpais National Monument. Sequencing showed that fungal classes varied from cave to cave. *Geomyces* and *Pseudogymnoascus* genera were identified in soil and guano from some of the caves. They found that ideal microclimate conditions

Fig. 8.7. Collecting samples for fungal testing in an El Malpais National Monument Cave. © Kenneth Ingham. Reproduced by permission of Kenneth Ingham. Permission to reuse must be obtained from the rights holder.

for the growth of *P. destructans* were present in parts of five caves for at least part of the year, which suggests that bats that hibernate at El Malpais might be at risk of WNS as it continues its westward spread. Tests for the presence of *P. destructans* identified close relatives, but not *P. destructans*. Fungal sequences from swabs of 10 bats recovered 134 fungal genera, with 21 found in higher numbers (>100 sequences) in at least one sample.

Before WNS, the most widely studied fungus in caves was *Histoplasma capsulatum*, the causative agent of histoplasmosis. The fungus is most commonly found in deposits of bat or bird guano both in and out of caves [36]. It occurs worldwide but is endemic in many areas including the Ohio River Valley in the United States. Histoplasmosis is usually a mild disease of the respiratory tract, but it can occasionally be more serious and even fatal. Any cave with significant organic inputs of bat or bird guano poses a risk for cavers of contracting histoplamosis.

Cave visitation and inputs of organic matter influence the mycoflora of caves. Griffin et al. [37] studied the aeromicrobiology of Carlsbad Cavern. Ubiquitous genera like *Cladosporium* and *Alternaria* were prevalent near the natural entrance, and dropped off with depth into the cave. *Penicillium* and *Aspergillus* were abundant in the Lunch Room, suggesting significant anthropogenic contamination.

Most fungal taxa commonly reported from caves are ubiquitous saprotrophs associated with soils, plant material, or insects. Vanderwolf et al. [28] found that a core group of fungi are widespread in karst caves along with rarely isolated species, a pattern common in both surface and cave environments. Saprotrophs can quickly colonize and grow on organic matter, and may mask the growth of oligotrophic cave fungi. The low nutrient, stable, low temperatures in caves favor communities of oligotrophic, cold-tolerant fungi [28].

8.4.2 Protozoa and algae

Protozoa and algae are single-celled eukaryotes. Protozoa function as primary consumers, and are themselves food for higher trophic levels of organisms. Protozoa are widespread in water, sediment, soil, moss, the guts of animals, and any temperate, humid environment. Most protozoa are free-living, and a few are parasitic and can cause disease. Algae and cyanobacteria are photoautotrophic, and are generally limited to entrance areas or around artificial lights in show caves as a component of lampenflora or lamp flora. Refer to Chapter 12 in this volume for more information. Some can survive and grow as chemoorganotrophs and may be a component of aquatic plankton, along with protozoa and mesofauna.

The first study of protozoa in caves was reported in 1845, and since then hundreds of species have been described. Generally what is found is a subset of what is found on the surface. A study of the distribution of testate amoeba associated with speleothems was done in lava caves from Europe, South America, Africa, and Australia by González

Fig. 8.8. Opal formations in Algar do Carvão, on the Azorean island of Terceira. © Kenneth Ingham. Reproduced by permission of Kenneth Ingham. Permission to reuse must be obtained from the rights holder.

López et al. [38]. They report a fairly homogeneous distribution of testate amoebae regardless of the external environment of the cave. Species were not exclusive to caves. The testate amoebae have a role in the silica cycle, particularly in the formation and deposition of biogenic amorphous opal found in some lava caves including Algar do Carvão [38, 39], a volcanic chimney on the island of Terceira in the Azores (▸ Fig. 8.8).

Pentecost [40] studied the phototrophic flora from a small lava cave near Laki, Iceland. Five cyanobacteria, two that fix nitrogen, and nine diatoms, some previously reported from caves, were identified. Two of five lichens fixed nitrogen, suggesting that nitrogen was limiting in the lava cave environment. The lava cave phototrophic community showed some similarities with those from karst caves. He concluded that environmental variables, particularly light and high humidity, were important factors in selection.

8.5 Bacteria and Archaea

8.5.1 Methods

A revolution in microbiology occurred with the introduction of culture-independent 16S rDNA methodology. Standard cultivation techniques with high-nutrients and body temperature incubation have met with limited success when applied to environmental microbes, including microorganisms from caves [41, 42]. Culture-independent, molecular, and phylogenetic techniques allow us to reveal the diversity present in many different environments [43], resulting in the detection of many novel microbial species and new phyla. Metabolically active microbial communities have been detected in nearly every environment studied, however extreme, including deep-sea thermal vents, within rock cores, and in caves [e.g. 44].

Early studies of lava cave microbial mats utilized culture-dependent methods for isolation of chemoorganotropic microorganisms associated with mats or “slimes.” Stoner and Howarth [45] first described fungi and aerobic bacteria from mats in Hawaiian lava caves. White and brown slimes were identified as possible sites of nutrient cycling, particularly nitrogen. The authors used high-nutrient, traditional culture media because at the time cave microbes were considered simply a subset of surface microbes washed into underlying caves. Although the surface may be a source of microbes entering caves, most should be unable to adapt to the extreme cave environment, and isolation from the surface and from other caves allows for adaptation and the evolution of novel taxa as observed by Brown et al. [46] and Saw et al. [44]. Although only a small fraction of the total microbial diversity can be grown on traditional microbiological media, cultures are still essential to characterize microbial physiology. Flow cytometry can be used to isolate individual cells. We also now understand that microbes in nature seldom occur in isolation, but rather as consortia or communities in biofilms. Culturing has improved with the use of lower nutrient media that more closely resemble environmental conditions and use of co-cultures with other bacteria [47].

These improvements will allow us to cultivate more of the existing microbial diversity as identified by molecular phylogenies. Results from culture studies and studies using molecular approaches are strikingly different, with only a small fraction of the total diversity grown on traditional microbiological media [41, 42]. Culture-based and molecular studies should be seen as complimentary, not competitive [48]. Molecular approaches did not capture all of the microbial diversity because many cultivated isolates were not detected by the molecular approach. Many cave studies were done in the age of Sanger sequencing, and high costs limited the number of genetic sequences that could be generated. Next generation sequencing techniques reveal a much greater extent of the diversity of a given habitat.

8.5.2 Microbes in volcanic environments

To understand the microbial diversity in lava caves, a good knowledge of the microorganisms present on the surface on volcanic rock deposits and in soils is important. Volcanic landscapes are oligotrophic environments where chemolithoautotrophy may be important as microbes seek to utilize reduced compounds, such as iron, sulfur, and manganese in the basalt as energy sources. Conditions may be similar to that on early Earth, and may help us interpret microbial adaptations, evolution, and exobiology. What mechanisms do the microbes use to colonize and alter host bedrock? What are they using as sources of nitrogen and phosphorus? Are mechanisms for getting energy and interactions with metals and minerals unique in lava caves or similar to those from other chemolithoautotrophic environments?

Acidobacteria, Alphaproteobacteria and Gammaproteobacteria, Actinobacteria, Firmicutes, and Cyanobacteria predominate in surface volcanic deposits in Hawai'i, with composition controlled by local differences in environments and the type of volcanic deposits [49]. Janssen [50] reviews studies of the microbial diversity of soils in many environments. Soils are dominated by Proteobacteria, averaging 39% of soil bacteria, with Acidobacteria, Actinobacteria, Verrucomicrobia, Bacteroidetes, Chloroflexi, Planctomycetes, Gemmatimonadetes, and Firmicutes. These findings show the need for a careful comparison of microbial populations from surface soils that are available to colonize the underlying caves. Using these studies as a backdrop, we can begin to build up knowledge of lava cave microbial diversity to compare to these surface environments.

8.5.3 Effects of mat color on microbial diversity

Northup et al. [22] compare white and yellow microbial mats and selected secondary minerals chosen based on color from four lava caves in three different locations: tropical and semiarid caves in Hawai'i; temperate caves in the Azores; and semiarid caves in New Mexico. Mats are more extensive in areas with greater rainfall. They found 15 phyla across all white and yellow mats from all three locations. All mats have Actinobacteria, Alpha-, Beta-, Delta-, and Gammaproteobacteria, Acidobacteria, and all but one sample had Nitrospirae. The number of phyla per cave ranged from five to eleven, with slightly greater diversity at the phylum level in yellow compared to white mats. Microbes from the secondary mineral deposits are more diverse at the OTU level, but share most of the same 15 phyla across samples. The most striking difference is that Actinobacteria dominated in mats, but are only found in two of the six mineral samples.

Another conclusion from the Northup et al. [22] study is that caves, in general, contain a core set of bacterial phyla. Six of 11 studies from carbonate caves (2006–2011) reviewed found Actinobacteria, Proteobacteria (Alpha-, Beta-, Delta-, and Gamma-), Acidobacteria, Verrucomicrobia, Planctomycetes, Nitrospirae, and Bacteroidetes (see Table 3 in reference [22]). Three studies of mats lacked Actinobacteria: Frasassi Caves in Italy [51], although found in biovermiculations and water from Frasassi [52], Movile Cave in Romania [53], and two caves in the United States: Lower Kane Cave and Cesspool Cave [54]. These caves are sulfur-based caves, which may account for the microbial diversity differences; moreover, these sulfur caves are also the only ones with Epsilonproteobacteria. The more recent the study the more bacterial phyla were reported, probably a reflection of improved sequencing technology and lower costs that result in more sequences per sample and per study. Newer studies of lava caves using next generation sequencing reveal many rarer phyla, including several candidate phyla.

8.5.4 Impact of location on microbial diversity

Hathaway et al. [27] compared bacterial diversity of dominant white and yellow microbial mats using clone sequences from lava caves from two widely separated oceanic archipelagos that form over a volcanic hotspot (Hawai'i) and a spreading center (the Azores). The two islands, the Big Island, Hawai'i, and Terceira, Azores, are widely separated, and the climates range from semiarid to tropical in Hawai'i, and temperate in the Azores. The authors tested the concept of "everything is everywhere: but the environment selects" [55, 56] by asking if lava cave microbial communities are similar in physically distant environments, or do local influences in each lava cave result in unique microbial communities. If microbial diversity follows the same scaling rules for other organisms, then researchers would expect to find endemism and spatial patterning in alpha diversity (i.e. local scale, such as mean species diversity or habitats) and beta diversity (i.e. differentiation among those habitats) resulting in gamma (i.e. total) diversity [57]. The authors [27] predicted they would find novel organisms with 16S rRNA gene sequences having <97% sequence identity with known bacteria due to isolation and unique selective regimes at each location.

Both white and yellow mats from four caves on each island were examined with SEM, chemical analysis of lava rock, and DNA extraction for construction of 16S rDNA clone libraries for phylogenetic and community analysis [27]. SEM showed an abundance of coccoid morphologies covered with pili or knobby appendages, smooth and segmented filaments, rods on or embedded in biofilm, beads-on-a-string, and reticulated filaments in samples from both islands. Statistically, the only significantly different element in the bulk chemical analysis of basalt from either location was a higher concentration of copper from the Big Island. DNA analysis of 1,370 full-length, nonchimeric 16S rDNA sequences produced from Sanger sequencing of clones, showed that the full extent of diversity was not sampled.

Of the 609 OTUs defined at 97%, only 34 OTUs (5% of the total) formed with sequences from both islands [27]. Taxonomic richness was higher in sampled caves on Hawai'i than on Terceira. Principle coordinate analysis (PCoA) by island showed that only 13.7% of the variability in microbial diversity was explained by geographic location and the concentrations of nitrogen, organic carbon, and copper. Rainfall, especially in Hawai'i, accounted for another 10.8% of the variability. PCoA by cave showed a strong influence of geography. Actinobacteria dominated the clones recovered from Hawaiian samples with 16% of the clone diversity recovered, but Acidobacteria dominated on Terceira, with 21% of the clones recovered. Alphaproteobacteria comprised 13% and 15% of the diversity in Hawai'i in Terceira, respectively.

Bacterial sequences with no known phyla affiliation were 14% of the total in Hawai'i and 12% in Terceira [27]. A nearest neighbor comparison among caves in Hawai'i and Terceira with other caves in the ARB database [58] show an overlap of only three OTUs (0.22% of the total). This high level of alpha diversity means that bacterial communities in each cave are different, and mat color is not a predictive factor.

They concluded that the geographical location is the dominant factor in explaining patterns of microbial diversity, with some contribution from nutrients.

Most microorganisms in subsurface environments grow as biofilms or individual or groups of cells attached to rock surfaces. Rocks and minerals are not homogeneous and can influence microbial diversity at the microniche level [59, 60] Components of the stratum may be toxic or beneficial to microbial growth by providing mineral nutrients, pH buffering, or other advantageous conditions. Given the long evolutionary history of microbes in geologic time, Jones and Bennett [60] hypothesize that each mineral surface is specifically altered by the best-adapted and most comprehensive microbial community that can use the mineral surface to the greatest advantage. Their experimental study under oligotrophic conditions concluded that community structure can be differentiated at the level of individual mineral grains supporting the unique microbial consortium best adapted to those conditions.

8.5.5 Microbial endemism

High levels of endemism among troglobites and stygobites are the norm. Culver et al. [61] found that about 30% of cave-adapted species in caves from the United States are found in only a single cave, and that trend can be extended to bacterial diversity. Although Hathaway et al. [27] were far from sampling total diversity, less than 5% of the OTUs found in lava caves occur in other caves or in other volcanic environments. They recovered 254 and 179 singletons in Terceira and Hawai'i, respectively. Sequences were more likely to be related to sequences from the same cave or the same island than between islands. If microbial distribution is ubiquitous, then they would expect a higher percentage of shared OTUs between the two archipelagos. Novel bacteria found in all sites indicate a need for conservation of caves as sources of novel bacteria and as simplified natural ecosystems for study of larger ecological questions.

Porca et al. [62] compared 16S rDNA gene sequences from yellow microbial communities on the walls of karst caves from Spain, the Czech Republic, and Slovenia. Those findings were compared to those of the yellow mats from the Azores and Hawaiian microbial mats [22]. A common core of microorganisms exists with 60% percent of the bacterial OTUs forming three common cores comprised of Actinobacteria and two gammaproteobacterial groups. Seven percent were common to two of the sites, and the remaining 35% were site-specific. Several estimates of diversity showed the greatest diversity in caves from Slovenia, then Czech Republic, and then Spain. There was no leveling off of diversity below the phylum level. Some of the Gammaproteobacteria are related to species that produce yellow pigments. The overall similarities of these and results from other studies reflect the similarity of limestone caves in terms of environmental conditions, geochemistry, and availability of organic matter to support microbial growth. This community may represent a group of true troglobitic microorganisms. Interestingly, from the lava cave microbial mats of Northup et al. [22], several of the core OTUs in the Porca et al. [62] study were similar, further supporting the idea of indigenous subsurface microorganisms.

8.5.6 Other lava cave microbiology

Roots are a very prominent feature of many lava caves (▸ Fig. 8.1) and have been studied extensively in lava caves in Hawai'i and elsewhere as invertebrate habitats [12, 63–68]; however, very little work has been done on microbial communities associated with roots in lava caves [68, 69]. Unsurprisingly, the levels of organic carbon and nitrogen are high in the root zone compared to other areas in a lava cave. Roots may act as a conduit for microorganisms and nutrients from the surface such that microbial communities associated with roots in lava caves would differ from those on nonroot habitats within the same lava cave [70]. The amount of nitrogen and organic carbon correlated with microbial community composition on roots, wet and dry ceilings, and corresponding floor zones. More sequences affiliated with Alphaproteobacteria and Bacteroidetes were retrieved from the roots compared to other samples, and fewer clones of Actinobacteria were recovered from the roots than from adjacent wet ceilings compared to dry ceiling and floor. Her findings support the role of roots as conduits of nutrients and microbes into lava caves.

Brown et al. [46] report on the phylogenetic diversity of microorganisms in a microbial mat in the entrance of a small, 100-year-old lava cave in Kilauea Crater, Hawai'i Volcanoes National Park, Hawai'i. High bacterial diversity was retrieved, with 31 of 53 phylotypes novel at or above the genus level. There was some overlap with bacteria from nearby lava caves, but most were closely related to sequences previously obtained from a hypersaline mat from an evaporative lake at Guerrero Negro, Mexico. Although these assemblages are from very different environments, the authors propose an inherent community structure in microbial mats. They also isolated the second recorded species in the genus *Gloeobacter* (*G. kilaueensis*) from a purple biofilm in the same cave in Kilauea crater [44]. *Gloeobacter* is believed to be an early ancestor of Cyanobacteria that diverged before the evolution of thylakoid membranes. Genome sequencing established *G. kilaueensis* as a new species which will provide insights into cyanobacterial evolution, including the shift from anoxic to oxygenic photosynthesis.

8.5.7 Nitrogen cycling

One of the first studies of nitrogen cycling in lava caves was conducted by Hathaway et al. [70] in 11 Tercerian lava caves. The presence of the *nifH* gene (essential for nitrogen fixation) and the *amoA* gene (essential for ammonia oxidation) was examined from microbial mats within lava caves underlying different surface land use regimes. Based on a greater diversity of *amoA* genes from caves underlying forested terrain, ammonia-oxidizing bacterial communities were affected by land use, but the distribution and diversity of *nifH* genetic sequences were evenly split between forest and pasture surface terrains. There was very little overlap between these two community types, which suggested substantially different communities. From a new, unpublished set of results (done by Jason Kimble), the *nifH*, *amoA*, and *nxrB* (essential for nitrite oxidation) genes are recovered to a much greater extent in oceanic and continental lava caves than in deeper carbonate caves.

Studies in both lava and carbonate caves suggest that nitrogen is often limiting and phosphorus is even more limiting for microbial growth. The role of extracellular enzyme activities in scavenging critically needed nitrogen and phosphorus for cave microorganisms in both lava and carbonate caves is just being studied (Noelle Martinez, unpublished results).

8.6 Human Impacts and Conservation

An unfortunate aspect of microbiology in caves is water pollution from fecal contamination. Lava caves have the potential to transport contaminants because they act as leaky pipes that transport significant water during heavy rains and floods. Most sewage enters groundwater *via* crevices in the lava or from local flooding. Halliday [71] conducted a study of solid waste and fecal pollution in Kaumana Cave in Hawai'i. Several passages in the cave are blocked by refuse, and one skylight is named Garbage Puka. They concluded that groundwater is currently not at risk, but that may change as standards for water quality keep getting more stringent.

Cave microorganisms are understudied and vulnerable to human impacts [72]. In addition to protecting the cave, its water inputs and the surface above the cave, people must modify their behavior in visiting and living above caves. Negative impacts from human activities include both direct and indirect threats: physical impacts, nutrient enrichment, imported exotic microbes, and threats from cave restoration efforts. Special concerns arise because most microbes in caves are invisible to visitors: out of sight, out of mind, despite the fact that many microbial mats in lava caves are extensive and visible. Visitors do not recognize the mats as alive and write their names in the mats, thereby eliminating or altering huge numbers of microorganisms. The success of recommendations for protecting cave microbiota rests on the value of microbes to cavers, scientists, and landowners and in our ability to educate people.

8.7 Microbial Morphologies

To help researchers beginning the study of microbes in lava caves, we present a photo survey of macroscopic colonies and SEM images of microbial mats and secondary minerals from lava caves.

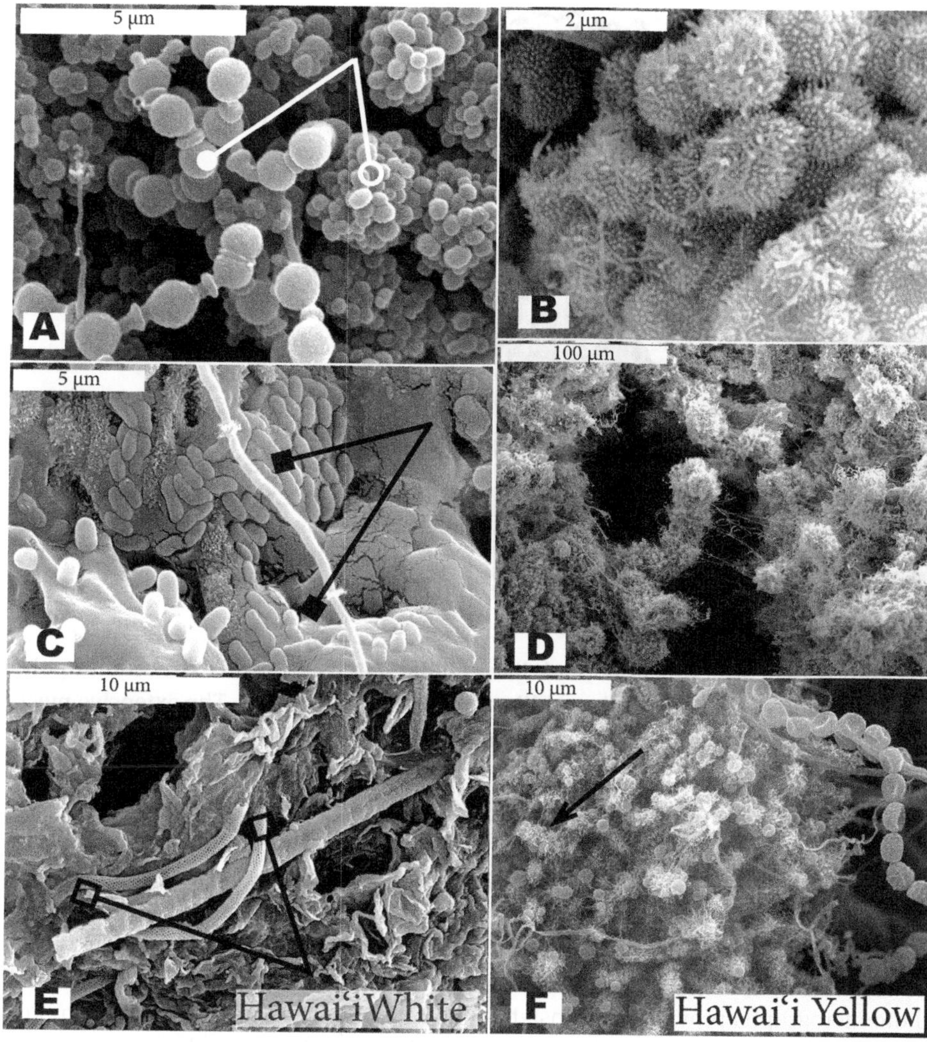

Fig. 8.9. Scanning electron micrographs (SEM) of white (left column) and yellow (right column) microbial mats from Hawaiian lava caves. Beads-on-a-string morphology is indicated by a line ending in a solid circle (a), while example cocci are indicated by a line ending in an open circle (a). Cocci with knobby appendages (b). Ordered rows of rods are indicated by a line ending in solid square (c). Note the biofilm in which the rods are embedded (c). Fuzzy colonies with thread-like extensions (d). The reticulated filaments are indicated by a line ending in an open square (e). Example thread-like appendages are indicated by an arrow (f). Photomicrographs by D. Northup, M. Spilde, and A. Kooser. Used with permission.

8.7.1 Microbial mats

Examination of lava cave microbial mat samples using SEM reveals an amazing diversity of microbial shapes with coccoid and rod-shaped morphologies as well as many

Fig. 8.10. Copper-containing speleothem in a Hawaiian lava cave. © Kenneth Ingham. Reproduced by permission of Kenneth Ingham. Permission to reuse must be obtained from the rights holder.

pili/fimbrae-like structures that look like cellular extensions and extracellular polymeric substances that have a thread-like appearance (▸ Fig. 8.9). The extensive mesh of appendages from cells is likely instrumental in hydrophobicity of microbial mats during wetter periods. Hydrophobicity may also be associated with exospore production by the microbial mat inhabitants [73].

One of the more interesting morphologies observed in lava cave mats appears to be rumpled biofilm covered with rows of rods. Other biofilm structures resemble sheets that have had holes poked in them, some creating geometric patterns. Some of the shapes observed overlap with those seen from carbonate caves and include the reticulated filaments [74, 75] and the beads-on-a-string [76]. Lava cave microbial mat SEM photomicrographs provide great insights into microbial mat structure, but also provide an opportunity for interpretive materials for cave visitors as the images reveal that the walls truly are alive.

8.7.2 Microbes masquerading as minerals

Lava caves contain a wide range of secondary minerals [11, 16, 39] that may contain interesting microbial communities. The possibility exists that these microbial communities may participate in the formation or dissolution of some of the secondary mineral deposits. This work has implications for life detection on Mars and other extraterrestrial bodies [22, 23] because of the discovery of lava caves on Mars and other bodies. Léveillé and associates [77–80] have explored the geomicrobiology of some of these secondary mineral deposits in lava caves from New Mexico, California, Hawai'i, and Azores, included copper- and iron-rich deposits and deposits that resemble reworked soil.

Copper-containing deposits in Hawaiian lava caves (▶ Fig. 8.10) have proven to be particularly fascinating due to the toxicity of copper for many life forms, but the resistance to copper by extremophiles [81]. Next generation sequencing from the copper deposits reveals large numbers of sequences affiliated with the *Acidomicrobium* genus in the Actinobacteria from a tropical lava cave and in the *Acidobacterium* genus in the Acidobacteria from a semiarid lava cave, both in Hawai'i. Genera present in these deposits from very different parts of the island suggest that the sulfur metabolism may be one metabolic strategy used by the bacteria in these copper deposits.

Besides the iron speleothems and the iron-oxide hexagons described in Northup et al. [22, 23], we have discovered that some lava caves appear to have a smooth iron oxide biofilm between microbial colonies (▶ Fig. 8.11). This morphology contrasts strongly with the iron-rich speleothems in not forming the more traditional red iron oxide muds or crusts. De los Ríos et al. [82] suggest that the iron-rich deposits found in a Terceiran lava cave are formed through the actions of bacteria such as *Leptothrix* and *Gallionella* spp.

Deposits that resemble reworked soil that forms on the walls of lava caves have been observed in gold, gray, butterscotch, and cream colors (▶ Fig. 8.12). They often form biovermiculation-like patterns that were originally observed in sulfur caves [52, 83, 84]. Preliminary results from 16S rRNA gene sequencing of these deposits from an Azorean sample document the presence of Actinobacteria, Alpha-, Beta-, Gamma-, and Deltaproteobacteria, Acidobacteria, Nitrospirae, Gemmatimonadetes, Bacteroidetes, Chlamydiae, and Candidate Divisions OD1, OP10 (renamed Armatimonadetes), and TM7. Future work will utilize next-generation sequencing to compare these deposits to traditional microbial mat samples from oceanic and continental lava caves. Another type of secondary mineral that is frequently observed in lava caves is a gold-colored deposit that forms veins, as well as hard, chunky deposits (▶ Fig. 8.13). These deposits have been observed to contain more usual microbial morphologies, as well as more unusual filamentous morphologies [22, 23]. Limited clone sequencing has documented the presence of Actinobacteria, Alpha-, Beta-, Gamma-, and Deltaproteobacteria, Acidobacteria, Chloroflexi, Nitrospirae, Gemmatimonadetes, Planctomycetes, and Bacteroidetes, with not all phyla being present in both locations. Studies of these different secondary mineral deposits can be particularly helpful in refining our life detection strategies to be used on extraterrestrial bodies.

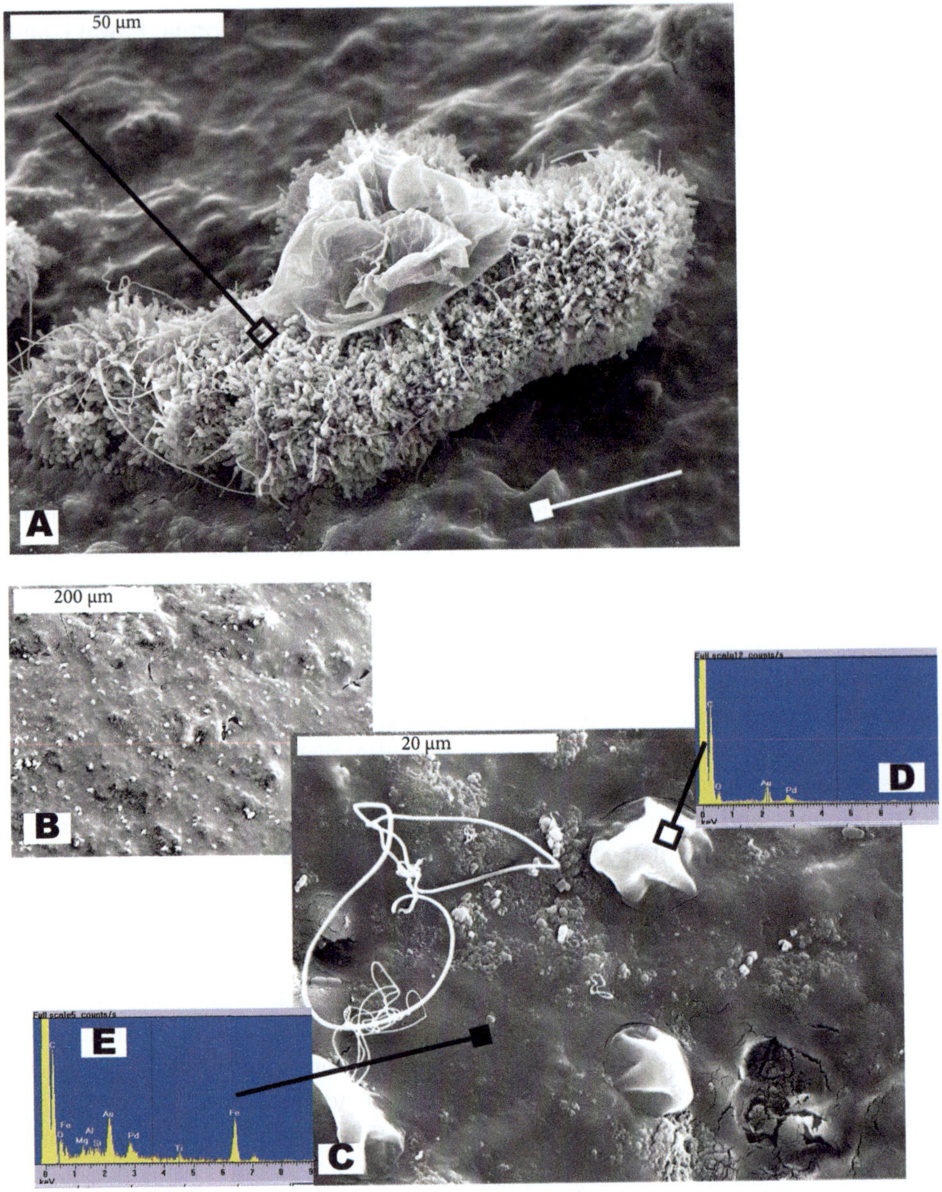

Fig. 8.11. Scanning electron micrographs of iron oxide biofilms with colonies erupting from the biofilm. Fully erupted colony (black line with an open square at end) atop an iron oxide biofilm (white line with solid white square at the end) (a). Overview of iron oxide biofilm with unerupted colonies (b). Close-up of unerupted colonies (c) showing the electron dispersive spectra with high carbon peak (d), and iron and oxygen peaks (e). Photomicrographs by D. Northup & M. Spilde. Used with permission.

Fig. 8.12. Butterscotch-colored organic ooze that resembles reworked soils coats this lava speleothem. A few microbial colonies are growing on top of the organic ooze. © Kenneth Ingham. Reproduced by permission of Kenneth Ingham. Permission to reuse must be obtained from the rights holder.

Fig. 8.13. Gold-colored deposits in a New Mexican lava cave. © Kenneth Ingham. Reproduced by permission of Kenneth Ingham. Permission to reuse must be obtained from the rights holder.

8.8 Astrobiology

Astrobiology is an interdisciplinary field that focuses on the study of the origin, evolution, and distribution of life in the universe, both on and beyond Earth, and how to detect life. As Carlton Allen of the United States National Aeronautical and Space Administration (NASA) said (personal communication to the authors), he has to be able to recognize a sulfur-based ecosystem if it is active or if it has been inactive for four billion years. On Earth, we are most familiar with the abundant life we observe on the surface, but Gold [10] speculates that there is as much microbial biomass below the surface as all the biomass above ground, being limited only by rising temperature with depth. Life on other planets will not likely be sentient green men, but will probably be microorganisms found below the surface, possibly in caves, and using sulfur as an energy source. Boston et al. [85] suggest that caves would be excellent sites to search for life or evidence of former life.

8.8.1 Life detection strategies

Caves have been recognized as potential model systems for us to study sulfur-based ecosystems. If we identify biomarkers, which are used as evidence left behind by living organisms, then we can identify past and current life on other planets. Searching for biomarkers in caves also helps us understand the contribution of microbes from biokarst (landscape level impacts of organisms) to speleogenesis.

Testing hypotheses for life on Mars involves searching early Precambrian deposits on Earth for evidence of life and the environmental conditions that allowed life to evolve and survive on early Earth. A second approach is to look for Mars analogs on Earth, which are places where Mars-like conditions currently exist, like the dry valleys in Antarctica and subsurface environments, particularly those with iron, which might serve as an energy source for chemolithoautotrophic bacteria. The subsurface can be accessed through drilling or *via* caves. As previously discussed, Northup et al. [22, 23] describe microbial mats and secondary mineral deposits that are nonbiological in appearance to determine if any patterns are detectable in the environmental conditions in which microorganisms are found. Such studies can help to guide how we detect life on extraterrestrial bodies. Basaltic glass forms whenever lava flows cool rapidly, and are widespread in the crusts of many rocky solar system bodies. Izawa et al. [86] found that basaltic glass on Earth is rapidly colonized by microbes when liquid water is available and microbial activity leaves very characteristic bioalteration textures that are preserved over geologic time, up to ~3.5 billion years on Earth. These tubular bioalterations can serve as high-value targets for astrobiological studies of samples from Mars, Europa, and other icy satellites.

Extant microbial growths and resulting lithified fossils from caves have been compared by Boston et al. [9] to identify characteristic biosignature suites that could be

Fig. 8.14. Snottites, acidic, pendant biofilms, in Cueva de Villa Luz, Mexico. © Kenneth Ingham. Reproduced by permission of Kenneth Ingham. Permission to reuse must be obtained from the rights holder.

applied to detection of life on other planets. They report on morphological fossils known as U-loops, and compare them to extant snottites in sulfur caves (▸ Fig. 8.14). Nonacidic snottites have also been observed in lava caves in Iceland (▸ Fig. 8.15). Other markers are of manganese and iron oxides, and calcite and sulfur minerals that can be recognized both by microscopic and unaided eye observations. Comparisons of isotopic ratios of sulfur and carbon, genetic data, unusual concentrations of minerals, and mineral forms can support morphological observations.

As magma cools, minerals crystallize out in a predictable sequence known as Bowens Reaction Series [13]. Olivine is the first to crystallize out of molten magma at close to 1,800 °C. It is a ferromagnesium mineral with very low silica content (<45%). Large areas of Mars are covered by basaltic rock from lava flows, with clear evidence of lava caves [80]. Life in caves on Mars would be protected from solar radiation, and the caves may retain water longer than on the surface [87]. Consequently, studies of lava cave microbial communities and their characteristics have many applications within the realm of astrobiology.

Bacteria capable of respiring olivine from the rock-ice interface of a lava cave from the Oregon Cascades were isolated by Popa et al. [88]. This is the first time an organism with such a metabolic ability has been reported. The lava-ice interface is a Mars analog environment. The isolate, *Pseudomonas* sp. HerB, is a mesophilic Gammaproteobacteria capable of chemolithoautotrophic iron oxidation using olivine as a source of energy under low oxygen conditions and near neutral pH which offsets abiotic iron

Fig. 8.15. Nonacidic snottite-like structures in a lava cave in Iceland.

oxidation. The authors speculate that this organism's metabolic abilities of would allow it to live in near-surface, icy volcanic environments like those found on Mars in the present or recent past, making this type of physiology a prime candidate for the search for life on other planets, and potentially an important primary producer in extraterrestrial ecosystems [88].

8.9 Conclusions and Future Opportunities

Microbiology has traditionally explained patterns of microbial biogeography using ecological concepts, such as "everything is everywhere: but the environment selects" [55, 56]. This claim is based on the concept that microbial distribution is ubiquitous; microorganisms are cosmopolitan in their distribution, but ecological conditions determine what grows. Recent discoveries of geographical patterns in microbial distribution are changing our fundamental assumptions and show that microbial biogeography is much more complicated than historically thought. Lava caves, which have a worldwide distribution and a variety of abiotic and biotic parameters, can provide an outstanding laboratory in which to test these hypotheses. Massively parallel next generation sequencing provides the means to capture the diversity that resides in these lava caves, in order to make these comparisons. Metagenomics will provide the tools to begin the study of how these microorganisms make their living in these caves, where chemolithoautotrophy and chemoheterotrophy are both viable life styles.

Comparisons of bacterial phylogeny are showing the presence of an indigenous subsurface microbiota from limestone and lava caves around the world. Many of the bacteria identified are unique, and may represent cave-adapted microorganisms. Geomicrobiological effects on cave formation, enlargement, and mineral precipitation have clear microbial contributions [5–7]. The potential for useful natural products from this unique biodiversity is only beginning to be studied [89]. Caves are an extreme habitat and the cave microbiome has great potential as a novel resource for drug discovery [90, 91]. Antimicrobials are one example of secondary metabolites produced by microorganisms to reduce competition, but Zheng [92] has broadened our understanding of the role of antimicrobials as toxins, ionophores, bioregulators, and signal molecules produced by "metabolically talented" microbes. Natural products are the most promising source of novel antibiotics and cave environments have great potential for the development of new bioactive microbial metabolites.

Lava caves provide accessible and simplified natural ecosystems for the study of life in extreme environments [93]. There is a need for additional studies of microbes in caves, particularly from lava caves, and to document the total diversity.

Acknowledgments

Our studies of lava caves have spanned the last two decades and have involved an immense number of students and colleagues without whom we would not be writing about the microbial communities of lava caves. We especially thank Dr. Kenneth Ingham for the use of his compelling lava cave photographs, and Mr. Michael Spilde for his equally compelling scanning electron micrographs and his teaching and coaching in scanning electron microscopy. Our colleagues in the Azores, Drs. Dapkevicius, Borges, and Gabriel inspired us to expand our studies of lava caves and provided a wealth of opportunities to explore the Azorean underworld. Our colleagues in Hawai'i, Drs. Fred Stone, Steve Smith, and Frank Howarth have generously shared their knowledge, time, and expertise concerning the Hawaiian lava caves. Drs. Nick and Sue White introduced us to the amazing world of Australian lava caves. Many friends have helped with fieldwork in the caves in these locations. We especially thank Árni and Gunnhildur Stefánsdóttir for their wonderful introduction to Iceland lava caves and their generous support of the Icelandic work. We thank members of the Subsurface Life in Mineral Environments (SLiME), Jenny Hathaway, Jessica Snider, Monica Moya, Matthew Garcia, Matthew Medina, and Ara Kooser, who contributed significantly to research that is featured in this chapter.

This work would not have been possible without the generous funding provided by Fundação para a Ciência e Tecnologia (FCT, Portugal), under the project PTDC/AMB/70801/2006 (Maria de Lurdes N. E. Dapkevicius, P. I.) and the Undergraduate Opportunities Program, Museum of Southwestern Biology (NSF-DEB 0731350, Joseph Cook, P. I.). This work was also supported by the Cave Conservancy of the Virginias, the Graduate Research Allocation Committee at UNM Biology, UNM Biology Grove Scholarship, the Student Research Allocation Committee at UNM, the National Speleological Society, the National Speleological Foundation, the New Mexico Space Grant Consortium, the New Mexico Alliance for Minority Participation Program, the New Mexico Geological Society, T & E Inc., the National Park Service through the Colorado Plateau Cooperative Ecosystems Study Unit, and Kenneth Ingham Consulting. We very much appreciate the support and collecting permits provided by the staff of the National Park Service in El Malpais National Monument, Lava Beds National Monument, and Hawai'i Volcanoes National Park, and several private land owners, including the Cave Conservancy of Hawai'i. We acknowledge support from the UNM Molecular Biology Facility, which is supported by NIH grant number P20GM103452. We gratefully thank the editor and an anonymous reviewer whose comments improved the manuscript.

References

[1] Poulson TL, Lavoie KH. The trophic basis of subsurface ecosystems. Chapter 12. In: Wilkens H, Culver DC, Humphreys WF, eds. Ecosystems of the World 30: Subterranean Ecosystems. Elsevier, 2000, 231–249.

[2] Howarth FG. High-stress subterranean habitats and evolutionary change in cave-inhabiting arthropods. Am Nat, 1993, 141, S65–S77.

[3] List of longest lava caves. Accessed July 12, 2013, at www.caverbob.com/lava.htm

[4] Holcomb RT. Eruptive History and Long-Term Behavior of Kilauea Volcano. United States Geological Survey Professional Paper, 1987, 1350, 261–350.

[5] Northup DE, Lavoie KH. Geomicrobiology of caves: A review. Geomicrobiol J, 2001, 18, 199–222.

[6] Barton HA, Northup DE. Geomicrobiology in cave environments: Past, current and future perspectives. J Cave Karst Stud, 2007, 69, 163–178.

[7] Lavoie KH, Northup DE, Barton HA. Chapter 1. Microbial-mineral interactions; Geomicrobiology in caves: A review. In: Jain SK, Kahn AA, Rai MK, eds. Geomicrobiology. Science Publisher, USA /CRC Press, 2010, 1–46.

[8] Ehrlich HL, Newman, DK. Geomicrobiology. 5th edn. Boca Raton, FL, Taylor and Francis Group, CRC Press LLC, 2009.

[9] Boston PJ, Spilde MN, Northup DE, et al. Cave biosignature suites: Microbes, minerals and Mars. Astrobiol J, 2001, 1, 57–70.

[10] Gold T. The deep, hot biosphere. Proc Nat Acad Sci USA, 1992, 89, 6045–6049.

[11] Palmer AN. Cave Geology. Dayton, Ohio, USA, Cave Books, 2007.

[12] Howarth FG. Lava cave ecosystems as a study site. In: Mueller-Dembois D, Bridges KW, Carson HL, eds. Island Ecosystems: Biology in Selected Hawaiian Communities. US/IBP Synth. Vol. 15. Hutchinson Ross, 1981, 222–230.

[13] Le Maitre RW, Streckeisen A, Zanetting A, Le Bas MJ, Bonin B, eds. Igneous Rocks: A Classification and Glossary of Terms: Recommendations of the International Union of Geological Sciences Subcommission on the Systematics of Igneous Rocks. 2nd edn. Cambridge University Press, 2004.

[14] Kempe S. Principles of pyroduct (lava tunnel) formation. In: White WB, ed. Proceedings of the 15th International Congress of Speleology, 2009, 1, 668–674.
[15] Rogers BW, Mosch CJ. Snowballs in the underground; lava-tube deposits and morphology: Natural history of El Malpais National Monument, New Mexico Bureau of Mines and Mineral Resources Bulletin, 1997, 156, 83–90.
[16] White WB. Secondary minerals in volcanic caves: data from Hawai'i. J Cave Karst Stud, 2010, 72, 75–85.
[17] Hill CA, Forti P. Cave Minerals of the World. 2nd edn. Huntsville, Alabama, USA, National Speleological Society, 1997.
[18] Tuttle MD, Stevenson DE. Variation in the cave environment and its biological implications. In: Zuber R, Chester J, Rhodes D, eds. National Cave Management Symposium Proceedings, Big Sky Montana. Albuquerque: Adobe Press, 1977, 108–121.
[19] Howarth FG. Cavernicoles in lava caves on the island of Hawai'i. Science, 1972, 175, 325–326.
[20] Snider J, Goin C, Miller RV, Boston PJ, Northup DE. Ultraviolet radiation sensitivity in cave bacteria: Evidence of adaptation to the subsurface? Int J Speleol, 2009, 3, 13–22.
[21] Northup DE, Connolly A, Trent A, et al. The nature of bacterial communities in Four Windows Cave, El Malpais National Monument, New Mexico, USA. AMCS Bull, 2008, 19, 119–125.
[22] Northup DE, Melim LA, Spilde MN, et al. Lava cave microbial communities within mats and secondary mineral deposits: Implications for life detection on other planets. Astrobiology, 2011, 11, 1–18.
[23] Northup DE, Hathaway JM, Snider JR, et al. Life in Earth's lava caves: Implications for life detection on other planets. In: Hanslmeier A, Kempe S, Seckbach J, eds. Dordrecht, Springer, Life on Earth and Other Planetary Bodies, 2012, 459–484.
[24] Garcia MG, Moya M, Spilde MN, Stone FD, Northup DE. Discovering new diversity in Hawaiian lava tube microbial mats. In: White WB, ed. Proc. 15th Int Cong Speleol, 2009, 1, 364–369.
[25] Moya M, Garcia MG, Spilde MN, Northup DE. Composition of bacterial mats in El Malpais National Monument, New Mexico, USA: Comparisons and contrasts with bacterial communities in Hawai'i lava caves. In: White WB, ed. Proc 15th Int Congr Speleol, 2009, 2, 709–713.
[26] Donachie SP, Hou S, Lee KS, et al. The Hawaiian Archipelago: A microbial diversity hotspot. Microb Ecol, 2004, 48, 509–520.
[27] Hathaway JJM, Garcia MG, Moya-Balash M, et al. Comparison of bacterial diversity in Azorean and Hawaiian lava cave microbial mats. Geomicrobiol J, 2014, 3, 205–220.
[28] Vanderwolf KL, Malloch D, McAlpine DF, Forbes GJ. A world review of fungi, yeasts, and slime molds in caves. Int J Speleol, 2013, 42, 77–96.
[29] Caumartin V. Review of the microbiology of underground environments. Natl Speleol Soc Bull, 1963, 25, 1–14.
[30] Cryan PM, Meteyer CU, Boyles JG, Blehert DS. White-nose syndrome in bats: illuminating the darkness. BMC Biol, 2013, 11, DOI: 10.1186/1741-7007-11-47.
[31] Puechmaille SJ, Wibbelt G, Korn V, et al. Pan-European distribution of white-nose syndrome fungus (*Geomyces destructans*) not associated with mass mortality. PLoS ONE, 2011, 6, e19167.
[32] Zukal J, Bandouchova H, Bartonicka T, et al. White-Nose Syndrome fungus: A generalist pathogen of hibernating bats. PLoS ONE, 2014, 9, e97224.
[33] Alves DMCC, Terrible LC, Brito D. The potential impact of White-Nose Syndrome on the conservation status of North American Bats. PLoS ONE, 2014, 9, e107395.
[34] McNab, BK. Evolutionary alternatives in the physiological ecology of bats. In: Kunz TH, ed. Ecology of Bats. Plenum Press, New York, USA, 1982, 151–200.
[35] Northup DE, Buecher DE. Characterization of roost microclimate and bat health in bat hibernacula with microbial sampling for the presence of *Geomyces* spp. in the lava caves

of El Malpais National Monument. Final Report for CPCESU Cooperative Agreements. J7190100019/UNM-55 & P11AT10728/UNM-70/, 2012.

[36] Ashford DA, Hajjeh RA, Kelley MF, Kaufman L, Hutwagner L, McNeil MM. Outbreak of histoplasmosis among cavers attending the National Speleological Society Annual Convention, Texas, 1994. Am J Trop Med Hyg, 1999, 60, 899–903.

[37] Griffin DW, Gray MA, Lyles M, Northup D. The transport of non-indigenous microorganisms into caves by human visitation: A case study at Carlsbad Caverns National Park. Geomicrobiol J, 2014, 31, 175–185.

[38] González López L, Vidal-Romaní JR, López Galindo MJ, Vaqueíro Rodríguez M, Sanjurjo Sánchez J. First data on testate amoebae in speleothems of caves in igneous rocks. Cadernos Lab. Xeolóxico de Laxe Coruña, 2013, 37, 37–56.

[39] Forti P. Genetic processes of cave minerals in volcanic environments: An overview. J Cave Karst Stud, 2005, 76, 3–13.

[40] Pentecost A. Flora of a small lava cave near Laki, Iceland. Cave Karst Sci, 2011, 39, 34–36.

[41] Amann RI, Ludwig W, Schleife KH. Phylogenetic identification and *in situ* detection of individual microbial cells without cultivation. Microbiol Rev, 1995, 59, 143–169.

[42] Hugenholtz P, Goebel BM, Pace NR. Impact of culture-independent studies on the emerging phylogenetic view of bacterial diversity. J Bacteriol, 1998, 180, 4765–4774.

[43] Pace NR. A molecular view of microbial diversity and the biosphere. Science, 1997, 276, 734–740.

[44] Saw JHW, Schatz M, Brown MV, et al. Cultivation and complete genome sequencing of *Gloeobacter kilaueensis* sp. nov., from a lava cave in Kīlauea Caldera, Hawai'i. PLoS ONE, 2013, 8, e76376.

[45] Stoner MF, Howarth FG. Community structure and niche differentiation in Hawaiian lava tubes. In: Mueller-Dembois D, Bridges KW, Carson HL, eds. Island Ecosystems: Biology in Selected Hawaiian Communities. US/IBP Synth. Vol. 15. Hutchinson Ross, 318–336.

[46] Brown MV, Donachie SP, Bissett A, Foster J. Deep evolutionary diversity in Hawaiian lava cave Cyanobacterial microbial assemblage. Meeting abstract. Astrobiology, 2012, 7, 505.

[47] Stewart EJ. Growing unculturable bacteria. J Bacteriol, 2012, 194, 41–51.

[48] Donachie SP, Foster JS, Brown MV. Culture clash: challenging the dogma of microbial diversity. ISME J, 2007, 1, 97–102.

[49] Gomez-Alvarez V, King GM, Nusslein K. Comparative bacterial diversity in recent Hawaiian volcanic deposits of different ages. FEMS Microbiol Ecol 2007, 60, 60–73.

[50] Janssen PH. Identifying the dominant soil bacterial taxa in libraries of 16S rRNA and 16S rRNA genes. Appl Environ Microbiol, 2006, 72, 1719–1728.

[51] Macalady JL, Dattagupta S, Schaperdoth I, et al. Niche differentiation among sulfur-oxidizing bacterial populations in cave waters. ISME J, 2008, 2, 590–601.

[52] Jones DS, Lyon EH, Macalady JL. Geomicrobiology of biovermiculations from the Frasassi Cave System, Italy. J Cave Karst Stud, 2008, 70, 78–93.

[53] Chen Y, Wu L, Boden R, et al. Life without light: microbial diversity and evidence of sulfur- and ammonia based chemolithotrophy in Movile Cave. ISME J, 2009, 3, 1093–1104.

[54] Porter ML, Engel AS, Kane TC, Kinkle BK. Productivity-diversity relationships from chemolithoautotrophically based sulfidic karst systems. Int J Speleol, 2009, 38, 27–40.

[55] Baas-Becking LGM. Geobiologie of Inleiding tot de Milieukunde. The Hague, the Netherlands, WP. Van Stockum & Zoon, 1934.

[56] O'Malley MA. 'Everything is everywhere: but the environment selects': ubiquitous distribution and ecological determinism in microbial biogeography. Stud Hist Phil Bio Biomed Sci, 2008, 39, 314–325.

[57] Whittaker RH. Evolution and measurement of species diversity. Taxon, 1972, 21, 213–251.

[58] Ludwig W, Strunk O, Westram R, et al. ARB: a software environment for sequence data. Nucl Acids Res, 2004, 32, 1363–1371.

[59] Barton HA, Taylor NM, Kreate MP, Springer AC, Oehrle SA, Bertog JL. The impact of host rock geochemistry on bacterial community structure in oligotrophic cave environments. Int J Speleol, 2007, 36, 93–104.

[60] Jones AA, Bennett PC. Mineral microniches control the diversity of subsurface microbial populations. Geomicrobiol J, 2014, 31, 246–261.

[61] Culver DC, Master LL, Christman MC, Hobbs HH. Obligate cave fauna of the 48 contiguous United States, Conserv Biol, 2000, 14, 386–401.

[62] Porca E, Jurado V, ŽGur-Bertok D, Saiz-Jimenez C, Pašić L. Comparative analysis of yellow microbial communities growing on the walls of geographically distinct caves indicates a common core or microorganisms in their formation. FEMS Microbiol Ecol, 2012, 81, 255–266.

[63] Howarth FG. Bioclimatic and geologic factors governing the evolution and distribution of Hawaiian cave insects. Entomologia Generalis, 1982, 8, 17–26.

[64] Howarth FG, James SA, McDowell W, Preston DJ, Imada CT. Identification of roots in lava tube caves using molecular techniques: implications for conservation of cave arthropod faunas. J Insect Cons, 2007, 11, 251–261.

[65] Jasinska EJ, Knott B, McComb AJ. Root Mats in ground water: A fauna-rich cave habitat. J N Am Benthol Soc, 1996, 15, 508–519.

[66] Stone FD, Howarth FG, Hoch H, Asche M. Root communities in lava tubes. In: Culver DC, White WB, eds. Encyclopedia of Caves. Burlington, MA, USA, Elsevier Academic Press, 2004, 477–484.

[67] Wessel A, Hoch H, Asche M, et al. Founder effects initiated rapid species radiation in Hawaiian cave planthoppers. Proc Natl Acad Sci USA, 2013, 110, 9391–9396.

[68] Snider, JR. Comparison of Microbial Communities on Roots, Ceilings and Floors of Two Lava Tube Caves in New Mexico. Unpublished Master's Thesis. Albuquerque, NM. 2010.

[69] Snider JR, Moya M, Garcia MG, Spilde MN, Northup DE. Identification of the microbial communities associated with roots in lava tubes in New Mexico and Hawai'i. 15th Proc Int Congr Speleol, 2009, 2, 718–723.

[70] Hathaway JJM, Sinsabaugh RL, Dapkevicius MLNE, Northup DE. Diversity of ammonia oxidation (*amoA*) and nitrogen fixation (*nifH*) genes in lava caves of Terceira, Azores, Portugal. Geomicrobiol J, 2014, 31, 221–235.

[71] Halliday WR. Raw sewage and solid waste dumps in lava caves of Hawai'i Island. J. Cave Karst Stud, 2003, 69, 68–75.

[72] Northup DE. Managing microbial communities in caves. In: van Beyen PE, ed. Karst Management. Berlin, Springer, 2011, 225–240.

[73] Senger CM, Crawford RL. Biological inventory: Mount St. Helens Cave basalt flow area: final report: unpublished report prepared for the Gifford Pinchot National Forest, Mount St. Helens National Volcanic Monument, St. Helens Ranger District, 1984.

[74] Melim LA, Northup DE, Spilde MN, Jones B, Boston PJ, Bixby RJ. 2008. Reticulated filaments in cave pool speleothems: microbe or mineral? J Cave Karst Stud, 2008, 70, 135–141.

[75] Miller AZ, Hernández-Mariné M, Jurado V, et al. 2012. Enigmatic reticulated filaments in subsurface granite. Environ Microbiol Rep, 2012, 4, 596–603.

[76] Northup DE, Dahm CN, Melim LA, et al. Evidence for geomicrobiological investigations in Guadalupe caves. J Cave Karst Stud, 2000, 67, 80–90.

[77] Léveillé RJ, Fyfe WS, Longstaffe FJ. Geomicrobiology of carbonate-silicate microbialites from Hawaiian basaltic sea caves. Chem Geol, 2000, 169, 339–355.

[78] Léveillé RJ, Longstaffe FJ, Fyfe WS. Kerolite in carbonate-rich speleothems and microbial deposits from basaltic caves, Kauai, Hawai'i. Clays Clay Min, 2002, 50, 514–524.

[79] Léveillé RJ, Longstaffe FJ, Fyfe WS. An isotopic and geochemical study of carbonate-clay mineralization in basaltic caves: abiotic versus microbial processes. Geobiol, 2007, 5, 235–249.

[80] Léveillé RJ, Datta S. Lava caves and basaltic caves as astrobiological targets on Earth and Mars: A review. Planet Space Sci, 2010, 58, 590–601.

[81] Orell A, Navarro CA, Arancibia R, Mobarec JC, Jerez CA. Life in blue: copper resistance mechanisms of bacteria and archaea used in industrial biomining of minerals. Biotechnol Adv, 2010, 28, 839–848.

[82] de los Ríos A, Bustillo MA, Ascaso C, Carvalho MR. Biocontructions in ochreous speleothems from lava tubes on Terceira Island (Azores). Sed Geol, 2011, 236, 117–128.

[83] Boston PJ, Curnutt J, Gomez E, Schubert K, Strader B. Patterned growth in extreme environments. In: EES Press. Proc Third IEEE Internat Conf Space Mission Challenges for Info Tech, 2009, 221–226.

[84] Hose LD, Palmer AN, Palmer MV, Northup DE, Boston PJ, DuChene HR. Microbiology and geochemistry in a hydrogen-sulphide-rich karst environment. Chem Geol, 2010, 169, 399–423.

[85] Boston PJ, Ivanov MV, McKay CP. On the possibility of chemosynthetic ecosystems in subsurface habitats on Mars. Icarus, 1992, 95, 300–308.

[86] Izawa MRM, Banerjee NR, Flemming RL, Bridge NJ, Schultz C. Basaltic glass as a habitat for microbial life: implications for astrobiology and planetary exploration. Planet Space Sci, 2010, 58, 583–591.

[87] Westall F, Brack A, Hoffmann BA, et al. An ESA study for the search for life on Mars. Planet Space Sci, 2000, 48, 181–202.

[88] Popa R, Smith AR, Popa R, Boone J, Fisk M. Olivine-respiring bacteria isolated from the rock-ice interface in a lava-tube cave, a Mars analog experiment. Astrobiology, 2012, 12, 9–18.

[89] Montano ET, Henderson LO. Studies of antibiotic production by cave bacteria. In: Cheeptham N, ed. Cave Microbiomes: A Novel Resource for Drug Discovery. Springer Briefs, Springer, 2013, 109–130.

[90] Pelaez F. The historical delivery of antibiotics from microbial natural products—can history repeat? BioChem Parmacol, 2006, doi:10.1016/j.bcp.2005.10.010.

[91] Gabriel CR, Northup DE. Microbial Ecology: Caves as an extreme habitat. In: Cheeptham N, ed. Cave Microbiomes: A Novel Resource for Drug Discovery. Springer Briefs, Springer, 2013, 85–108.

[92] Zhang L. Integrated approach for discovery of novel drugs from microbial natural products. In Zhang L, Demain AL, eds. Natural Products. Drug Discovery and Therapeutic Medicine. Totowa, NJ, Humana, 2005, 30–55.

[93] Northup DE, Welbourn WC. Life in the twilight zone—lava-tube ecology. New Mexico Bureau of Mines & Mineral Resources Bulletin, 1997, 156, 69–81.

Audrey Paterson and Annette Summers Engel

9 Predicting bacterial diversity in caves associated with sulfuric acid speleogenesis

Abstract: Most caves and karst features form by carbonic acid dissolution. A minor percentage of actively forming caves and karst systems globally also form from sulfuric acid dissolution, whereby hydrogen sulfide is oxidized to sulfuric acid in the process known as sulfuric acid speleogenesis (SAS). Microorganisms colonize sulfidic karst systems, and prior research has almost exclusively focused on surveying the diversity of microbial assemblages in modern systems. The goal for most studies has been to identify the uniqueness of certain populations or species rather than on evaluating metabolic activities and potential microbial effects on cave or karst processes. This chapter reviews our current knowledge of microbial diversity from these systems. Community composition reflects habitat types, such as cave-wall biofilms or subaqueous microbial mats. The relative abundances of major taxonomic groups correlate with geochemical conditions that define community niche space. Similar diversity and niche space patterns result in a predictive model that diminishes the need to conduct additional diversity-based studies in the future. Rather, future research focused on understanding and quantifying microbial functions, such as from experimental omics investigations, single-cell genomics, and RNA-based transcriptional studies, will be more important because findings would more appropriately elucidate microbial roles in SAS and other geological processes.

9.1 Introduction

Cave systems formed from sulfuric acid speleogenesis (SAS) represent an estimated 5% [1] to 10% [2] of caves worldwide, including the Guadalupe Mountains of New Mexico, USA [3–8], Kraushöhle in Austria [9], caves within Şălitrari Mountain in the Cerna Valley of Romania [10], Provalata Cave in Republic of Macedonia [11], the Kugitangtou caves in Turkmenistan [12], and caves in the Italian regions of Umbria and Marche [13, 14], Tuscany [15], and Sardinia [16]. A few active cave systems provide insight into SAS and biogeochemical processes, such as Lower Kane Cave in Wyoming, USA [17–19], Cueva de Villa Luz in Mexico [20, 21], Movile Cave in Romania [22–24], the Frasassi caves and Acquasanta Terme in Italy [13, 14, 25–30], Steamboat Springs and Glenwood Caverns in Colorado, USA [31], as well as the Edwards karst aquifer in Texas, USA [32–35]. The concept of SAS has a long history [36–38], but Egemeier [39] introduced the chemistry of SAS to a wider audience following his dissertation research [17]. He suggested subaerial, abiotic oxidation of H_2S in the fluids that rise to the shallow subsurface and surface [2], followed by the replacement of carbonate with gypsum (Equation 9.1). The removal of gypsum by waters undersaturated with respect to gypsum increases void space.

$$H_2SO_4 + CaCO_3 + H_2O \rightarrow CaSO_4 \cdot 2H_2O + CO_2 \tag{9.1}$$

Hydrogen sulfide can be an abiotic product of thermochemical sulfate reduction or a biological product of microbial dissimilatory sulfate reduction, sulfur disproportionation, or decomposition of sulfur-bearing organic matter. In karst, sulfidic water from oil field brine or hydrothermal activity rises or upwells into the shallow subsurface and mixes with nonsulfidic groundwater [28, 40, 41]. Rising fluid is diagnostic of hypogenic cave formation [42, 43], and water that discharges into cave passages or circulates in the aquifer containing hydrogen sulfide often has elevated concentrations of other gases like methane and carbon dioxide.

Despite early recognition that microbes actively colonize caves, and that microbes produce acidity that dissolves carbonates, including from the oxidation of reduced sulfur compounds [44–46], demonstration that microbes play a role in SAS occurred decades later after experimental studies in Lower Kane Cave [18, 19, 47, 48], the Frasassi caves [13, 14, 27, 42], and the Edwards Aquifer [33, 34, 49]. The effective rate of carbonate dissolution in these systems due to microbial activity, which focuses acidity at the water-rock surface colonized by microbes, can average ~20 mg carbonate lost per square centimeter of carbonate exposed per year [34, 42]. This is because taxonomically diverse groups of Bacteria and Archaea are capable of oxidizing sulfide and other reduced sulfur compounds at faster rates than abiotic reactions, even using nitrate as an electron acceptor instead of oxygen [33, 34, 48, 50]. Other key metabolic processes within the sulfur cycle are linked to organic matter decomposition, intra- or extracellular elemental sulfur precipitation, and storage of inorganic and organic sulfur compounds [51–53]. These processes can also produce acidity and cause carbonate dissolution at the microbe-rock-water interface, although acidity is likely to be as carbonic and/or organic acids.

Of the known and readily accessible sulfidic cave systems, microbial colonization results in morphologically distinct biofilms prominently seen on cave walls, rocks, sediments, or as benthic or floating mats in lakes and streams. Cave-wall microbial communities from sulfidic caves have been studied extensively because of their conspicuous morphologies. Representative sulfidic karst studies with microbial community descriptions and contextual geochemistry are listed in ▸ Tab. 9.1. Veils or webs of mucous that hang, drip, and ooze from subaerial surfaces are referred to as snottites [20, 21]. A photograph of these features can be found in Chapter 8, ▸ Fig. 8.14. Snotties have solutions with low pH values between 0 and 2.5 and are associated with elemental sulfur precipitation and mineralized crusts of gypsum and other minerals [7, 21, 25, 54, 55]. Chapter 2 of this book includes a case study focused on describing active microbial communities and metabolisms from snottites in sulfidic caves from Italy. Relict SAS caves have similarly shaped features, referred to as biothems, such as U-loops, and they may indicate past microbial life [56, 57] (▸ Fig. 9.1 (a)). Some microbial mats float on the surface of water, such as in the Movile Cave, and form

Table 9.1. Compiled summary data used in the analysis of 16S rRNA gene sequences from sulfidic karst and nonkarst systems. The samples are organized by interface (e.g. air-rock cave-wall biofilms, air-water interface with floating mats, benthic sediments and microbial mats, aquifer well water floating filaments and dislodged biofilms from water-rock interface, cenote open water and water-rock interface, and mixed interfaces from nonkarst terrestrial springs).

Interface	System Location	▸ Fig. 9.2 & 9.3 ID No.	Sample Description	Temp (°C)	pH	H_2S (g) ppmv	CO_2 (g) ppmv	Mineral Substrate	Ref.
Air-Rock Interface Cave Wall Biofilms	Frasassi cave system, Italy	1	Snottite PC1 cave-wall snottite		0–1.5	0.2–2	1200	Microcrystalline gypsum	[55]
		2	Snottite RS2 cave-wall snottite		0–1.5	24	3000	Microcrystalline gypsum with S_0	
		3	Cave wall biofilm "RS24," hanging 3 cm long (metagenomics data indicates ~17% Archaea)	13	0–2.5	8–24	3300	Gypsum & S_0 crust	[84]

Interface	System Location		Sample Description	Temp (°C)	pH	Dissolved Sulfide (µmol/L)	Dissolved Oxygen (µmol/L)	Conductivity (µS/cm)	Ref.
Air-Water Interface Floating Mats	Movile Cave, Romania	4	50 mL water + ~1 g floating mat	20	7.5	200–300			[23]
		5	Floating biofilms (3–4 mm thick) at the air-water interface of air bells in submerged passages; airbells have 7–10% O_2 (g), 2–3.5% CO_2, 1–2% CH_4	20.8	7.3	240	< 0.1	1443	[24]

Table 9.1. (cont.) Compiled summary data used in the analysis of 16S rRNA gene sequences from sulfidic karst and nonkarst systems. The samples are organized by interface (e.g. air-rock cave-wall biofilms, air-water interface with floating mats, benthic sediments and microbial mats, aquifer well water floating filaments and dislodged biofilms from water-rock interface, cenote open water and water-rock interface, and mixed interfaces from nonkarst terrestrial springs).

Interface	System Location		Sample Description	Temp (°C)	pH	Dissolved Sulfide (μmol/L)	Dissolved Oxygen (μmol/L)	Conductivity (μS/cm)	Ref.
Cave Benthic Sediments and Microbial Mats	Frasassi cave system, Italy	6	Grotto Sulfurea; thick white subaqueous filamentous mats; 5–10 mm thick gray-black at sediment interface, white-beige gelatinous interior, yellowish patches with white streamers	13.6	7.5	188	6.8	1870	[24]
		7	Grotto Sulfurea cave stream; sample "GS02-WM," a "feathery" biofilm attached to limestone	~13.5	7.3–7.4			1200 to 3500	[27, 86]
		8	Sample GS02-zEL, a "cottony" biofilm on the surface of fine dark sediment in slow-flow margins of streambed						
	Acquasante Terme, Italy	9	Grotto Nuova del Rio Garrafo site AS1; sample; AS07-7 streamer; H_2S (g) = 100 ppmv	44.1	6.37	801	3.06	10,600	[29]
		10	Grotto Nuova del Rio Garrafo site AS1 AS05-3 streamer	40.4	6.38			8400	
	Cesspool Cave, Virginia, USA	11	White filaments, resurgent spring, dark zone; mats	12–13		450–660			[79]
	Cesspool Cave, Virginia, USA	12	White filaments (<1–2 mm) in dark zone near back pool spring orifice	12.2	6.8	1470		600	[24]
	Glenwood Springs, Colorado, USA	13	Microbial filament samples	49.6	6.39	48.5	1.4	29,300	[31]

Table 9.1. (cont.) Compiled summary data used in the analysis of 16S rRNA gene sequences from sulfidic karst and nonkarst systems. The samples are organized by interface (e.g. air-rock cave-wall biofilms, air-water interface with floating mats, benthic sediments and microbial mats, aquifer well water floating filaments and dislodged biofilms from water-rock interface, cenote open water and water-rock interface, and mixed interfaces from nonkarst terrestrial springs).

Interface	System Location		Sample Description	Temp (°C)	pH	Dissolved Sulfide (µmol/L)	Dissolved Oxygen (µmol/L)	Conductivity (µS/cm)	Ref.
Cave Benthic Sediments and Microbial Mats	Parker Cave, Sulphur River, Kentucky, USA	14	Phantom Waterfall and Sulphur river Macroscopic microbial mat composed of white filaments; cave floor pH = 0.13; ceilings description indicates acidic layer of slime	12	6.4–6.7	Waterfall, 323–618; Confluence, 1.5			[40]
	Lower Kane Cave, Wyoming, USA	15	Fissure Spring; LKC04_FS124f; white filaments	22	7.30	39.7	< 0.2	580	[18, 87]
		16	LKC04_FS127f; white filaments						
		17	Upper Spring; LKC04_US190f; subaqueous white filaments collected 1 m downstream of spring orifice						
		18	LKC11_191G; benthic gray sediment-interface mat						
		19	LKC11_191S; short white filaments	21.3	7.39	35.3	< 0.2	577	
		20	LKC11_191L; long white filaments						
		21	LKC04_US195f; white filament bundles						
		22	LKC04_US198f; white filament bundles						
		23	LKC04_US203f; white filament bundles						
		24	LKC04_US203w; white mats						
		25	LKC04_US203y; yellow-white mats						
		26	LKC04_US203g; gray filaments						
		27	Lower Spring; LKC_04_LS248f; white filaments	22.1	7.39	39.4	< 0.2	575	
		28	LKC_04_LS248y; yellow mats						

Table 9.1. (cont.) Compiled summary data used in the analysis of 16S rRNA gene sequences from sulfidic karst and nonkarst systems. The samples are organized by interface (e.g. air-rock cave-wall biofilms, air-water interface with floating mats, benthic sediments and microbial mats, aquifer well water floating filaments and dislodged biofilms from water-rock interface, cenote open water and water-rock interface, and mixed interfaces from nonkarst terrestrial springs).

Interface	System Location		Sample Description	Temp (°C)	pH	Dissolved Sulfide (μmol/L)	Dissolved Oxygen (μmol/L)	Conductivity (μS/cm)	Ref.
Aquifer Well Water, Floating Filaments and Dislodged Biofilms from Water-Rock Interface	Edwards Aquifer, Texas, USA	29	"Comal Springs"; collected at spring	23.3	7.14	0.0	166.9	658	[49]
		30	"Girl Scout Shallow"; purged well	26.4	7.32	BDL	BDL	600	
		31	"Girl Scout Deep"; purged well	24.8	7.38	BDL	BDL	574	
		32	"LCRA Shallow"; purged well	25.3	7.41	11.2	3.75	722	
		33	"LCRA Deep"; purged well	26.2	7.3	26.5	3.12	595	
		34	"Paradise Alley Shallow"; purged well	26.3	7.17	59.1	2.5	1570	
		35	"Paradise Alley Deep"; purged well	25.7	7.0	231.2	3.43	2640	
		36	"Aquarena well"; purged well	21.7	6.89	2588	0	19,700	
Cenote Open Water and Water-Rock Interface Microbial Mats	Sistema Zacaton, Mexico	37	Mat, 273 m; delicate striated mat	30.1	6.53			761	[41, 74]
		38	Mat, 195 m; water chemistry meas. @198 m	30.1	6.54			761	
		39	Mat, 114 m; water chemistry meas. @150 m	30.0	6.52			760	
		40	Mat, 100 m; water chemistry meas. @90 m	30.0	6.72			767	
		41	Mat, 58 m; water chemistry meas. @60 m	30.0	6.72			767	
		42	Mat, 32 m; water chemistry meas. @30 m	30.0	6.73			767	
		43	Mat, 20 m; thick, complex photosynthetic mat; water chemistry meas. @22 m	30.0	6.58	10	0	767	
		44	Mat, 10 m; thick, complex photosynthetic mat						
		45	Mat, 7 m; thick, complex photosynthetic mat						
		46	Mat, 4 m; thick, complex photosynthetic mat; top 5 m clear blue water turn milky sulfur clouds in sun						
		47	Sinkhole 105 m diameter, 318 m deep Mat, 1 m; thick, complex photosynthetic mat (water chemistry meas. @surface)	30.1	6.67	10	0	759	

Table 9.1. (cont.) Compiled summary data used in the analysis of 16S rRNA gene sequences from sulfidic karst and nonkarst systems. The samples are organized by interface (e.g. air-rock cave-wall biofilms, air-water interface with floating mats, benthic sediments and microbial mats, aquifer well water floating filaments and dislodged biofilms from water-rock interface, cenote open water and water-rock interface, and mixed interfaces from nonkarst terrestrial springs).

Interface	System Location		Sample Description	Temp (°C)	pH	Dissolved Sulfide (μmol/L)	Dissolved Oxygen (μmol/L)	Conductivity (μS/cm)	Ref.
Terrestrial Surface Sulfidic Springs; Mixed Interfaces	Rattlesnake Spring, Oklahoma, USA	48	Mat at spring orifice	20.7	6.85	181.8	4.4	7140	
		49	Mat at 1.6 m drop; thick white mat on carbonate	20.7	7.16	118	90.9	7600	[99]
		50	5.5 m transect; water depth shallows to less than 5 cm	21.0	7.48	19.7	149	7610	
	Richfield Springs, New York, USA	51	Water and thin white filamentous mats collected from concrete/stone basin of springhead	7.5	7.07	142.31	29.68	2068	[77]
	Sharon Springs, New York, USA	52	White Sulphur Spring site, water and white filamentous mats	8.5	7.09	94.19	16.25	2463	[77]
	Palmetto Springs, Texas, USA	53	Thick white microbial mats with some green material collected on concrete surface and floating in the pool "Artesian Well and Pond"	28.6	8.75	94.48	16.25	647	[77]
	South Canyon Hot Springs, Colorado, USA	54	White filaments mixed with sediments at orifice	45.3	7.18	40.49	16.56	1775	[77]
	Great Sulfur Spring, Lake Eerie, Michigan, USA	55	Great Sulfur Spring; pumped water with blackish particulates from the throat of the spring	11.3	6.3	31.2	21.88	2590	[103]
		56	Great Sulfur Spring mat; White filament bundles in water column attached to rocks; shallow pool edge	14–17			87.5–143.7		

Table 9.1. (cont.) Compiled summary data used in the analysis of 16S rRNA gene sequences from sulfidic karst and nonkarst systems. The samples are organized by interface (e.g. air-rock cave-wall biofilms, air-water interface with floating mats, benthic sediments and microbial mats, aquifer well water floating filaments and dislodged biofilms from water-rock interface, cenote open water and water-rock interface, and mixed interfaces from nonkarst terrestrial springs).

Interface	System Location		Sample Description	Temp (°C)	pH	Dissolved Sulfide (μmol/L)	Dissolved Oxygen (μmol/L)	Conductivity (μS/cm)	Ref.
Terrestrial Surface Sulfidic Springs; Mixed Interfaces	Zodletone Springs, Oklahoma, USA	57	Spring source			800–1000	< 0.1		[104]
		58	Mat underlying the water flowing (15 m) from the spring source, collected as cores						
	Icelandic Hot Springs	59	"Sulfur mat" white filaments at outer surface, inner gelatinous dark gray mat; covered in sulfur	67	6.7	352			[105]
		60	"Chloroflexus" mat with long pink streamers on gray filaments; covered in sulfur precipitate	65–70	8.3	32			

paper-thin floating biofilms that often have entrained bubbles (▶ Fig. 9.1 (b)). Subaqueous microbial growth found in cave streams and springs in active sulfidic caves includes biofilms and mats of filaments, streamers, stringers, webs, veils, or filamentous masses, with other different descriptive terms (e.g. short, long, gelatinous, feathery, cottony) and colors (e.g. white, gray, brown, yellow) [2–4, 6, 21, 25, 40, 44, 58–64] (▶ Fig. 9.1 c–e). Mineralized features dominated by elemental sulfur in relict SAS caves, such as Lechuguilla Cave in New Mexico, might also be due to past subaqueous microbial growth [65] (▶ Fig. 9.1 (f)). Such mineralized structures in SAS caves have been considered biosignatures, and can be used as analogs for the detection of life processes on Mars and other planetary bodies [66, 67].

Through the oxidation of reduced sulfur compounds or other electron donors, microbes gain energy for chemolithoautotrophic primary productivity [22, 24]. Chemolithoautotrophically based cave ecosystems demonstrate higher ecosystem productivity [24] and can have relatively high biological species richness and diversity compared to nonsulfidic karst systems [68, 69] based on research done from Movile Cave [22–24], the Frasassi caves [24, 63], from submarine caves with sulfidic springs near Cape Palinuro in Italy [70], and the Edwards Aquifer in Texas [35]. There is the potential for other sulfidic, chemolithoautotrophically based ecosystems to be identified with continued exploration of sulfidic karst aquifers. In 2006, Ayyalon Cave in Israel, associated with the Yarqon-Tanninim aquifer, was discovered to have a sulfidic, chemolithoautotrophically based ecosystem with four endemic stygobitic crustacean species and four troglobitic terrestrial invertebrates [71, 72]. Another area where more sulfidic karst ecosystem work could be done is where there is sulfidic groundwater, such as in Florida [73], deep water-filled sinkholes or cenotes [41, 74], and sulfidic springs in karst areas [75–78].

In the past, microbiological studies from sulfidic caves focused on basic microbial diversity and not on microbial metabolism or function. In general, molecular genetics studies, sometimes linked to culture-based work, have been done on 16S rRNA gene sequences obtained from many active sulfidic caves associated with subaerially exposed corrosion crusts or mineral deposits on cave walls, subaqueous sediments, or floating mats to characterize the taxonomic diversity of Bacteria and Archaea in these habitats and systems [21, 25, 31, 62, 79]. However, linkages between taxonomic microbial diversity and functional diversity, which is done by using different methods, have not been developed. Essentially, the community of researchers studying the microbiology of caves is still focused on answering, "who's there?" [80]. But, gene surveys alone do not define geomicrobial impacts to karst formation processes, such as SAS and the production of localized acidity that leads to carbonate rock dissolution. Moreover, gene surveys do not define how (and if) microbes affect the formation of subaerial mineral crusts and precipitation of elemental sulfur and other minerals.

For this chapter, we review the current knowledge of active sulfidic karst microbial diversity based on 16S rRNA gene sequence surveys and use the data to differentiate communities based on habitat type and geochemical conditions that define niche

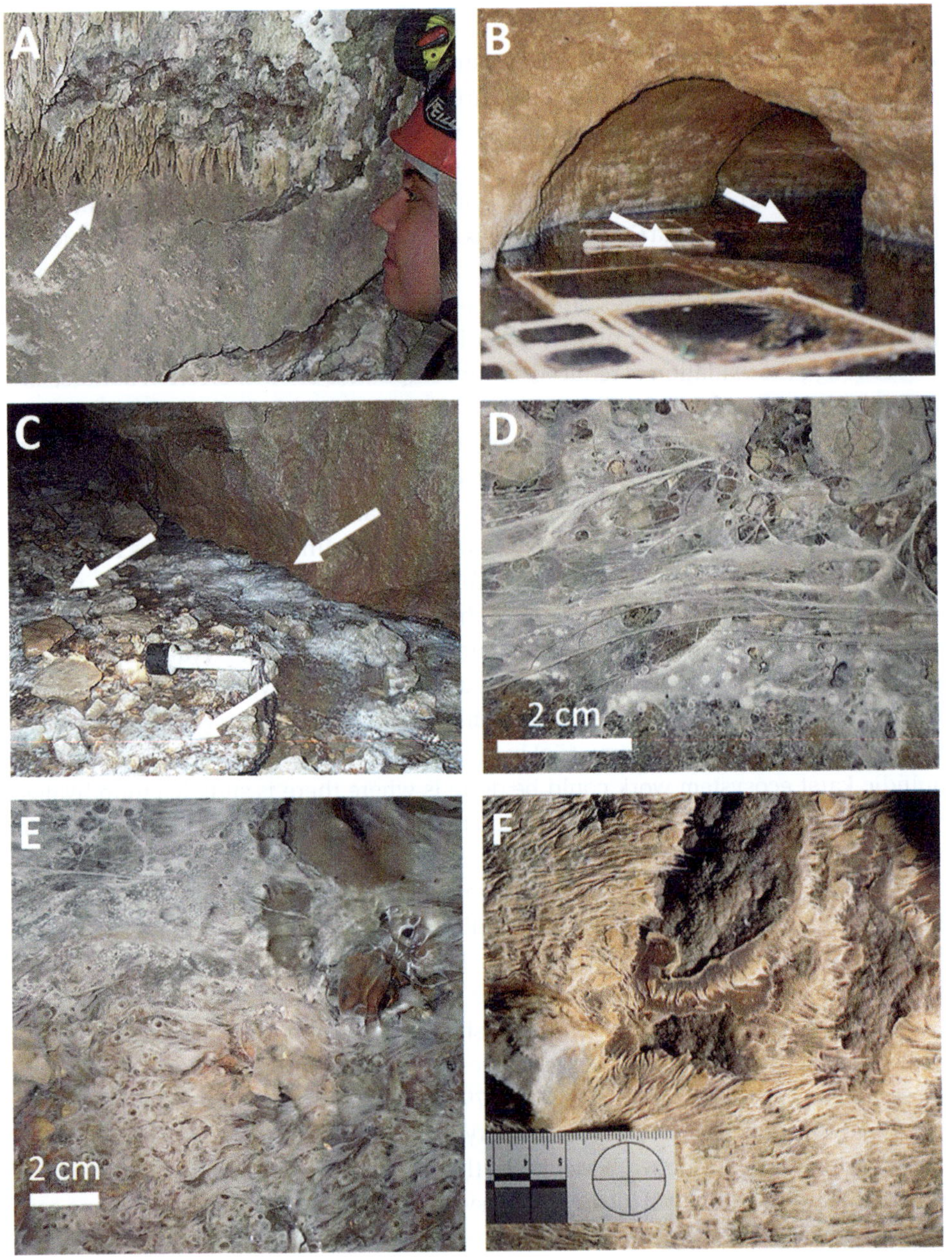
A
B
C
D
2 cm
E
2 cm
F

◂ **Fig. 9.1.** Photographs from active sulfidic caves and relict caves formed by SAS. (a) Mineralized cave-wall biothems from Spirit Mountain Cave, Wyoming, USA. (b) Floating mats (arrows) in a Movile Cave, Romania, airbell. Floating tubes are ~10 cm on a side, for scale. (c) White filamentous streamers (arrows) in stream passage from Big Sulphur cave, Kentucky, USA; flashlight is 10 cm long. (d) Long filaments and webs on carbonate cobbles in Lower Kane Cave, Wyoming, USA. (e) Thick filamentous microbial mats with webs and yellow patches in Lower Kane Cave, Wyoming, USA. (f) Sulfur-containing, mineralized filamentous features on the floor of Lechuguilla Cave, New Mexico, USA, from the Far East MAB Survey. Photographs (a)–(e) taken by Annette Summers Engel. Photograph (f) taken by Kathryn and Harvey DuChene, used with permission.

space. The concept of ecological niche space is fundamental to interpreting how microbial groups and specific taxa are spatially distributed in an environment, as well as to interpreting how groups have and will respond functionally if niches change [81]. A synthesis of available bacterial gene sequences from sulfidic karst systems has not been done previously. Therefore, our motivation was to identify whether or not we know the answer to "who's there?" by now, and after nearly two decades worth of molecular genetics research from sulfidic caves (the history of which is reviewed in Chapter 1 of this book). Our statistical analyses suggest that presence/absence and abundance patterns of specific bacterial groups strongly relate to niche space that reflects habitat suitability. With limited uncertainty, it may be possible to predict which bacterial taxonomic groups might colonize certain types of sulfidic karst systems. It was not possible to assess archaeal gene sequence data because so few studies have been done. Future research could continue to address "who's there?" among Archaea in sulfidic karst systems. Most importantly, this chapter demonstrates that future research on the microbiology of sulfidic caves should turn to experiments that address functional, biochemical, and geomicrobiological diversity instead of purely gene sequence diversity. These future findings will also elucidate what processes may have been strongly influenced by microbes in ancient or relict SAS systems, which could guide biosignature detection and analyses for planetary analog investigations.

9.2 Compilation of Bacterial Diversity Data

Our current understanding of the microbial diversity and metabolic activity of active SAS caves is based on investigations of 16S rRNA gene surveys to generate gene phylogenies [21, 27, 40, 47, 78, 82, 83]. Depending on gene sequence similarities between the unknown and previously characterized microbes, putative metabolic function can be conservatively interpreted. But, usually, a number of other methods are combined with 16S rRNA gene sequence surveys to assist in the evaluation of ecological or geochemical context for microbiological samples, including fluorescence in situ hybridization (FISH) to quantify cell abundances of specific groups [27, 29, 82], functional gene sequence phylogenetics [23], metagenomics [30, 54, 84], experimen-

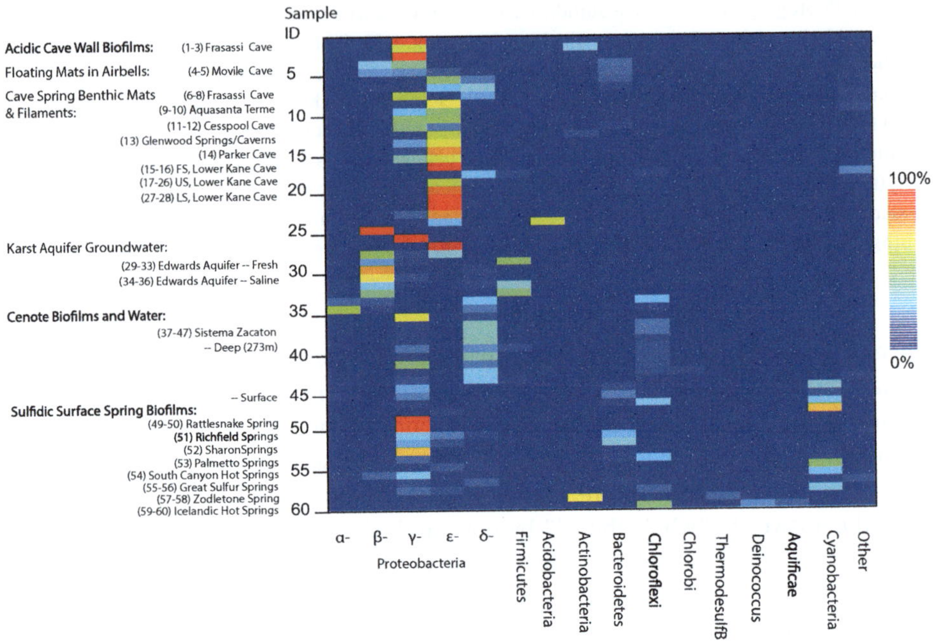

Fig. 9.2. Normalized relative abundances of bacterial diversity from cave-wall biofilms, microbial mats in cave streams, and sulfidic aquifer water samples listed in ▸ Tab. 9.1. Analyses were conducted using the computer program PAST (version 2.14) [88] (http://folk.uio.no/ohammer/past/).

tal primary and heterotrophic productivity rate determinations [24, 79], isotope systematics [18, 21, 33, 59, 83, 85, 86], and from geochemical methods, including X-ray absorption spectroscopy [85, 87].

For this review, we examined bacterial 16S rRNA genes retrieved from 60 samples from sulfidic caves, karst aquifers, cenotes, and sulfidic springs upwelling at the terrestrial surface (▸ Tab. 9.1). We attempted to be as inclusive as possible, even though original studies had different motivations and analytical methods. Our analyses included bacterial diversity data from other sulfidic groundwater or aquatic systems that have similar aqueous geochemistry to sulfidic karst systems, such as from sulfidic surface springs. This was done in an attempt to demonstrate the novelty or ubiquity of microbial populations in sulfidic karst systems. There were many more studies from nonkarst systems than we could have included in the analyses, but nonkarst sample selection was based on which research relied on comparable methods to earlier karst studies. Some data were not used because the study did not provide sufficient geochemical information or gene sequence information (i.e. GenBank accession numbers). Where possible, we include relevant geochemical or other contextual information to help with habitat condition interpretations (▸ Tab. 9.1).

For the dataset, we used the published (i.e. not reanalyzed) taxonomic data, including assigned bacterial phyla and proteobacterial classes from 16S rRNA gene se-

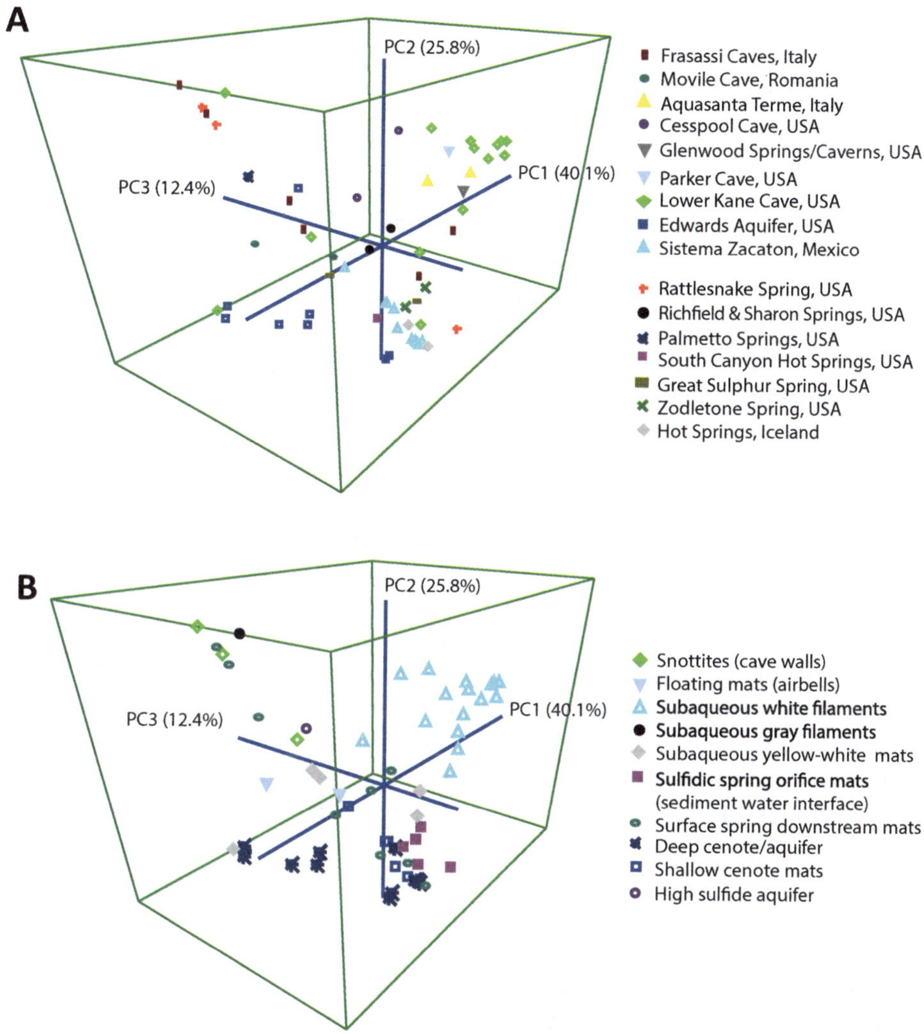

Fig. 9.3. Principal component analysis (PCA) of normalized abundance data for bacterial phyla and proteobacterial classes summarized in ▸ Fig. 9.2. Two sets of symbols and legends are noted, with (a) showing sample locations and (b) showing common attributes related to microbial habitats and biofilm morphological properties (i.e. morphotypes). Analyses were conducted using the computer program PAST (version 2.14) [88] (http://folk.uio.no/ohammer/past/).

quence clone surveys or pyrosequencing surveys (Figures 9.2 and 9.3). For ▸ Fig. 9.2, taxa are represented in a summary matrix of bacterial 16S rRNA gene relative abundances if the group represented at least 10% of the total community in one or more of the environmental samples. To generate the principal components analyses (PCA) for ▸ Fig. 9.3, taxa were included when the group represented at least 5% of the total community in one or more samples; this excluded rarer taxonomic groups from the analyses and emphasized the major taxonomic groups shared among the samples. Data were organized by geographic location (▶ Fig. 9.3 (a)) and environmental sample types (i.e. cave-wall biofilms, filamentous mats), habitat type, biofilm or mat morphotype, and geochemistry (▶ Fig. 9.3 (b)).

9.3 Controls on Bacterial Diversity in Sulfidic Karst

Sulfidic karst bacterial communities have relatively low diversity, and generally less than five taxonomic groups comprise the majority of sequence affiliations (▶ Fig. 9.2). Most of the retrieved diversity is among groups associated with putative reduced sulfur compound oxidation (e.g. Gammaproteobacteria, Epsilonproteobacteria, or Betaproteobacteria) and sulfate reduction or sulfur disproportionation (e.g. Deltaproteobacteria). The presence/absence and relative proportions of these proteobacterial classes in sulfidic karst systems may distinguish the microbial communities and associated metabolic processes from those in nonsulfidic karst. Specifically, from the systems evaluated here, Gammaproteobacteria are represented by the genera *Thiobacillus*, *Beggiatoa*, and *Thiothrix*, which have metabolic functions associated with sulfur cycling [89, 90]. These taxa are rarely retrieved in nonsulfidic karst systems [91]. Similarly, although Epsilonproteobacteria inhabit a number of sulfidic environments, such as deep-sea vent communities, contaminated aquifers, and SAS caves, Epsilonproteobacteria are also almost never retrieved in 16S rRNA gene surveys from nonsulfidic karst systems [91]. There are indications that some epsilonproteobacterial groups are biogeographically congruent to sulfidic caves and karst systems, while other groups are only known from single locations [24, 77, 78, 82]. The metabolic pathway(s) for reduced sulfur oxidation among Epsilonproteobacteria are poorly understood [92, 93], but recent metagenomics studies help to define sulfur, nitrogen, and hydrogen metabolisms for several key taxa that have been identified from sulfidic caves, such as *Sulfurovum*-like and *Sulfuricurvum*-like spp. [30].

The relative abundances of Gammaproteobacteria and Epsilonproteobacteria also play an important role in describing variance among the samples analyzed (▶ Fig. 9.3). Community diversity in SAS caves and karst systems is strongly linked to specific habitat, such as acidic cave-wall snottites or microbial mats from cave streams (▶ Fig. 9.3 (b)). Gammaproteobacteria dominate (60–100% in some samples) snottites from the Frasassi caves, belonging exclusively to the genus *Acidithiobacillus* spp. [formerly reported as *Thiobacillus* in earlier studies; 21, 94]. But, Actinobacteria also have high relative abundances based on 16S rRNA gene surveys (0–40% of the community). Snottites from Cueva de Villa Luz and Lower Kane Cave also have relatively high *Acidithiobacillus* spp. abundances and Actinobacteria [21, 83], which may suggest that snottite and cave-wall geochemical conditions select for similar bacterial communities, regardless of geography. Actinobacteria are rarely retrieved from subaqueous samples in sulfidic karst (▸ Fig. 9.2), but diverse genera of Actinobacteria are found in other, nonsulfidic cave-wall systems [95, 96].

Gammaproteobacteria also dominate low-diversity microbial mats from water-sediment interface habitats in springs and streams in active sulfidic karst systems. From studies in Lower Kane Cave and the Frasassi caves, variable abundances of sulfur-oxidizing genera within the Gammaproteobacteria correlate to changing dissolved oxygen concentrations downstream caused by ingassing, aeration, or mixing sulfidic and freshwater fluids. Some gammaproteobacterial taxa can represent up to 97% of the total community in higher dissolved oxygen locations compared to lower oxygen conditions that are dominated by Epsilonproteobacteria (▸ Fig. 9.2). The distribution of Gammaproteobacteria and Epsilonproteobacteria in cave stream microbial mats based on geochemical conditions has been noted previously [18, 47, 86]. From the PCA analysis (▸ Fig. 9.3), PC1 describes 40% of the variance among bacterial communities, and is represented by inversely correlated loading coefficients between Gammaproteobacteria (-0.46) and Epsilonproteobacteria (0.87). PC2 explains 25% of variance, with positively related loading coefficients for Gammaproteobacteria (0.81) and Epsilonproteobacteria (0.35).

Deltaproteobacteria represent the most diverse taxonomic group of known sulfidogenic or dissimilatory sulfate- and sulfur-reducing microbes [97, 98]. Among the reviewed samples (▸ Tab. 9.1), Deltaproteobacteria rarely strictly dominated microbial communities, but are retrieved in greatest relative abundances (~20–40% of total bacterial diversity) from Sistema Zacaton cenote mats (deeper than 50 m) and benthic mats at the sediment interface in the Lower Kane and Frasassi caves (sometimes described as gray-black sediments [24, 87]) and from sulfidic spring orifices (▸ Tab. 9.1) [24, 87]. Deltaproteobacteria are also retrieved from Edwards Aquifer samples, and in sulfidic surface spring sediment samples from Zodletone Spring, Oklahoma (~17%), Richfield Spring, New York (~13%), orifice samples from Great Sulfur Springs, Michigan (~10%), South Canyon Hot Springs, Colorado (~8%), and Sharon Springs, New York (~6%). Deltaproteobacteria were a minor component (<1–5%) of subaqueous white-yellow webs or white filamentous web samples collected above the sediment interface in Lower Kane Cave spring samples. Deltaproteobacteria are not detected or retrieved in large abundances from cave-wall snottites or long white filaments or streamers suspended in water columns. As a result, Deltaproteobacteria contribute notably to the loadings on PC2 and PC3 (with loading coefficients of 0.18 and 0.27, respectively), but generally contribute less to the total variance among samples compared to the putative sulfur-oxidizing bacterial taxa, Epsilonproteobacteria, Gammaproteobacteria, and Betaproteobacteria.

9.4 Predicting the Distribution of Sulfur Bacteria in Sulfidic Karst

Niche space, defined by habitat conditions and available metabolic substrates, is clearly distinguished from the PCA (► Fig. 9.3). Epsilonproteobacterial-dominated long filaments and streamers collected from above the sediment-water interface in low oxygen niches at spring orifices are separated in ordination space from epsilonproteobacterial-dominated and gammaproteobacterial-dominated bundles and webs from cave streams with increasing oxygen availability. These samples are also separated in ordination space from mixed proteobacterial communities from floating mats on the water surface in airbells. Moreover, regardless of geographic proximity (► Fig. 9.3 (a)), bacterial communities from each of these niches defined by habitat-type and geochemical gradient (which dictates potential bacterial metabolisms) cluster tightly with each other in PCA space (► Fig. 9.3 (b)). Previous studies of bacterial diversity from terrestrial, surface sulfidic springs (not necessarily from karst) [77, 99] suggest that taxonomic and metabolic diversity among Gammaproteobacteria and Epsilonproteobacteria is due to niche partitioning that results in the colonization of geochemical gradients by certain groups. From comparing genetic and geochemical data for SAS karst and nonkarst systems with more than one mat morphotype, and from multiple samples collected across geochemical gradients in one system with shared taxonomic groups, such as a flowing cave stream with high to low sulfide concentrations, PCA evidence indicates that SAS karst bacterial diversity is also controlled by niche partitioning, which is likely ubiquitous in sulfidic cave and karst aquifers systems (► Fig. 9.3). Consequently, clear patterns of bacterial presence/absence can be predicted from the previously characterized SAS karst systems. Essentially, from the extensive research already done for two decades, we may already know "who's there?" in sulfidic karst systems. There may not be a need in the future for basic 16S rRNA gene sequencing surveys, and in the absence of other molecular investigations such as metagenomics, in newly discovered sulfidic systems, unless biogeography hypotheses are being tested [81].

9.5 Conclusions

We make these conclusions while recognizing that the taxonomic classifications presented in this chapter are quite broad (i.e. phylum- and class-level identification). The 16S rRNA gene, as a bacterial biomarker, is widely used to provide insight into taxonomic profiles and bacterial niche differentiation without laboratory cultivation and isolation [100]. However, one of the limitations of using 16S rRNA gene-based analyses is that taxonomic assignments, particularly at the phylum level, allude to, but do not strictly provide, direct links to microbial physiology [101]. Generalizations about metabolic function based on taxonomic affiliations must be made carefully to understand drivers of, and responses to, microbial ecological niche differentiation. Taxonomic assignments based on 16S rRNA genes do not provide direct indications of metabolic function, particularly for groups that we do not understand from fundamental taxonomic and functional diversity perspectives. Therefore, putative function based on taxonomic identity may be wrong, and will not be known until culturing approaches are successful and interdisciplinary molecular techniques, such as metagenomics and metatranscriptomics based on RNA, uncover functional diversity. Confirmation of the genetic basis for metabolic activities, which has been done extensively from pure cultures, has not been done yet from most sulfidic karst settings. Nevertheless, future research in these directions take attention away from "who's there?" and emphasize "what are they doing?" through experimentation and assessment of the impacts of microbial functional diversity in karst. After all, we argue that knowing which microbes occupy a habitat is not as critical as knowing what the microbes are capable of doing. With this knowledge about microbial function in niche space, we can then begin to speculate about how microbial activities proceeded in relict SAS karst systems. This would be important future work because, currently, SAS processes are still considered predominately abiotic in nature [1, 9, 43, 102] despite mounting evidence to the contrary [19, 27, 33, 34, 48, 50, 86].

Acknowledgments

The research was partially supported by the Jones Endowment for Aqueous Geochemistry at the University of Tennessee-Knoxville. Some of the results used in this chapter were generated from two grants from the National Science Foundation (DEB-0640835 and EAR-0844364) and a grant from the Louisiana Board of Regents Pilot fund for New Research Program [NSF(2010)-PFUND-174].

References

[1] Palmer AN. Sulfuric acid caves: Morphology and evolution. In: Frumkin A (vol. ed.), Shroder J (chief ed.). Treatise on Geomorphology. Vol. 6. San Diego, CA: Academic Press, 2013, 241–257.

[2] Palmer AN. Origin and morphology of limestone caves. GSA Bull, 1991, 103, 1–21.

[3] Hill CA. H_2S-related porosity and sulfuric acid oil-field karst. In: Budd DA, Saller AH, Harris PM, eds. Unconformities in carbonate strata: Their recognition and significance of associated porosity: AAPG Memoir, 1995, 301–306.

[4] Hill CA. Overview of the geologic history of cave development in the Guadalupe Mountains, New Mexico. J Cave Karst Stud, 2000, 62, 60–71.

[5] Jagnow DH, Hill CA, Davis DG, DuChene HR. History of the sulfuric acid speleogenesis in the Guadalupe Mountains, New Mexico. J Cave Karst Stud, 2000, 62, 54–59.

[6] Palmer AN, Palmer MV. Hydrochemical interpretation of cave patterns in the Guadalupe Mountains, New Mexico. J Cave Karst Stud, 2000, 62, 91–108.

[7] Polyak VJ, Provencio PP. Summary of the timing of sulfuric acid speleogenesis for the Guadalupe caves based on ages of alunite. J Cave Karst Stud, 2000, 62, 72–74.
[8] Northup DE, Barns SM, Yu LE, et al. Diverse microbial communities inhabiting ferromanganese deposits in Lechuguilla and Spider Caves. Environ Microbiol, 2003, 5, 1071–1086.
[9] Plan L, Tschegg C, De Waele J, Spötl C. Corrosion morphology and cave wall alteration in an Alpine sulfuric acid cave (Kraushöhle, Austria). Geomorphology, 2012, 169–170, 45–54.
[10] Puşcaş CM, Onac BP, Tămaş T. The mineral assemblage of caves within Şălitrari Mountain (Cerna Valley, SW Romania): depositional environment and speleogenetic implications. Carbonate Evaporite, 2010, 25, 107–115.
[11] Temovski M, Audra P, Mihevc A, et al. Hypogenic origin of Provalata Cave, Republic of Macedonia: a distinct case of successive thermal carbonic and sulfuric acid speleogenesis. Int J Speleol, 2013, 42, 235–246.
[12] Maltsev V, Korshunov V. Geochemistry of fluorite and related features of the Kugitangtou Ridge Caves, Turkmenistan. J Cave Karst Stud, 1998, 60, 151–155.
[13] Galdenzi S, Menichetti M. Occurrence of hypogenic caves in a karst region: Examples from central Italy. Environ Geol, 1995, 26, 39–47.
[14] Galdenzi S, Cocchioni M, Morichetti L, Amici V, Scuri S. Sulfidic ground-water chemistry in the Frasassi Caves, Italy. J Cave Karst Stud, 2008, 70, 94–107.
[15] Piccini L, De Waele J, Galli E, Polyak VJ, Bernasconi SM, Asmerom Y. Sulphuric acid speleogenesis and landscape evolution: Montecchio Cave, Albegna river valley (Southern Tuscany, Italy). Geomorphology, 2015, 229, 134–143.
[16] De Waele J, Forti P, Naseddu A. Speleogenesis of an exhumed hydrothermal sulphuric acid karst in Cambrian carbonates (Mount san Giovanni, Sardinia). Earth Surf Proc Land, 2013, 38, 1369–1379.
[17] Egemeier SJ. Cavern development by thermal waters. Nat Speleol Soc Bull, 1981, 43, 31–51.
[18] Engel AS, Porter ML, Stern LA, Quinlan S, Bennett PC. Bacterial diversity and ecosystem function of filamentous microbial mats from aphotic (cave) sulfidic springs dominated by chemolithoautotrophic '*Epsilonproteobacteria*'. FEMS Microbiol Ecol, 2004, 51, 31–53.
[19] Engel AS, Stern LA, Bennett PC. Microbial contributions to cave formation: New insights into sulfuric acid speleogenesis. Geology, 2004, 32, 369–372.
[20] Hose LD, Pisarowicz JA. Cueva de Villa Luz, Tabasco, Mexico: Reconnaissance study of an active sulfur spring cave. J Cave Karst Stud, 1999, 61, 13–21.
[21] Hose LD, Palmer AN, Palmer MV, Northup DE, Boston PJ, DuChene HR. Microbiology and geochemistry in a hydrogen-sulphide-rich karst environment. Chem Geol, 2000, 169, 399–423.
[22] Sarbu SM, Kane TC, Kinkle BK. A chemoautotrophically based cave ecosystem. Science, 1996, 272, 1953–1955.
[23] Chen Y, Wu L, Boden R, et al. Life without light: microbial diversity and evidence of sulfur- and ammonium-based chemolithotrophy in Movile Cave. ISME J, 2009, 3, 1093–1104.
[24] Porter ML, Engel AS, Kane TC, Kinkle BK. Productivity-diversity relationships from chemolithoautotrophically-based sulfidic karst systems. Int J Speleol, 2009, 38, 27–40.
[25] Vlasceanu L, Sarbu SM, Engel AS, Kinkle BK. Acidic cave-wall biofilms located in the Frasassi Gorge, Italy. Geomicrobiol J, 2000, 17, 125–139.
[26] Galdenzi S, Maruoka T. Gypsum deposits in the Frasassi Caves, central Italy. J Cave Karst Stud, 2003, 65, 111–125.
[27] Macalady JL, Lyon EH, Koffman B, et al. Dominant microbial populations in limestone-corroding stream biofilms, Frasassi Cave System, Italy. Appl Environ Microb, 2006, 72, 5596–5609.
[28] Galdenzi S, Cocchioni F, Filipponi G, et al. The sulfidic thermal caves of Acquasanta Terme (Central Italy). J Cave Karst Stud, 2010, 72, 43–58.

[29] Jones DS, Tobler DJ, Schaperdoth I, Mainiero M, Macalady JL. Community structure of subsurface biofilms in the thermal sulfidic caves of Acquasanta Terme, Italy. Appl Environ Microb, 2010, 76, 5902–5910.

[30] Hamilton TL, Jones DS, Schaperdoth I, Macalady JL. Metagenomic insights into S(0) precipitation in a terrestrial subsurface lithoautotrophic ecosystem. Front Microbiol, 2014, 5, 756.

[31] Barton HA, Luiszer F. Microbial metabolic structure in a sulfidic cave hot spring: potential mechanisms of biospeleogenesis. J Cave Karst Stud, 2005, 67, 28–38.

[32] Longley G. The Edwards Aquifer: Earth's most diverse groundwater ecosystem? Int J Speleol, 1981, 11, 123–128.

[33] Engel AS, Randall KW. Experimental evidence for microbially mediated carbonate dissolution from the saline water zone of the Edwards Aquifer, Central Texas. Geomicrobiol J, 2011, 28, 313–327.

[34] Gray CJ, Engel AS. Microbial diversity and impact on carbonate geochemistry across a changing geochemical gradient in a karst aquifer. ISME J, 2013, 7, 325–337.

[35] Hutchins BT, Schwartz BF, Engel AS. Environmental controls on organic matter production and transport across surface-subsurface geochemical boundaries in the Edwards Aquifer, Texas, USA. Acta Carsologica, 2013, 42/2–3, 245–259.

[36] Principi P. Fenomeni di idrologia sotterranea nei dintorni di Triponzo (Umbria). Grotte d'Ital, 1931, 5, 1–4.

[37] Eraso A. Sobre la intensification de la accion corrosiva del aqua en la genesis de las cavernas, en presencia de iones oxidables. Proceedings of the 4th International Congress of Speleology in Yugoslavia, 1965. Ljubljana: Federation Yougoslave de Speleologie, 1968, 79–89.

[38] Moorehouse D. Cave development via the sulfuric acid reaction. Nat Speleol Soc Bull, 1968, 30, 1–10.

[39] Egemeier SJ. Cavern development by thermal waters with a possible bearing on ore deposition. Stanford, CA: Stanford University; 1973.

[40] Angert ER, Northup DE, Reysenbach A-L, Peek AS, Goebel BM, Pace NR. Molecular phylogentic analysis of a bacterial community in Sulphur River, Parker Cave, Kentucky. Am Mineral, 1998, 83, 1583–1592.

[41] Sahl JW, Gary MO, Harris JK, Spear JR. A comparative molecular analysis of water-filled limestone sinkholes in north-eastern Mexico. Environ Microbiol, 2011, 13, 226–240.

[42] Forti P, Galdenzi S, Sarbu SM. The hypogenic caves: a powerful tool for the study of seeps and their environmental effects. Cont Shelf Res, 2002, 22, 2373–2386.

[43] Klimchouk A. Morphogenesis of hypogenic caves. Geomorphology, 2009, 106, 100–117.

[44] Smyk B, Drzal M. Research on the influence of microorganisms on the development of karst phenomena. Geog Polonica, 1964, 2, 57–60.

[45] Caumartin V. Review of the microbiology of underground environments. Nat Speleol Soc Bull, 1963, 25, 1–14.

[46] Caumartin V. La corrosion microbienne dans un reseau karstique. Proceedings of the 4th International Congress of Speleology in Yugoslavia, 1965. Ljubljana: Federation Yougoslave de Speleologie, 1968, 31–38.

[47] Engel AS, Meisinger DB, Porter ML, et al. Linking phylogenetic and functional diversity to nutrient spiraling in microbial mats from Lower Kane Cave (USA). ISME J, 2010, 4, 98–110.

[48] Steinhauer ES, Omelon CR, Bennett PC. Limestone corrosion by neutrophilic sulfur-oxidizing bacteria: A coupled microbe-mineral system. Geomicrobiol J, 2010, 27, 723–738.

[49] Gray C. Influences of microbial diversity on carbonate geochemistry across a transition from fresh to saline water in the Edwards Aquifer, Texas. Baton Rouge: Louisiana State University, 2010.

[50] Jones AA, Bennett PC. Mineral microniches control the diversity of subsurface microbial populations. Geomicrobiol J, 2014, 31, 246–261.
[51] Prange A, Arzberger I, Engemann C, et al. In situ analysis of sulfur in the sulfur globules of phototrophic sulfur bacteria by X-ray absorption near edge spectroscopy. Biochim Biophys-Acta, 1999, 1428, 446–454.
[52] Grabowski A, Nercessian O, Fayolle F, Blanchet D, Jeanthon C. Microbial diversity in production waters of a low-temperature biodegraded oil reservoir. FEMS Microbiol Ecol, 2005, 54, 427–443.
[53] Dahl C, Prange A. Bacterial sulfur globules: occurrence, structure and metabolism. In: Shively J, ed. Microbial Sulfur Metabolism. Springer, Berlin/Heidelberg, 2006, 21–51.
[54] Jones DS, Schaperdoth I, Macalady JL. Metagenomic evidence for sulfide oxidation in extremely acidic cave biofilms. Geomicrobiol J, 2014, 31, 194–204.
[55] Macalady JL, Jones DS, Lyon EH. Extremely acidic, pendulous cave wall biofilms from the Frasassi cave system, Italy. Environ Microbiol, 2007, 9, 1402–1414.
[56] Queen JM, Melim LA. Biothems: Biologically influenced speleothems in caves of the Guadalupe Mountains, New Mexico, USA. 57th Field Conference, Caves and Karst of Southeastern New Mexico. New Mexico Geological Society Guidebook, 2006, 167–174.
[57] Melim LA, Liescheidt R, Northup DE, Spilde MN, Boston PJ, Queen JM. A biosignature suite from cave pool precipitates, Cottonwood Cave, New Mexico. Astrobiology, 2009, 9, 907–917.
[58] Hubbard DA, Herman JS, Bell PE. Speleogenesis in a travertine scarp: observations of sulfide oxidation in the subsurface. In: Herman JS, Hubbard DA, eds. Travertine-Marl: Stream Deposits in Virginia. Charlottesville: Department of Mines, Minerals and Energy, Division of Mineral Resources, 1990, 177–184.
[59] Bottrell SH, Smart PL, Whitaker F, Raiswell R. Geochemistry and isotope systematics of sulphur in the mixing zone of Bahamian blue holes. Appl Geochem, 1991, 6, 97–103.
[60] Brigmon RL, Martin HW, Morris TL, Bitton G, Zam SG. Biogeochemical ecology of *Thiothrix* spp. In underwater limestone caves. Geomicrobiol J, 1994, 12, 141–159.
[61] Sarbu SM, Kinkle BK, Vlascenau L, Kane TC, Popa R. Microbiological characterization of a sulfide-rich groundwater ecosystem. Geomicrobiol J, 1994, 12, 175–182.
[62] Vlasceanu L, Popa R, Kinkle BK. Characterization of *Thiobacillus thioparus* LV43 and its distribution in a chemoautotrophically based groundwater ecosystem. Appl Environ Microb, 1997, 63, 3123–3127.
[63] Sarbu SM, Galdenzi, S, Manichetti, M, Gentile, G. Geology and biology of Grotte di Frasassi (Frasassi Caves) in Central Italy, an ecological multi-disciplinary study of a hypogenic underground karst system. In: Wilkens H, Culver DC, Humphreys S, eds. Ecosystems of the World: Subterranean Ecosystems. Oxford, Elsevier Science, 2000, 361–381.
[64] Engel AS, Porter ML, Kinkle BK, Kane TC. Ecological assessment and geological significance of microbial communities from Cesspool Cave, Virginia. Geomicrobiol J, 2001, 18, 259–274.
[65] Davis DG. Extraordinary features of Lechuguilla Cave, Guadalupe Mountains, New Mexico. J Cave Karst Stud, 2000, 62, 147–157.
[66] Boston PJ, Hose LD, Northup DE, Spilde MN. The microbial communities of sulfur caves: A newly appreciated geologically driven system on Earth and potential model for Mars. In: Harmon RS, Wicks C, eds. Perspectives on Karst Geomorphology, Hydrology, and Geochemistry—A tribute volume to Derek C Ford and William B White, Geological Society of America Special Paper 404. Boulder, CO, Geological Society of America, 2006, 331–344.
[67] Boston PJ, Spilde MN, Northup DE, Rosales-Lagarde L, Stafford K. Subsurface sulfur systems: Production and preservation of distinctive biogenic signatures in sulfur, iron, manganese, and carbonate cave systems. Geochim Cosmochim Acta, 2006, 70, A60.
[68] Culver DC, Pipan T. The Biology of Caves and Other Subterranean Habitats. Oxford, Oxford University Press, 2009, 256p.

[69] Engel AS. Observations on the biodiversity of sulfidic karst habitats. J Cave Karst Stud, 2007, 69, 187–206.

[70] Southward AJ, Kennicutt MC, Herrera-Alcalà J, et al. On the biology of submarine caves with sulphur springs: Appraisal of $^{13}C/^{12}C$ ratios as a guide to trophic relations. J Mar Biol Assoc UK, 1996, 76, 265–285.

[71] Por FD. Ophel: a groundwater biome based on chemoautotrophic resources. The global significance of the Ayyalon Cave finds, Israel. Hydrobiologia, 2007, 592, 1–10.

[72] Por FD, Dimentman C, Frumkin A, Naaman I. Animal life in the chemoautotrophic ecosystem of the hypogenic groundwater cave of Ayyalon (Israel): A summing up. Sci Res, 2013, 5, 7–13.

[73] Opsahl SP, Chanton JP. Isotopic evidence for methane-based chemosynthesis in the Upper Floridan aquifer food web. Oecologia, 2006, 150, 89–96.

[74] Sahl JW, Fairfield N, Harris JK, Wettergreen D, Stone WC, Spear JR. Novel microbial diversity retrieved by autonomous robotic exploration of the world's deepest vertical phreatic sinkhole. Astrobiology, 2010, 10, 201–213.

[75] Porter ML, Engel AS. Diversity of uncultured *Epsilonproteobacteria* from terrestrial sulfidic caves and springs. Appl Environ Microb, 2008, 74, 4973–4977.

[76] Iurkiewicz AA, Stevanovic ZP. Reconnaissance study of active sulfide springs and cave systems in the southern part of the Sulaimani Governorate (NE Iraq). Carbonate Evaporite, 2010, 25, 203–216.

[77] Headd B, Engel AS. Biogeographic congruency among bacterial communities from terrestrial sulfidic springs. Front Microbiol, 2014, 5, 473.

[78] Rossmassler K, Engel AS, Twing KI, Hanson TE, Campbell BJ. Drivers of epsilonproteobacterial community composition in sulfidic caves and springs. FEMS Microbiol Ecol, 2012, 79, 421–432.

[79] Engel AS, Porter ML, Kinkle BK, Kane TC. Ecological assessment and geological significance of microbial communities from Cesspool Cave, Virginia. Geomicrobiol J, 2001, 18, 259–274.

[80] Engel AS, Northup DE. Caves and karst as model systems for advancing the microbial sciences. In: Martin J, White WB, eds. Frontiers in Karst Research, Special Publication 13. Leesburg, Virginia, Karst Waters Institute, 2008, 37–48.

[81] Green JL, Bohannan BJM, Whitaker RJ. Microbial biogeography: from taxonomy to traits. Science, 2008, 320, 1039–1043.

[82] Engel AS, Lee NM, Porter ML, Stern LA, Bennett PC, Wagner M. Filamentous "*Epsilonproteobacteria*" dominate microbial mats from sulfidic cave springs. Appl Environ Microbiol, 2003, 69, 5503–5511.

[83] Meisinger DB, Zimmermann J, Ludwig W, et al. In situ detection of novel Acidobacteria in microbial mats from a chemolithoautotrophically based cave ecosystem (Lower Kane Cave, WY, USA). Environ Microbiol, 2007, 9, 1523–1534.

[84] Jones DS, Albrecht HL, Dawson KS, et al. Community genomic analysis of an extremely acidophilic sulfur-oxidizing biofilm. ISME J, 2012, 6, 158–170.

[85] Engel AS, Lichtenberg H, Prange A, Hormes J. Speciation of sulfur from filamentous microbial mats from sulfidic cave springs using X-ray absorption near-edge spectroscopy. FEMS Microbiol Lett, 2007, 269, 54–62.

[86] Macalady JL, Dattagupta S, Schaperdoth I, Jones DS, Druschel GK, Eastman D. Niche differentiation among sulfur-oxidizing bacterial populations in cave waters. ISME J, 2008, 2, 590–601.

[87] Paterson A. The distribution and diversity of functional gene pathways controlling sulfur speciation in Lower Kane Cave, Wyoming [Masters]. Baton Rouge, Louisiana State University, 2012.

[88] Hammer O, Harper DAT, Ryan PD. PAST: Paleontological statistics software package for education and data analysis. Palaeont Electron. 2.14 ed2001, 4, 9 pp.

[89] Friedrich CG, Rother D, Bardischewsky F, Quentmeier A, Fischer J. Oxidation of reduced inorganic sulfur compounds by bacteria: emergence of a common mechanism? Appl Environ Microbiol, 2001, 67, 2873–2882.

[90] Friedrich C, Bardischewsky F, Rother D, Quentmeier A, Fischer J. Prokaryotic sulfur oxidation. Curr Op Microbiol, 2005, 8, 253–259.

[91] Lee NM, Meisinger DB, Aubrecht R, et al. Caves and karst environments. In: Bell EM, ed. Life at Extremes; Environments, Organisms and Strategies for Survival, Oxfordshire, UK: CAB International, 2012, 320–344.

[92] Sievert SM, Hugler M, Taylor CD, Wirsen CO. Sulfur oxidation at deep-sea hydrothermal vents. In: Dahl C, Friedrich C, eds. Microbial Sulfur Metabolisms: Springer, 2008, 238–258.

[93] Yamamoto M, Nakagawa S, Shimamura S, Takai K, Horikoshi K. Molecular characterization of inorganic sulfur-compound metabolism in the deep-sea epsilonproteobacterium *Sulfurovum* sp. NBC37-1. Environ Microbiol, 2010, 12, 1–10.

[94] Kelly DP, Wood AP. Reclassification of some species of *Thiobacillus* to the newly designated genera *Acidithiobacillus* gen. nov., *Halothiobacillus* gen. nov. and *Thermithiobacillus* gen. nov. Int J Syst Evol Microbiol, 2000, 50, 511–516.

[95] Pašić L, Kovče B, Sket B, Herzog-Velikonja B. Diversity of microbial communities colonizing the walls of a karstic cave in Slovenia. FEMS Microbiol Ecol, 2010, 71, 50–60.

[96] Cuezva S, Fernandez-Cortes A, Porca E, et al. The biogeochemical role of Actinobacteria in Altamira Cave, Spain. FEMS Microbiol Ecol, 2012, 81, 281–290.

[97] Rabus R, Hansen T, Widdel F. Dissimilatory sulfate- and sulfur-reducing prokaryotes. In: Dworkin M, Falkow S, Rosenberg E, Schleifer K-H, Stackebrandt E, eds. The Prokaryotes. Springer New York, 2006, 659–768.

[98] Muyzer G, Stams A, JM. The ecology and biotechnology of sulphate-reducing bacteria. Nature Rev Microbiol, 2008, 6, 441–454.

[99] Headd B, Engel AS. Evidence for niche partitioning revealed by the distribution of sulfur oxidation genes collected from areas of a terrestrial sulfidic spring with differing geochemical conditions. Appl Environ Microbiol, 2013, 79, 1171–1182.

[100] Fabrice A, Didier R. Exploring microbial diversity using 16S rRNA high-throughput methods. J Comput Sci Syst Biol, 2009, 2, 74–92.

[101] Langille MGI, Zaneveld J, Caporaso JG, et al. Predictive functional profiling of microbial communities using 16S rRNA marker gene sequences. Nat Biotech, 2013, 31, 814–821.

[102] De Waele J, Plan L, Audra P. Recent developments in surface and subsurface karst geomorphology: An introduction. Geomorphology, 2009, 106, 1–8.

[103] Chaudhary A, Haack SK, Duris JW, Marsh TL. Bacterial and archaeal phylogenetic diversity of a cold sulfur-rich spring on the shoreline of Lake Erie, Michigan. Appl Environ Microbiol, 2009, 75, 5025–5036.

[104] Elshahed MS, Senko JM, Najar FZ, et al. Bacterial diversity and sulfur cycling in a mesophilic sulfide-rich spring. Appl Environ Microbiol, 2003, 69, 5609–5621.

[105] Skirnisdottir S, Hreggvidsson GO, Hjörleifsdottir S, et al. Influence of sulfide and temperature on species composition and community structure of hot spring microbial mats. Appl Environ Microbiol, 2000, 66, 2835–2841.

Deepak Kumaresan, Alexandra M. Hillebrand-Voiculescu, Daniela Wischer, Jason Stephenson, Yin Chen, and J. Colin Murrell

10 Microbial Life in Unusual Cave Ecosystems Sustained by Chemosynthetic Primary Production

Abstract: Life in hypogenic cave ecosystems, such as Movile Cave in Romania and Ayyalon Cave in Israel, is sustained by *in situ* chemosynthetic primary production, which is analogous to the ecosystem energy at deep-sea hydrothermal vent ecosystems. Movile Cave is completely isolated from the surface and hydrogen sulfide and methane gases are the primary energy sources for microorganisms at the base of the ecosystem. Invertebrates, many of which are endemic to Movile Cave and surrounding aquifers, are isotopically lighter in both carbon and nitrogen than surface organisms because they derive nutrition from chemolithoautotrophic primary producers within the cave. Recent microbiological studies in Movile Cave focus on microbial nutrient cycling, particularly sulfur and carbon, using both cultivation-dependent and -independent techniques. Here, we review the past work on functional diversity of microbes involved in sulfur, carbon, and nitrogen cycling in Movile Cave and possible directions for future research for Movile Cave and Ayyalon Cave where there has been limited study.

10.1 Introduction

Microbial life in caves encounters extreme environments with high limitations on nutrient and energy availability. Cave ecosystems are characterized by limited light, nearly constant air and water temperatures, and relative humidity at near saturation. Usually the formation of caves results from the seepage of meteoric surface waters into carbonate rock structures, and processes are entirely controlled by water, air, gravity, and fauna from the surface [1]. Similarly, the biological communities within these cave ecosystems are dependent on the flow of nutrients and energy from the surface [2, 3]. A small percentage of the world's caves are of hypogenic origin, whereby energy needed to dissolve the rock and to support the biological communities inhabiting the caves is supplied by ascending fluids and gases [3]. Geochemical characteristics of hypogenic caves differ depending on the origin of the rising waters, the type of host rock, and the temperature and composition of the released gases. From a microbiological perspective, the most interesting hypogenic caves are those with inputs of gases like hydrogen sulfide (H_2S) and methane (CH_4). These gases provide energy sources for microbial communities [2].

Geologist Christian Lascu discovered Movile Cave in 1986 when an artificial shaft dug for geological investigation created access to the narrow cave passages. The cave is located near the town of Mangalia in south-eastern Romania, a few kilometers from the Black Sea (43° 49′ 44″ N; 28° 33′ 32″ E). Since the discovery of Movile Cave, studies provide evidence that the ecosystem is isolated from the surface. Radioactive artificial nuclides ^{90}Sr and ^{137}Cs, which were released as a result of the 1986 Chernobyl nuclear accident, have been found in relatively high concentrations in soil and in lakes surrounding Movile Cave, as well as in the Black Sea and in sediments of other caves, but not within Movile Cave itself [4].

Life within the Movile Cave ecosystem is sustained entirely by chemosynthesis from chemolithoautotrophic bacteria and archaea and methylotrophic bacteria [4]. The cave ecosystem is similar to some deep-sea hydrothermal vents [5], although vent environments are not completely independent from inputs of photosynthetically derived nutrients produced higher in the water column [6–8]. Within Movile Cave, however, the chemolithoautotrophs derive energy exclusively from the oxidation of inorganic compounds (e.g. H_2S, H_2) or CH_4 at the base of the food web, with heterotrophic bacteria, protists, and remarkably diverse invertebrates across several trophic levels, including worms, insects, spiders, and crustaceans [9]. Paleogeographical evidence suggests that some animal species were trapped in Movile Cave as early as 5.5 million years ago [10, 11]. Over time, these animals have adapted to life in the dark. Thirty-three out of 48 invertebrate species observed are endemic to Movile Cave [4].

Ayyalon Cave, located about 21 km from the Mediterranean Sea near the city of Ramla, Israel, was discovered in 2006 and is a similar cave ecosystem to Movile Cave [12]. Chemosynthetic primary production in this ecosystem was confirmed by carbon isotope analysis from the carcass of the troglobitic scorpion *Akravis rachanani* Levy 2007 [13]. In comparison to deep-sea hydrothermal vents, these caves offer easier access to interesting model ecosystems to study food-web interactions that are primarily driven by chemosynthetic primary production and microbial biomass.

In this chapter, we focus on the microbial functional diversity of the Movile Cave ecosystem, particularly those microbes involved in carbon, sulfur, and nitrogen cycling. It should be noted that, to date, microbiological studies in the Movile Cave ecosystem primarily focus on the microbial mat floating on the cave water and information on microbial populations from Ayyalon Cave is limited.

10.2 Cave Formation and Features

10.2.1 Movile Cave

Movile Cave formed from two major corrosion processes: condensation corrosion by carbon dioxide (CO_2) and acid corrosion by sulfuric acid (H_2SO_4) [14]. A detailed discussion of these processes are summarized by Kumaresan et al. [15]. The upper passages of the cave (approximately 200 m long) are completely dry as a consequence of the lack of water infiltrating from the surface, but the lower level (approximately 40 m long) is partly flooded by thermal waters that contain 0.3 mM dissolved H_2S, 0.2 mM

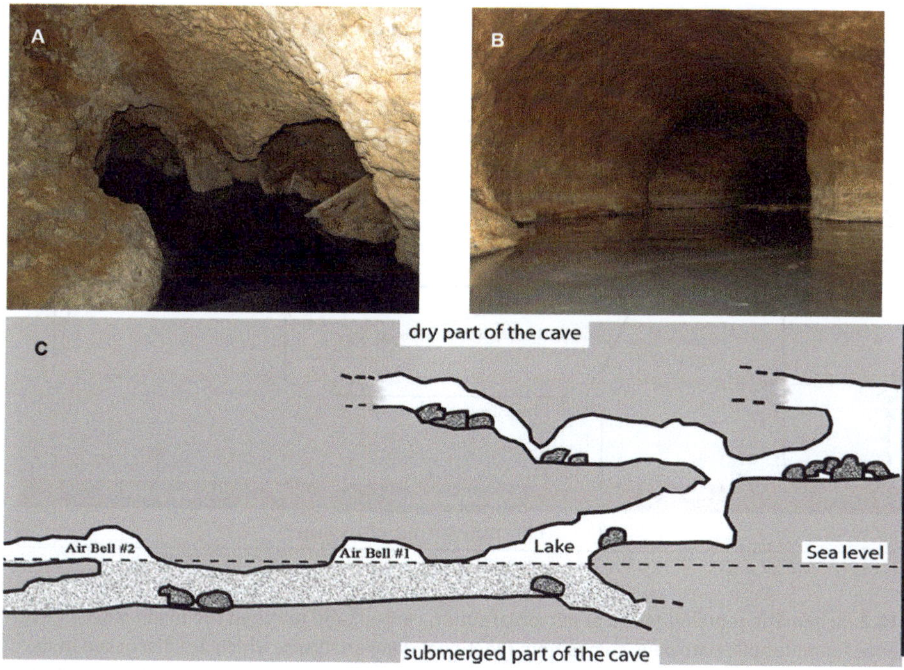

Fig. 10.1. Movile Cave photographs (a& b) and cross-section of Movile Cave (c) (used with permission, from the Ph. D. thesis of Daniel Muschiol).

dissolved CH_4, and 0.3 mM ammonium (NH_4^+) [14]. The physiochemical properties of the water are not affected by seasonal climatic changes (Serban Sarbu, personal communication). The water is at a constant pH of 7.4 at 22–24 °C due to the buffering capacity of the carbonate bedrock and 5 L/s water flow rate [14]. Dissolved oxygen concentration ranges between 9 and 16 µM at the water surface and decreases to less than 1 µM below the first few centimeters of the surface; anoxic conditions continue at depth [9].

The air temperature in the lower cave passage ranges from 20.7 °C to 20.9 °C [14] and the relative humidity is between 98% and 100% [9]. The cave air in the upper dry level contains 20–21% O_2 and 1–2% CO_2, while the airbells (i.e. air pockets) (▸ Fig. 10.1) in the lower level contain only 7–10% O_2, up to 2.5% CO_2, and 1–2% (v/v) methane. Hydrogen sulfide is detected in proximity to the air-water interface but not in the upper level [9]. The airbells create an active redox interface on the surface of the water where microorganisms in floating microbial mats oxidize reduced sulfur compounds, CH_4, and NH_4^+ using oxygen from the atmosphere (▸ Fig. 10.2). Extensive microbial mats composed of bacteria, fungi, and protozoa float on the water surface and also grow on the limestone walls of the cave [16]. Macrofauna in Movile Cave appear in proximity to the microbial mats within the airbells, and the upper dry nonsulfidic cave passages are devoid of macrofauna [3].

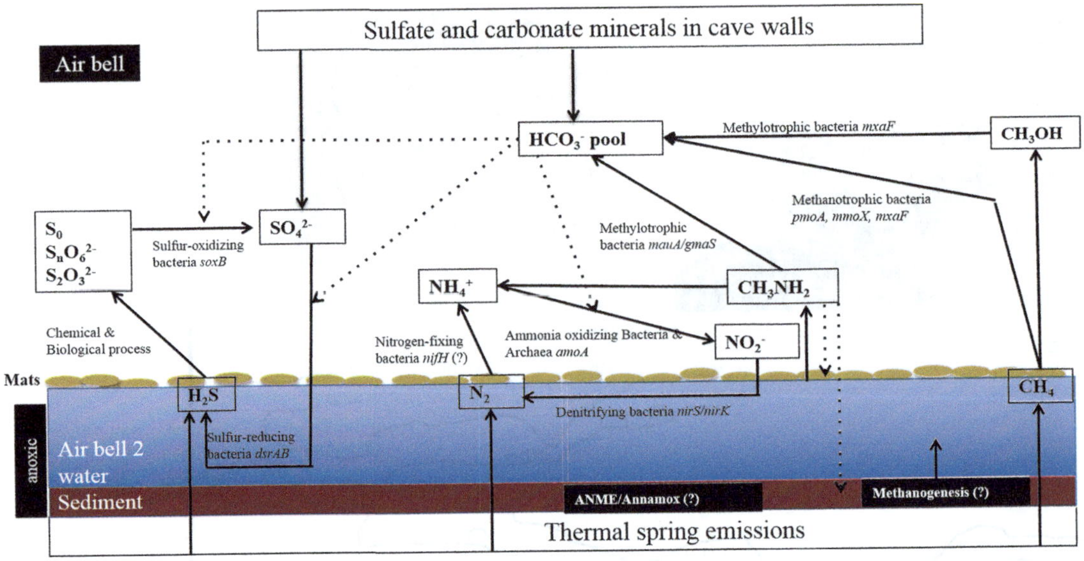

Fig. 10.2. Schematic representation of microbial sulfur, carbon, and nitrogen cycling in Movile Cave. Evidence for metabolic pathways comes from functional gene analyses, which are discussed in detail in the text. For sulfur cycling, *soxB* encodes the SoxB component of the periplasmic thiosulfate-oxidizing Sox enzyme complex, and *dsrAB* encode for the α- and β-subunits of the dissimilatory sulfite reductase enzyme. For carbon cycling, *pmoA* encodes the α-subunit of the particulate methane monooxygenase enzyme, *mmoX* encodes the α-subunit of the hydroxylase of the soluble methane monooxygenase enzyme, and *mxaF* encodes the α-subunit of the methanol dehydrogenase enzyme. Also, *mauA* encodes the small subunit of the methylamine dehydrogenase enzyme, and *gmaS* encodes the gamma-glutamylmethylamide synthetase. For nitrogen cycling, *nifH* encodes the nitrogenase reductase subunit, *amoA* encodes the α-subunit of ammonia monooxygenase for both Bacteria and Archaea, and *nirS* and *nirK* encodes a copper and a cytochrome cd1-containing nitrite reductase enzyme, respectively. Text within black boxes refers to processes that currently require evidence for anaerobic methane oxidation (ANME) and anaerobic ammonia oxidation (ANNAMOX). figure adapted from [15].

10.2.2 Ayyalon Cave

Similar to Movile Cave, Ayyalon Cave was accidentally discovered at the bottom of a cement manufacturer's chalk quarry in a karst area. The region is known for underground voids created by carbonate dissolution in the artesian Yarkon-Taninim aquifer. Upwelling moderately saline (524 mg/L chloride), sulfidic (5 mg/L), and thermal (29 °C) waters intersect the freshwater table of the aquifer. This is known as the Ayyalon Salinity Anomaly, and the cave water has higher temperature than the rest of the aquifer. Where overlain by chalk, horizontal galleries which have developed in the Turonian limestone have round cupolas and smooth walls indicative of slow-moving water [17]. Ayyalon Cave is situated ~100 m beneath the surface and has 2.7 km of galleries in two levels. The upper level preserves the morphological features of all caverns in the area, but the lowest level is filled with debris piles and collapsed passages [18]. The concentration of dissolved oxygen in the water is too low to be detected, even if its presence is obvious. Ayyalon Cave has also been shown to be isolated from any surface inputs.

10.3 Microbial Life in a Chemolithoautotrophic Ecosystem

Early studies of the Movile Cave ecosystem by Sarbu and colleagues [4, 19] provide evidence for chemolithoautotrophic carbon fixation. Microbial mats were incubated with ^{14}C-bicarbonate and the incorporation of radioactive carbon into microbial lipids was measured. Activity of ribulose-1,5-bisphosphate carboxylase/oxygenase (RuBisCO), a key enzyme of the Calvin Cycle, was observed in homogenates of microbial mat samples, as well as in lysates of bacterial cells cultivated from cave water. The results support the hypothesis that chemosynthate is being produced *in situ* [19]. Stable isotope ratio analyses of carbon and nitrogen also provide conclusive evidence that Movile Cave is a self-sustained ecosystem dependent on chemolithoautotrophically fixed carbon (for detailed discussion, refer to [4]). Similar methods were used to evaluate the ecosystem from Ayyalon Cave comprised of a few endemic stygobitic and troglobitic species, including three crustacean species, two arachnid species, and one insect [12]. Characterization of microbial populations in Ayyalon Cave has been limited, and the only known microorganism identified to date floats on the water surface and covers the walls; it belongs to the genus *Beggiatoa*.

The following sections mainly focus on research that targets microbial life in the Movile Cave ecosystem. Details for each of the pathways are included in ▸ Fig. 10.2.

10.3.1 Methanotrophy and methylotrophy

Airbells within the Movile Cave ecosystem offer a favorable environment for aerobic methane-oxidizing bacteria (methanotrophs). Methanotrophs, using CH_4 as a sole carbon and energy source, are found within the phyla Proteobacteria and Verrucomicrobia. Over the last few decades, the biochemistry and molecular biology of methanotrophs have been well characterized [20]. Based on 16S rRNA and key metabolic genes in the methane oxidation pathway, our knowledge of the composition and distribution of methanotrophic communities and environmental regulation of methanotrophic activity in various environments has significantly advanced [21]. The functional genes *pmoA* and *mmoX* encode for the active-site subunit of particulate methane monooxygenase and soluble methane monooxygenase, respectively. Stable isotope probing (SIP) experiments using Movile Cave water and microbial mats incubated with $^{13}CH_4$ [22, 23] identified active methanotrophs belonging to both the Alpha-

and Gammaproteobacteria [24]. Strains of *Methylomonas*, *Methylococcus*, and *Methylocystis/Methylosinus* assimilated $^{13}CH_4$ according to 16S rRNA and metabolic gene analyses. Sequences of nonmethanotrophic bacteria and an alga *Ochromonasdanica*, based on 18S rRNA gene sequence analysis, were also retrieved from ^{13}C-labeled DNA (i.e. heavy DNA), which suggested the possibility of nonmethanotrophs cross-feeding on ^{13}C-labeled biomass or metabolites arising from the initial consumption of $^{13}CH_4$ by methanotrophs. The gene *mxaF* encodes for the active-site subunit of methanol dehydrogenase. Clone libraries of *mxaF* and 16S rRNA genes from the heavy DNA contained sequences similar to the extant methylotrophs *Methylophilus* and *Hyphomicrobium*. This indicated that organisms assimilated ^{13}C-methanol excreted by methanotrophs that metabolized $^{13}CH_4$.

Other methylotrophs utilize reduced carbon substrates that have no carbon-carbon bond (e.g. methanol and methylated amines) as their sole carbon and energy source [25]. Enumeration studies by Rohwerder et al. [26] revealed the presence of methylotrophs in the floating mat, at up to 10^6 colony forming units (CFU) per gram dry weight (gdw) of mat. It is possible that these methylotrophs can feed on methanol released by methanotrophs during CH_4 oxidation. Methylated amines (MA) are typically associated with saline ecosystems and are produced from degradation of glycine betaine and trimethylamine *N*-oxide, osmolytes found in marine organisms [27, 28]. Recent studies report new methylotrophs from a wide range of environments and identify genera involved in MA utilization that were not previously associated with this one-carbon compound [25, 29]. Interestingly, 16S rRNA gene surveys by Chen et al. [30] and Porter et al. [31] indicate the presence of obligate MA-utilizing methylotrophs *Methylotenera mobilis*, *Methylophilus* spp., and *Methylovorus* spp. in Movile Cave microbial mats.

Building on this preliminary evidence for the genetic potential for MA utilization in Movile Cave, recently Wischer et al. [32] used a time-course SIP experiment with ^{13}C-mono methylamine (CH_3NH_2) to show that *M. mobilis* was an active methylotroph utilizing CH_3NH_2 in Movile Cave mat samples. At later time points, 16S rRNA gene sequences related to the genera *Methylobacterium* and *Methylovorus* were also detected in ^{13}C-labeled fractions, in addition to the genera *Catellibacterium*, *Cupriavidus*, *Porphyrobacter*, and *Altereythrobacter*, none of which have been previously reported to be methylotrophs. Wischer et al. [32] also developed new sets of PCR primers to target the gene *gmaS* involved in an alternative methylated amine utilization pathway, via gamma-glutamylmethylamide (GMA) and/or *N*-methylglutamate [33–35].

10.3.2 Microbial metabolism of sulfur

Sulfur-oxidizing bacteria (SOB), a heterogeneous group of organisms sharing the ability to oxidize reduced inorganic sulfur compounds, are distributed within Bacteria and Archaea. The biochemistry and molecular biology of microbial sulfur oxidation has been extensively reviewed by Ghosh and Dam [36]. SOB were the main focus of early studies characterizing microbial diversity and activity in Movile Cave [16, 26, 37]. Sarbu and colleagues [16] identified *Thiothrix* and *Beggiatoa*-like filamentous bacteria in the floating mat. Vlasceanu et al. [37] isolated a strain of *Thiobacillus thioparus* from cave water samples and characterized this bacterium at both the physiological and molecular level. Using most probable number (MPN) enumeration, Rohwerder et al. [26] showed that SOB that could use sulfur, tetrathionate, and thiosulfate as energy sources existed at up to 10^7 CFU/g mat and that sulfate-reducers capable of reduction of sulfate to H_2S were also present in the microbial mat.

Cultivation-independent studies targeting bacterial 16S rRNA genes in microbial mat DNA revealed sequences related to the genera *Thiobacillus* from the Betaproteobacteria, *Thiovirga*, *Thiothrix*, and *Thioploca* from the Gammaproteobacteria, and *Sulfuricurvum* from the Epsilonproteobacteria [30]. Employing an identical approach, Porter et al. [31] reported similar SOB diversity results from the microbial mat DNA, including *Halothiobacillus* and *Thiomonas* from the Betaproteobacteria and Gammaproteobacteria, respectively, and *Sulfurospirillum* from the Epsilonproteobacteria. The *soxB* gene encodes the SoxB component of the periplasmic thiosulfate-oxidizing enzyme complex. Clone libraries retrieved *soxB* genes from Alpha-, Beta- and Gammaproteobacteria, some being closely related to the *soxB* genes of *Thiobacillus* and *Methylibium petroleiphilum* from the Betaproteobacteria, and *Thiothrix* and *Halothiobacillus* (at 65% identity) from the Gammaproteobacteria. By targeting RuBisCO, specifically the form I green type RuBisCO (*cbbL*) gene sequences, Chen et al. [30] detected sequences closely related (80–85% identity) to RuBisCO gene sequences of *Thiobacillus denitrificans* and *T. thioparus*. DNA-SIP experiments using ^{13}C-labeled bicarbonate showed that SOB from the Beta- and Gammaproteobacteria (e.g. *Thiobacillus*, *Thiovirga*, *Thiothrix*, and *Thioploca*) were particularly active in assimilating CO_2 [30].

Rohwerder and colleagues [26] detected an extremely acidophilic SOB, which is interesting considering that the pH of the water in Movile Cave is neutral. A study by Sarbu et al. [38] in Frasassi caves in Italy, comprised of 23 km of limestone cave passages formed by sulfuric acid speleogenesis, reported pH values of 3.8–4.5 measured from the surface of microbial mats covering limestone walls. They suggested that not all the sulfuric acid produced by SOB was immediately buffered. See Chapter 2 for more information about the Frasassi caves. It was possible that similar processes could lead to ecological niches occupied by acidophilic bacteria in Movile Cave.

Rohwerder et al. [26] also reported activity of facultatively anaerobic SOB in Movile Cave that were capable of using nitrate (NO_3^-) rather than oxygen as an alternative electron acceptor, and that were present at the same levels as obligately aerobic SOB (i.e. 10^7 CFU/gdw of mat). The results suggested that both groups could contribute substantially to the biomass produced. Chen et al. [30] detected 16S rRNA gene sequences related to *Sulfuricurvum* spp. (Epsilonproteobacteria) from DNA extracted from Movile

Cave microbial mat, which can oxidize sulfur anaerobically [39]. This provided evidence for anaerobic sulfur oxidation, as demonstrated by Rohwerder et al. [26].

Although less abundant than SOB, sulfate-reducing bacteria (SRB) have been shown to be present in Movile Cave. They appear to belong to a higher trophic level, using the organic carbon released by SOB and other primary producers as the electron donor [26]. Sulfate-reducers in sulfidic caves appear to fall mainly within the Deltaproteobacteria [2]. Sequences related to members of the family *Desulfobulbaceae* have been found in Movile Cave in two independent 16S rRNA gene based clone library analyses [30, 31].

Endosymbiotic SOB living within invertebrates have been reported in various habitats, such as deep-sea hydrothermal vents and mangrove swamps [7, 40–42]. SOB benefit from sulfide, oxygen, and CO_2 from the host, and the bacteria in turn supply organic compounds to the host [43]. Recently, Flot et al. [44] reported the presence of *Thiothrix* epibionts associated with *Niphargus cf stygius* in Movile Cave. While the niphargids from Frasassi and Movile Caves were not closely related, their association pattern with *Thiothrix* was similar and the authors suggesting that *Niphargus–Thiothrix* association might be widespread in European groundwater ecosystems.

10.3.3 Nitrogen cycling

Studies targeting microbial nutrient cycling within the Movile Cave ecosystem have focused largely on carbon and sulfur, whereas microbial nitrogen cycling has received little attention. Ammonia- and nitrite-oxidizing bacteria were hypothesized as being important primary producers in the Movile Cave ecosystem, alongside methane- and sulfur-oxidizing bacteria [30]. Although ammonium concentrations in the cave waters are relatively high (0.2–0.3 mM), nitrate has not been detected [9]. This could be due to a rapid turnover of nitrate by either assimilatory nitrate reduction or denitrification. Facultatively anaerobic sulfur-oxidizers, such as *T. denitrificans*, can use NO_3^- as an alternative electron acceptor for respiration in oxygen-depleted conditions [45]. Therefore, sulfur oxidation linked to denitrification could be an important process in Movile Cave. Rohwerder et al. [26] detected high numbers of SOB in enrichments with thiosulfate and nitrate incubated under anoxic conditions, which supports the hypothesis. Chen et al. [30] also retrieved 16S rRNA gene sequences related to denitrifiers from the genus *Denitratisoma* (Betaproteobacteria) from Movile Cave mat sample DNA. Methylotrophs within the family Methylophilaceae, and some species of *Methylotenera*, require nitrate for growth on methanol [46]. Investigating the role of methylotrophy-linked to denitrification, alongside the role of methylated amines as a nitrogen source within the Movile Cave ecosystem, will yield more insight into the interactions between the carbon cycle and the nitrogen cycle in this ecosystem.

Microbial nitrogen (N_2) fixation may be another significant process of the nitrogen cycle in the Movile Cave ecosystem. The ability to fix N_2 is widespread among bacteria and archaea and many bacteria present in Movile Cave are known N_2-fixers (e.g. *Beggiatoa* and *Methylocystis*) [47–49]. Reduction of N_2 to NH_4^+ is highly energy-consuming and generally carried out during nitrogen-limited conditions [50]. Although there are relatively high NH_4^+ concentrations in the cave water, there may well be nitrogen-depleted niches within the microbial mats where N_2 fixation occurs. Process-based N_2-fixation measurements and assaying for *nif*H transcripts in cave mat samples could determine whether N_2 fixation occurs in this ecosystem.

10.4 Current Research and Future Perspectives

10.4.1 Isolates and whole genome sequence analysis

Efforts to isolate microbes and then sequence their genomes are integral in understanding the role of microbes in nutrient cycling, particularly in complex ecosystems such as Movile Cave. *Methylomonas* strain LWB was isolated from Movile Cave mat and water samples using CH_4 as the sole carbon source. The isolate's genome has been sequenced (Stephenson et al., unpublished research). Phylogenetic analysis of 16S rRNA and *pmoA* genes retrieved from the *Methylomonas* strain LWB genome reveals closer affiliation with other members of the genus *Methylomonas*, but the genome is divergent enough to suggest that this is a novel *Methylomonas* species. Genes encoding the sequence-divergent particulate methane monooxygenase (pXMO) enzyme system have been observed only in gammaproteobacterial methanotrophs [51]. The *pxmA* gene sequence from *Methylomonas* LWB clusters closely with the *pxmA* sequence of *Methylomonas* LW13 [52]. Pxm proteins closely cluster with proteins involved in ethane or ammonia oxidation rather than CH_4 oxidation, which has led to speculation that the pXMO enzyme might be capable of oxidizing a broader range of substrates [51]. Wischer et al. [32] isolated seven methylotrophic bacteria that represented six different genera from enrichments containing MMA as a sole carbon and nitrogen source. Eight non-methylotrophic MA-utilizing bacteria were also isolated from a separate enrichment with MMA as a nitrogen source and a mixture of sugars and carboxylic acids as the carbon source. Subsequently, genomes of facultative methylotrophs belonging to the genera *Catellibacterium* and *Mesorhizobium* were sequenced.

The resulting genomes can be effectively used to design comparative genomics and transcriptomics experiments on closely related isolates (i.e. strains *Methylomonas* LWB vs. *Methylomonas methanica* MC09) to understand how these organisms respond and adapt to different environmental niches. Also, it should be noted that experimental data resulting from cultivated model organisms are key to analyze and interpret any "meta-omics" study that attempts to understand microbial roles in biogeochemical cycling.

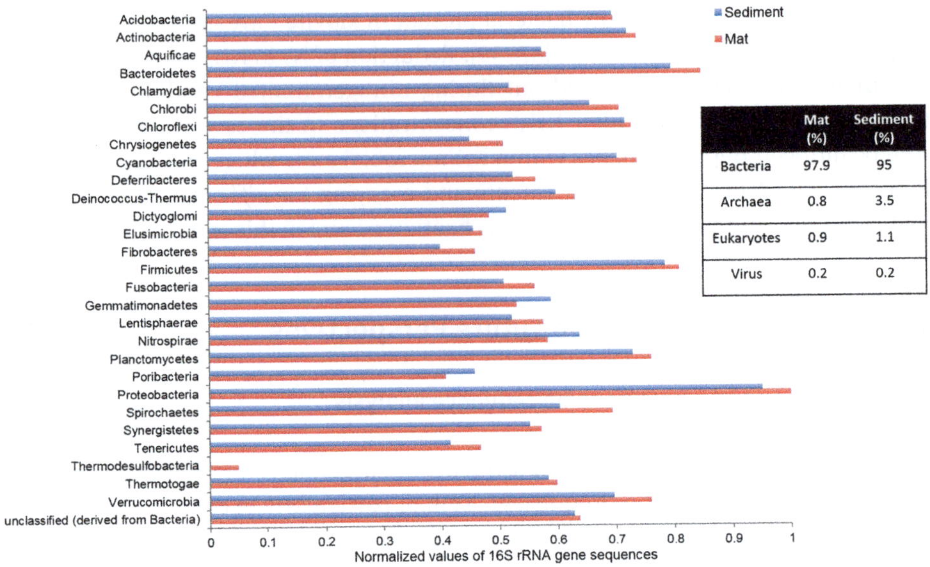

	Mat (%)	Sediment (%)
Bacteria	97.9	95
Archaea	0.8	3.5
Eukaryotes	0.9	1.1
Virus	0.2	0.2

Fig. 10.3. Histogram representing normalized values of 16S rRNA gene sequences from microbial mat and sediment metagenome sequences assigned to different bacterial phyla. The table (inset) represents the percentage of sequences assigned to different domains.

10.4.2 Microbial community composition analysis using metagenome sequences

Next-generation sequencing advances provide new and improved dimensions to high-throughput molecular microbial ecology studies. To determine microbial community composition in an ecosystem, amplicon and shot-gun metagenomic sequencing greatly provide us with a metabolic blue print of microbial systems [53]. Shot-gun metagenomic sequences of microbial mat and sediment samples (i.e. approximately 1 million sequences, each of 400 bp average read length) reveal widely diverse microbes (D. Kumaresan, unpublished results). Preliminary analysis of metagenomic data using the open web server platform MG-RAST [54] reveals that bacterial sequences dominate both microbial mat and sediment communities (▸ Fig. 10.3). In-depth analysis of metagenomic data and mining for key metabolic genes will yield better insight into the functional diversity and genetic potential of microbial communities in the Movile Cave ecosystem.

10.4.3 Archaeal communities

Archaeal communities are suggested to play an important role in nutrient recycling within cave ecosystems [55–57]. Over the years, the focus of research on microbial systems in Movile Cave has been on Bacteria, whereas no in-depth study has been performed on the contribution of Archaea to nutrient cycling. Chen et al. [30] revealed that the archaeal community in the Movile Cave, based on 16S rRNA gene libraries, possessed some Archaea that had also been originally retrieved by Takai and Horikoshi [58] from deep-sea hydrothermal vents. Metagenome sequence analyses on both Movile Cave mat and sediment DNA revealed gene sequences taxonomically annotated to Archaea, albeit in lower abundance when compared to bacterial communities (D. Kumaresan, unpublished results). Recently, Schirmack et al. [59] isolated *Methanobacterium movilense*, a novel hydrogenotrophic methanogen that could use H_2/CO_2, formate, 2-propanol, and 2-butanol as substrates. The same research group isolated *Methanosarcina spelaei* spp. nov., a methanogenic archaeon from the microbial mats in the cave water [63]. In situ measurements, particularly within the sediment profile, are required to assess the contribution of biogenic CH_4 production to the total CH_4 budget in Movile Cave. Moreover, it is essential to specifically target archaeal communities in future experiments to determine if they play a significant role in carbon, sulfur, and nitrogen cycling. In previous studies investigating the microbiology of Movile Cave, only mat and water samples were used. Sediment remains largely unexplored. While metagenomic sequences from sediment DNA reveals interesting microbes involved in nutrient cycling, it is essential to support these observations with process-based measurements.

10.5 Conclusions

Next-generation sequencing, a suite of other molecular ecology techniques, and a concerted effort to isolate novel organisms will improve our understanding of the functional diversity of microbial communities. The results allow us to study the contributions of different functional guilds in maintaining chemolithoautotrophic ecosystems. By using a combination of techniques, such as SIP and metagenomic sequencing of isotope-labeled DNA [60, 61], to target particular microbes involved in a specific function, we can reduce sequence information complexity and examine functional guilds in detail. For example, a composite genome of *M. mobilis* was compiled from the metagenomic sequences from Lake Washington sediment DNA [62]. Similar strategies can be used in Movile Cave to obtain genome information for microbes that are difficult to isolate and cultivate in the laboratory, particularly *Methylotenera* spp. that was abundantly represented in metagenome sequences and SIP incubations with ^{13}C-labeled MMA (D. Kumaresan, unpublished results; [32]). Although we do have a basic understanding of important functional guilds (e.g. SOB and methanotrophs), further research is required to appreciate the trophic interactions between the communities in Movile Cave. Whilst Ayyalon Cave shares similar features to Movile Cave, study on microbial diversity is limited and future research should focus to unravel the major microbial players and their interactions in sustaining this interesting ecosystem.

Acknowledgments

We acknowledge funding from the Natural Environment Research Council (NERC) to JCM (NE/G017956), the NERC and Earth and Life Systems Alliance (ELSA) for funding DK, a Warwick Postgraduate Research Fellowship and University of East Anglia-ELSA funding for DW and a NERC-CASE studentship for JS. We thank Serban Sarbu and Rich Boden for valuable discussions, and Vlad Voiculescu and Mihai Baciu for sampling. We also acknowledge GESS (Group for Underwater and Speleological Exploration, Bucharest, Romania) – Movile Cave custodian – for supporting our research, permission to use its research station in Mangalia, and technical assistance needed for sampling activities in the cave.

References

[1] Palmer AN. Origin and morphology of limestone caves. Geol Soc Am Bull, 1991, 103, 1–21.
[2] Engel AS. Observations on the biodiversity of sulfidic karst habitats. J Cave Karst Stud, 2007, 69, 187–206.
[3] Forti P, Galdenzi S, Sarbu SM. The hypogenic caves: a powerful tool for the study of seeps and their environmental effects. Cont Shelf Res, 2002, 22, 2373–2386.
[4] Sarbu SM, Kane TC, Kinkle BK. A chemoautotrophically based cave ecosystem. Science, 1996, 272, 1953–1955.
[5] Nakagawa S, Takai K. Deep-sea vent chemoautotrophs: diversity, biochemistry and ecological significance. FEMS Microbiol Ecol, 2008, 65, 1–14.
[6] Campbell KA. Hydrocarbon seep and hydrothermal vent paleoenvironments and paleontology: Past developments and future research directions. Palaeogeogr Palaeocl, 2006, 232, 362–407.
[7] Dubilier N, Bergin C, Lott C. Symbiotic diversity in marine animals: the art of harnessing chemosynthesis. Nat Rev Microbiol, 2008, 6, 725–740.
[8] Lutz RA, Kennish MJ. Ecology of deep-sea hydrothermal vent communities: A review. Rev Geophys, 1993, 31, 211–242.
[9] Sarbu SM. Movile Cave: a chemoautotrophically based groundwater ecosystem. In: Wilkens H, Culver DC, Humphreys WF, eds. Subterranean Ecosystems. Amsterdam, Elsevier, 2000, 319–343.
[10] Sarbu SM, Kane TC. A subterranean chemoautotrophically based ecosystem. Nat Speleol Soc Bull, 1995, 57, 91–98.
[11] Falniowski A, Szarowska M, Sirbu I, Hillebrand A, Baciu M. *Heleobia dobrogica* (Grossu & Negrea, 1989) and the estimated time of its isolation in a continental analogue of hydrothermal vents. Molluscan Res, 2008, 28, 165–170.
[12] Por F, Dimentman C, Frumkin A, Na'aman I. Animal life in the chemoautotrophic ecosystem of the hypogenic groundwater cave of Ayyalon (Israel): A summing up. Nat Sci, 2013, 5, 7–13.
[13] Levy G. The first trogbolite scorpion from Israel and a new chaetoid family (Arachnida: Scorpiones). Zoology in the Middle East, 2007, 40, 91.
[14] Sarbu SM, Lascu C. Condensation corrosion in Movile Cave, Romania. J Cave Karst Stud, 1997, 59, 99–102.
[15] Kumaresan DW, Stephenson J, Hillebrand-Voiculescu A, Murrell JC. Microbiology of Movile Cave—A chemolithoautotrophic ecosystem. Geomicrobiol J, 2014, 31, 186–193.

[16] Sarbu SM, Kinkle BK, Vlasceanu L, Kane TC, Popa R. Microbiological characterization of a sulfide-rich groundwater ecosystem. Geomicrobiol J, 1994, 12, 175–182.

[17] Frumkin AZY, Na'aman I, Tsatskinc A, Porat N, Vulfsone L. Karst depression over hypogenic hydrothermal karst in a carbonate terrain. Karst Geomorphology, 2013, 1–13.

[18] Na'aman I, Dimentman C, Frumkin A. Active hypogene speleogenesis in a regional karst aquifer: Ayyalon Cave, Israel. In: Klimchouck A, Sasowsky ID, Mylroie J, Engel SA, Engel AS, eds. Hypogene Cave Morphology, Special Publication 18, Leesburg, Virginia, Karst Waters Institute, 2014, 73–74.

[19] Sarbu SM, Vlasceanu L, Popa R, Sheridan P, Kinkle BK, Kane TC. Microbial mats in a thermomineral sulfurous cave. In: Stal LJ, Caumette P, eds. Microbial Mats: Structure, Development and Environmental Significance. Berlin, Springer-Verlag, 1994, 45–50.

[20] Trotsenko YA, Murrell JC. Metabolic aspects of aerobic obligate methanotrophy. In: Allen I, Laskin SS, Geoffrey MG, eds. Advances in Applied Microbiology, Academic Press, 2008, 183–229.

[21] McDonald IR, Bodrossy L, Chen Y, Murrell JC. Molecular ecology techniques for the study of aerobic methanotrophs. Appl Environ Microbiol, 2008, 74, 1305–1315.

[22] Dumont MG, Murrell JC. Stable isotope probing—linking microbial identity to function. Nat Rev Microbiol, 2005, 3, 499–504.

[23] Radajewski S, Ineson P, Parekh NR, Murrell JC. Stable-isotope probing as a tool in microbial ecology. Nature, 2000, 403, 646–649.

[24] Hutchens E, Radajewski S, Dumont MG, McDonald IR, Murrell JC. Analysis of methanotrophic bacteria in Movile Cave by stable isotope probing. Environ Microbiol, 2004, 6, 111–120.

[25] Chistoserdova L, Kalyuzhnaya MG, Lidstrom ME. The expanding world of methylotrophic metabolism. Annu Rev Microbiol, 2009, 63, 477–499.

[26] Rohwerder T, Sand W, Lascu C. Preliminary evidence for a sulphur cycle in Movile Cave, Romania. Acta Biotechnol, 2003, 23, 101–107.

[27] Barrett E, Kwan H. Bacterial reduction of trimethylamine oxide. Annu Rev Microbiol, 1985, 39, 131–149.

[28] Lin T-Y, Timasheff SN. Why do some organisms use a urea-methylamine mixture as osmolyte? Thermodynamic compensation of urea and trimethylamine N-oxide interactions with protein. Biochemistry, 1994, 33, 12695–12701.

[29] Chistoserdova L. Modularity of methylotrophy, revisited. Environ Microbiol, 2011, 13, 2603–2622.

[30] Chen Y, Wu L, Boden R, et al. Life without light: microbial diversity and evidence of sulfur- and ammonium-based chemolithotrophy in Movile Cave. ISME J, 2009, 3, 1093–1104.

[31] Porter ML, Engel AS, Kane TC, Kinkle BK. Productivity-diversity relationships from chemolithoautotrophically based sulfidic karst systems. Int J Speleol, 2009, 38, 27–40.

[32] Wischer D, Kumaresan D, Johnston A, et al. Bacterial metabolism of methylated amines and identification of novel methylotrophs in Movile Cave. ISME J, 2015, 9, 195–206.

[33] Chen Y, McAleer KL, Murrell JC. Monomethylamine as a nitrogen source for a nonmethylotrophic bacterium, *Agrobacterium tumefaciens*. Appl Environ Microbiol, 2010, 76, 4102–4104.

[34] Chen Y, Scanlan J, Song L, et al. *gamma*-Glutamylmethylamide is an essential intermediate in the metabolism of methylamine by *Methylocella silvestris*. Appl Environ Microbiol, 2010, 76, 4530–4537.

[35] Latypova E, Yang S, Wang Y, et al. Genetics of the glutamate-mediated methylamine utilization pathway in the facultative methylotrophic beta-proteobacterium *Methyloversatilis universalis* FAM5. Mol Microbiol, 2010, 75, 426–439.

[36] Ghosh W, Dam B. Biochemistry and molecular biology of lithotrophic sulfur oxidation by taxonomically and ecologically diverse bacteria and archaea. FEMS Microbiol Rev, 2009, 33, 999–1043.

[37] Vlasceanu L, Popa R, Kinkle BK. Characterization of *Thiobacillus thioparus* LV43 and its distribution in a chemoautotrophically based groundwater ecosystem. Appl Environ Microbiol, 1997, 63, 3123–3127.

[38] Galdenzi S, Cocchioni M, Morichetti L, Amici V. Sulfidic ground water chemistry in the Frasassi Caves, Italy. J Cave Karst Stud, 2008, 70, 94–107.

[39] Kodama Y, Watanabe K. *Sulfuricurvum kujiense* gen. nov., sp. nov., a facultatively anaerobic, chemolithoautotrophic, sulfur-oxidizing bacterium isolated from an underground crude-oil storage cavity. Int J Syst Evol Microbiol, 2004, 54, 2297–2300.

[40] Distel DL. Evolution of chemoautotrophic endosymbioses in bivalves. Bioscience, 1998, 48, 277–286.

[41] Russell JA, Dubilier N, Rudgers JA. Nature's microbiome: introduction. Mol Ecol, 2014, 23, 1225–1237.

[42] Wood AP, Kelly DP. Isolation and physiological characterisation of *Thiobacillus thyasiris* sp. nov., a novel marine facultative autotroph and the putative symbiont of *Thyasira flexuosa*. Arch Microbiol, 1989, 152, 160–166.

[43] Dahl C, Prange A. Bacterial sulfur globules: Occurrence, structure and metabolism. In: Shively J, ed. Inclusions in Prokaryotes: Springer, Berlin Heidelberg, 2006, 21–51.

[44] Flot JF, Bauermeister J, Brad T, Hillebrand-Voiculescu A, Sarbu SM, Dattagupta S. *Niphargus-Thiothrix* associations may be widespread in sulphidic groundwater ecosystems: evidence from southeastern Romania. Mol Ecol, 2014, 23, 1405–1417.

[45] Claus G, Kutzner H. Physiology and kinetics of autotrophic denitrification by *Thiobacillus denitificans*. Appl Environ Microbiol, 1985, 22, 283–288.

[46] Kalyuhznaya MG, Martens-Habbena W, Wang T, et al. *Methylophilaceae* link methanol oxidation to denitrification in freshwater lake sediment as suggested by stable isotope probing and pure culture analysis. Environ Microbiol Reports, 2009, 1, 385–392.

[47] Murrell JC, Dalton H. Nitrogen fixation in obligate methanotrophs. J Gen Microbiol, 1983, 129, 3481–3486.

[48] Nelson D, Waterbury J, Jannasch H. Nitrogen fixation and nitrate utilization by marine and freshwater Beggiatoa. Arch Microbiol, 1982, 133, 172–177.

[49] Oakley CJ, Murrell JC. Cloning of nitrogenase structural genes from the obligate methanotroph *Methylococcus capsulatus* (Bath). FEMS Microbiol Lett, 1991, 62, 121–125.

[50] Postgate J. Biological nitrogen fixation. Nature, 1972, 226, 25–27.

[51] Tavormina PL, Orphan VJ, Kalyuzhnaya MG, Jetten MS, Klotz MG. A novel family of functional operons encoding methane/ammonia monooxygenase-related proteins in gammaproteobacterial methanotrophs. Environ Microbiol Rep, 2011, 3, 91–100.

[52] Auman A, Stoylar S, Costello A, Lidstrom ME. Molecular characterization of methanotrophic isolates from freshwater lake sediment. Appl Environ Microbiol, 2000, 66, 5259–5266.

[53] Gilbert JA, Dupont CL. Microbial metagenomics: beyond the genome. Ann Rev Mar Sci, 2011, 3, 347–371.

[54] Meyer F, Paarmann D, D'Souza M, et al. The metagenomics RAST server – a public resource for the automatic phylogenetic and functional analysis of metagenomes. BMC Bioinformatics, 2008, 9, 386.

[55] Chelius MK, Moore JC. Molecular phylogenetic analysis of Archaea and Bacteria in wind cave, South Dakota. Geomicrobiol J, 2004, 21, 123–134.

[56] Gonzalez JM, Portillo MC, Saiz-Jimenez C. Metabolically active *Crenarchaeota* in Altamira Cave. Naturwissenschaften, 2006, 93, 42–45.

[57] Northup DE, Barns SM, Yu LE, et al. Diverse microbial communities inhabiting ferromanganese deposits in Lechuguilla and Spider Caves. Environ Microbiol, 2003, 5, 1071–1086.

[58] Takai K, Horikoshi K. Genetic Diversity of Archaea in deep-sea hydrothermal vent environments. Genetics, 1999, 152, 1285–1297.

[59] Schirmack J, Mangelsdorf K, Ganzert L, Sand W, Hillebrand-Voiculescu A, Wagner D. *Methanobacterium movilense* sp. nov., a hydrogenotrophic, secondary-alcohol-utilizing methanogen from the anoxic sediment of a subsurface lake. Int J Syst Evol Microbiol, 2014, 64, 522–527.

[60] Chen Y, Dumont MG, Neufeld JD, et al. Revealing the uncultivated majority: combining DNA stable-isotope probing, multiple displacement amplification and metagenomic analyses of uncultivated *Methylocystis* in acidic peatlands. Environ Microbiol, 2008, 10, 2609–2622.

[61] Chen Y, Neufeld JD, Dumont MG, Friedrich MW, Murrell JC. Metagenomic analysis of isotopically enriched DNA. Methods Mol Biol, 2010, 668, 67–75.

[62] Kalyuzhnaya MG, Lapidus A, Ivanova N, et al. High-resolution metagenomics targets specific functional types in complex microbial communities. Nat Biotechnol, 2008, 26, 1029–1034.

[63] Ganzert L, Schirmack J, Alawi M, Mangelsdorf K, Sand W, Hillebrand-Voiculescu A, Wagner D. *Methanosarcina spelaei* sp. nov., a methanogenic archaeon isolated From a floating biofilm of a subsurface sulphurous lake in the Movile Cave, Romania. Int J Sys Evol Microbiol, 2014, 64, 3478–3484.

Cesareo Saiz-Jimenez

11 The Microbiology of Show Caves, Mines, Tunnels, and Tombs: Implications for Management and Conservation

Abstract: The subsurface is a highly complex and challenging environment to study and the knowledge of their microbial ecology is sparse. This chapter reviews the microbiology of show caves, tunnels, mines, and tombs, with an emphasis on the major microbial groups found in these types of underground habitats. The biogeochemical cycles identified from these systems are also described, as well as cave management and conservation approaches.

11.1 Introduction

Caves are windows in the Earth's crust that allow for the exploration of the subsurface. Microbial colonization of caves is an ancient and natural process, although often it is attributed to current cave managements and visits. During millennia, caves have maintained a delicate balance between subsurface microorganisms and cave-adapted fauna (i.e. troglobites) due to its isolation from the outside. Caves lack natural light, which excludes the growth of phototrophic microorganisms. Cave can also be oligotrophic because of relatively little or no organic carbon able to support heterotrophic microbial life. Some caves maintain relatively constant temperatures throughout the year and have extensive areas of mineral formations and osmophilic niches. These stressing conditions provide an extreme environment for most microorganisms [1]. Not only microorganisms in subterranean environments are worth of study, but also their interactions with minerals [2] and troglobites [3].

At the moment of a cavity's discovery, communication with the exterior is established and the cavity is subjected to the impact of terrestrial microbial and animal communities that have the potential to alter the subsurface ecosystem. Visitors entering into the cavity also contribute to this disruption because they bring abundant organic matter (e.g. hair, dander, clothing fibers, organic matter in the shoes, etc.) that can profoundly alter the ecosystem food web [4, 5]. In fact, a cave subjected to a strong input of organic carbon will suffer the replacement of oligotrophic microorganisms and troglobites by terrestrial microorganisms and animals that metabolize faster the exogenous organic carbon, which can shift the composition of autochthonous cave inhabitants.

Caves are highly complex and challenging environments to study. The most well-known caves are those formed on limestone and other calcareous rocks. Other types of caves have been rarely studied and are relatively unknown for the geomicrobiologists,

Fig. 11.1. Paleolithic bison from Altamira Cave, Spain [9, 32]. The bison has 170 cm in length and 130 cm height. Photo from Pedro Saura, Museo de Altamira, Spain.

such as the caves on ice, halite, or gypsum [6, 7]. In the last few years interest on lava caves on basalt is increasing [8]. Working and sampling in caves are not easy tasks, particularly caves housing rock art paintings where gaining entrance is restricted or prohibited to avoid damage to the Paleolithic art. Opportunities to obtain samples in these caves are very rare, and the few samples acquired are usually so small that only molecular techniques can be applied.

Consequently, in contrast to the literature describing the microbiology of surface terrestrial and marine environments, studies on the microbial ecology of subterranean environments overall are also sparse. It is also not surprising that in the past decade, most research on caves with rock art paintings have used culture-independent methods. Generally, the research has been limited to only a few caves where the focus is on the importance of a cavity from a geological, biological, or cultural point of view, including so-called show caves. For instance, extensive research in the last decade, such as in the Lascaux Cave in France and the Altamira Cave in Spain, has focused on rock art painting preservation from microbial attacks and human impacts (▸ Fig. 11.1) [9, 10]. Chapters 13 and 14, respectively, in this book describe research done in these caves. This chapter reviews the microbiology of show caves, tunnels, mines, and tombs, with an emphasis on the major microbial groups found in these types of underground habitats. The biogeochemical cycles identified from these systems are also described, as well as cave management and conservation approaches.

11.2 Major Groups of Microorganisms

Microbial biofilms in caves adopt different morphologies. Some forms are well-known, such as that of snottites hanging from cave ceilings, such as in the Frasassi caves, Italy, with a mucous consistency and being composed primarily of *Acidithiobacillus thiooxidans* [11]. The snottites are not from the tourist section of the Frasassi caves, but in the lower levels of the cave system associated with sulfidic streams. Another well-known example of different microbial structures includes the colorful microbial mats coating the walls of Spanish caves [9, 12] (▸ Fig. 11.2). Other biofilms are more bizarre, such as the sprout-like aggregates attached to the bed of a cave stream in Bosnia and Herzegovina [13]. The aggregate core consists of a member of a novel deep-branching lineage in the bacterial phylum Nitrospirae [13]. Recently, an acidophilic bacterial-archaeal-fungal ecosystem linked to the formation of ferruginous crusts and stalactites was described by Gherman et al. [14] in an abandoned pyrite mine.

The domains Archaea and Bacteria, and the kingdom Fungi, are generally well represented in show caves and other subterranean environments. Diversity of these microbial groups is preliminary based on 16S and 18S rRNA gene sequences. In addition, growth of members of the division Chlorophyta (green algae) and the class Bacillariophyceae (diatoms) is common in caves with artificial lighting, but these groups will not considered in this chapter. Chapter 12 focuses on growth of organisms around artificial lighting in caves. Among bacteria, the most abundant and/or frequently retrieved phyla found in caves are: Acidobacteria, Actinobacteria, Bacteroidetes, Cyanobacteria, Firmicutes, Nitrospirae, and Proteobacteria. However, none of the caves studied to date has been exhaustively sampled to verify whether a microbial group is truly absent from its ecosystem [15].

Fig. 11.2. Bacterial colonization on the walls of Altamira Cave, Spain [9, 12, 47].

11.2.1 Archaea

Archaea are ubiquitous in terrestrial and marine ecosystems [16, 17], including subsurface environments. Mesophilic Crenarchaeota (Thaumarchaeota) communities are reported from Lechuguilla [18] and Wind caves [19] (USA). Members of the Crenarchaeota have also been retrieved from steam vents and caves from volcanic national parks in the USA [20]. Thaumarchaeota can comprise a significant portion of an archaeal community, such as representing > 89% of the Archaea in a Portuguese underground aqueduct from the 16th century [21]. From diversity surveys and a metagenome sequence, Thaumarchaeota are the third most abundant phylum on calcite speleothems from the oligotrophic Kartchner Caverns in Arizona (USA) [22, 23]. As much as 13% of the Thaumarchaeota reads are also classified as *Nitrosopumilus maritimus*, an ammonia-oxidizing marine archaeon [23]. *N. maritimus* is involved in nitrogen and phosphorus cycling [24]. This microbe has also been found associated with phototrophic microorganisms on the walls of artificially illuminated caves (Valme Jurado, personal communication). In general, an interesting association between Archaea and Cyanobacteria has been reported in the Roman tomb of Servilia, in Spain [25]. Archaea closely related to Thaumarchaeota (*N. maritimus*) have also been identified from the oxygen-rich, surface layer of moonmilk from the cold alpine Hundalm Cave in Tyrol, Austria [26]. Archaea in the oxygen-depleted subsurface layers are more distantly related to members within the Euryarchaeota [26]. Euryarchaeota from the order Thermoplasmatales have been found from the sulfidic section of the Frasassi caves [27] and in an abandoned copper mine in the Czech Republic [28].

Biogenic methane production is performed by methanogenic Archaea in anoxic environments and methane consumption is mainly achieved by methanotrophic Proteobacteria [29]. The production of very low amounts of methane corresponds to a methanogenic community that represents only a minor fraction of the Archaea in Hundalm Cave moonmilk [26]. Although Movile Cave in Romania is not a show cave or artificial underground cavity, it was discovered through digging of an artificial shaft. Crenarchaeota and Euryarchaeota are present in Movile Cave based on 16S rRNA sequences [30]. However, no sequences related to known sulfur-oxidizing archaea, ammonia-oxidizing archaea, methanogens or anaerobic methane-oxidizing archaea have been detected [30]. A novel species of methanogenic archaea, *Methanobacterium movilense*, was isolated from the anoxic sediment of a subsurface lake in the Movile Cave [31].

11.2.2 Bacteria

11.2.2.1 Acidobacteria

Acidobacterial sequences are commonly retrieved in caves, but none have been isolated and cultured in the laboratory to date. Pioneering studies by Schabereiter-

Gurtner et al. [32, 33] of red pigment from a bison painting in Altamira Cave and from a horse painting in Tito Bustillo Cave, Spain, reveal complex and partly unknown bacterial communities with Acidobacteria. In another study of rock art paintings from two other Spanish caves, Acidobacteria are important to the community as the second most abundant group after Proteobacteria [34]. It is possible that the abundance of Acidobacteria reported in the Spanish caves might be biased by the sampling (i.e. concentrated mainly on red paintings) and/or by the low number of sequences studied.

An analysis of 3000 acidobacterial 16S rRNA gene sequences reveals that the phylum Acidobacteria is highly diverse. Zhou et al. [35] compare the distribution of bacteria in a Chinese cave with those of various Australian, Spanish, and US caves, and found the relative abundance of Acidobacteria was second to Proteobacteria. This is a common pattern based on 16S rRNA genes from soils [36]. However, a molecular survey on 60 caves carried out by Lee et al. [1] showed that Acidobacteria account for only ~3% of all bacterial groups.

The Acidobacteria phylum includes 26 subgroups, which were identified by Barns et al. [37] from subsurface sediments from uranium-contaminated sites. Acidobacteria subgroups 1, 3, 4, and 6 are the most abundant in terrestrial environments, including some cave samples. Primarily subgroups 1 to 11 (except subgroup 2) have been identified from caves [19, 38, 39]. For example, from the Altamira Cave in Spain, 12 samples mostly from areas colonized in the past by phototrophic microorganisms due to the use of artificial lighting for facilitating tourist visits reveal, from a total of 100 clones classified into 41 different operational taxonomic units, larger diversity within the Acidobacteria than previously shown [38]. Also, unexpected acidobacterial taxa include representatives from subgroups 3–7 and 9–11. Acidobacteria associated with phototrophic microorganisms thriving around lighting lamps in the catacombs of Saint Callixtus in Rome, Italy, shared sequences from Altamira Cave subgroups 3, 4, 6, 9, and 10, and subgroups 5, 7, and 11 were also identified from Altamira Cave samples [40]. An association may exist between Acidobacteria and the phototrophic microorganisms, whose growth in subterranean environments is triggered by artificial lighting. Acidobacteria from subgroup 5 dominate Wind Cave, South Dakota (USA), communities with Gammaproteobacteria [19]. Novel 16S and 23S rRNA gene sequences from subgroups 7 and 8 were retrieved from Lower Kane Cave, Wyoming, a cave undergoing sulfuric acid speleogenesis and not a show cave [39]. Acidobacterial cells were always found embedded in a matrix of epsilon- or gammaproteobacterial filaments and suggest perhaps a lifestyle based on heterotrophy or chemoorganotrophy. Soils from other nonshow caves, including the fumarolic ice caves on Mt. Erebus, an active volcano on Ross Island, Antarctica [6], contain distinct communities of very low to moderate diversity, with Acidobacteria being the only phylum detected in all the caves and a major component of the communities; no subgroup analysis was reported.

11.2.2.2 Actinobacteria

The subterranean environment is a favorable habitat for Actinobacteria [41, 42], including show caves and tombs from different geographical locations [23, 43–46]. Actinobacteria are present in colorful, small, and scattered microbial colonies with defined colony edges and in microbial mats that extensively cover cave walls and ceilings. At first glance, three colony types are easily distinguished: yellow, white, and gray (▸ Fig. 11.2). The question arises as to whether similarly colored colonies are comprised of the same bacteria or consortia, despite being spatially distributed and having different niches based on organic matter input and microclimatic parameters [9, 12].

Yellow colonies are found in European and North American limestone caves (▸ Fig. 11.2) [47], in gypsum karst (Angel Fernandez-Cortes, personal communication), and in lava tubes from different geographic regions [8]. In a comparative analysis between yellow colonies from Spanish, Slovenian, and Czechian caves, Porca et al. [47] conclude that colonies growing on the walls of geographically distant caves share bacteria that are morphologically similar and phylogenetically related. Actinobacterial Pseudonocardinae phylotypes, gammaproteobacterial Chromatiales, and the genus *Nitrospira* comprise the core of these communities, which is considered to be the result of similar environmental parameters in karst. But, difference in site-specific geochemistry and availability and nature of organic carbon also relate to each community also support diverse, site-specific phylotypes. Colony growth is triggered by organic carbon dissolved in dripping waters from the overlaying soil and rocks. As such, yellow colony communities are likely true cave dwellers. The yellow coloration could be a consequence of carotenoid production by members of Xanthomonadales and other yellow pigment-producing bacteria, such as members of Pseudonocardinae and species of the genus *Steroidobacter* [47].

White colonies have been found in lava caves [8], and Spanish [12, 43] and Italian limestone caves [42]. Yellow with white microbial mats coat walls and ceilings of lava caves from The Azores, and Hawai'i and New Mexico (USA), although other colors are also observed (pink, dark orange, etc.) [8]. Microbial mats from these three locations with different climatic regimes have similar compositions dominated by Acidobacteria, Alpha- and Gammaproteobacteria, Actinobacteria, and Cyanobacteria. White colonies in Altamira Cave are composed of Actinobacteria and Alphaproteobacteria, although when these microbial mats are located on iron (i.e. hematite), actinobacterial abundances increase three times [48]. In some caves and Italian Etruscan tombs, white colonies are dominated by the genera *Nocardia* and *Pseudonocardia* [43, 46]. Also in the Altamira Cave, gray colonies (▸ Fig. 11.3) are primarily composed of Actinobacteria, with the predominant microorganism probably representing an unknown species belonging to the order Nitriliruptoridae [45].

Other studies uncover diverse Actinobacteria in caves. Ortiz et al. [23] describe actinobacterial predominance from Kartchner Caverns stalactites, with the majority being members of the orders Actinomycetales, lower relative abundances of Acidimi-

Fig. 11.3. Gray colonizations on the wall of Altamira Cave, Spain [45].

crobiales, and Rubrobacterales. Extremely acidophilic snottites from the Frasassi cave system also have low abundances of the family Acidimicrobiaceae that belong to a monophyletic sister group to the genus *Ferrimicrobium* [11]. Rubrobacterales are well represented in subterranean environments. *Rubrobacter* spp. are associated with rosy discoloration of mural paintings from an Italian crypt and Austrian chapels [49–51] and have been isolated from Roman tombs in Spain [52]. 16S rRNA gene sequences have also been retrieved from Ardales Cave in Spain [43] and Etruscan tombs [46, 53].

The low, stable temperature and high relative humidity of most caves may have selected for Actinobacteria [54]. However, there are some exceptions, as reported for the Cave of Crystals, in Chihuahua, Mexico. This cave was opened up by mining activities, where gypsum crystals up to 15 m long were discovered. The cave is a natural but unique subterranean ecosystem with a nearly constant year-round temperature of 55 °C and 100% relative humidity. Two actinobacterial groups isolated from gypsum crystals and cave walls are closely related to the genus *Prauserella* [7]. Conversely, a 50 °C geothermal mine adit in Colorado (USA) did not contain Actinobacteria, but instead the most abundant phylum was Deinococcus-Thermus [55].

The identification of new actinobacterial genera and species from subterranean environments means that caves constitute an ecological niche with solid potential for biodiversity studies, particularly in the search of novel bacteria. From 2000 to 2009, 34 new actinobacteria were isolated and described from subterranean environments [44], as gathered from the *International Journal of Systematic and Evolutionary Microbiology*. From 2010 to 2013, 13 more species have been added. In the last 14 years, 47 novel species were described and validated, six species belong to *Agromyces*, and five to *Kribbella* and *Amycolatopsis*, which represent 34% of the total species identified. These three genera have a wide distribution in different subterranean environments, including caves, mines, catacombs, etc. Seven novel genera (*Catelliglobosispora*, *Ferrimicrobium*, *Fodinibacter*, *Fodinicola*, *Hoyosella*, *Knoellia*, and *Spelaeicoccus*) were erected. Other novel Actinobacteria have been described but not yet validated, such as *Microlunatus cavernae* and *Saccharopolyspora cavernae* [56, 57].

Caves also are an untapped source of novel bioactive compounds. In fact, Groth and Saiz-Jimenez [54] highlight the importance of caves as a new habitat from which Actinobacteria that produce bioactive substances could be isolated. A high number of streptomyces were isolated from an Italian cave (Grotta dei Cervi, Porto Badisco) [42]. Later, Herold et al. [58] report the production of cervimycins A–D by *Streptomyces tendae*, isolated from this same cave. These polyketide glycosides show activity against multi-drug-resistant staphylococci and vancomycin-resistant enterococci. Subramani and Aalbersberg [59] show that, by focusing on rare actinomycetes from underexplored natural habitats, it has been possible to isolate about 220 rare genera, from which 50 taxa produce 2500 bioactive compounds. Rarer actinomycetes have also been isolated [60], but at much lower frequencies than streptomycete strains isolated using by conventional methods [41, 42, 61].

11.2.2.3 Bacteroidetes

Members of the Bacteroidetes are widely present in European and North American caves [30, 47] and are less abundant in some crypts and tombs [46, 49]. Persistence of Bacteroidetes in caves might be related to growth preferences that require low temperatures, high salinity gradients, and darkness [62]. Bacteroidetes are the dominant operational taxonomic unit from a manganese-rich biofilm in Carter Saltpeter Cave, Tennessee (USA), a former cave mined for niter deposits [63] and also recently impacted from sewage waste [64]. Uranium mines also seem to be a favorable environment for this phylum [65]. Sources of Bacteroidetes in caves can be from top soil seepage through the epikarst. But, seepage can have human and animal fecal waste contamination. Bacteroidetes (order Bacteroidales) represent an important fraction of the gastrointestinal flora of many mammals. This group is often used as a fecal indicator taxon because of molecular genetics detection advantages over coliforms [62]. A fecal pollution source into caves can also be from guano, as suggested the study of the Neolithic paintings from the Magura Cave in Bulgaria [66, 67]. Similar paintings made with guano have also been found in Grotta dei Cervi in Italy (▸ Fig. 11.4), although the microbial communities from the paintings are uninvestigated due to sampling prohibition [42].

11.2.2.4 Cyanobacteria

Cave systems are heterogeneous and composed of different ecotones, defined as zones of transition between adjacent ecological systems [68]. Relatively strong changes in community structure and composition occur across ecotone boundaries because ad-

Fig. 11.4. Neolithic guano paintings in Grotta dei Cervi, Italy [42].

jacent communities meet their limits. No communities are more exempliary of this ecotone concept than phototrophic communities (i.e. cyanobacteria, algae, lichens, and lower plants) that colonize cave walls, ceilings, and ground sediments where there is some light. In most caves, phototrophic communities disappear unless artificial lighting is installed for visitors. The primary stress factors associated with colonization of a cave by Cyanobacteria include light availability, humidity, lack of nutrients, and temperature [69].

There is a tendency in most review papers to report separately on cyanobacteria, despite their associations with heterotrophic bacteria, fungi, algae, and bryophytes. In general, the diversity of Cyanobacteria in cave entrances, regardless of being show caves or artificial cavities, can be separated into different zones or ecotones [70–73]. For example, the cyanobacterial community in Kastria Cave, a show cave in Greece, is structured based on light gradients [71]. Mucilaginous temporary biofilms of *Aphanocapsa* spp., *Chroococcus* spp., *Eucapsis minor*, *Leptolyngbya gracillama*, *Leptolyngbya perelegans*, and *Pseudophormidium spelaeoides* prevail in the entrance zone, whereas *Scytonema julianum* and *Iphinoe spelaeobios* are present in the dimly lighted section of the Kastria Cave entrance. Dominant Cyanobacteria in some other caves also include *Spelaeopogon sommierii*, *Geitleria calcarea*, *Geitleria floridana*, *Herpyzonema pulverulentum*, and *Symponema cavernicolum* [72, 73]. From L'Aigua Cave (Water Cave), a popular tourist destination in Spain, species of chasmoendolithic growth in caves includes for the first time *Calothrix elenkinii*, *Gloeothece confluens*, *Hormothece cylindrocellulare*, *Aphanothece saxicola*, and *Pleurocapsa minor* [74]. *Iphinoe spelaeobios*, *Loriellopsis cavernicola*, *Toxopsis calypsus*, *Phormidium melanochroum*, and *Chalicogloea cavernicola* have been isolated from different Greekand Spanish show caves [74, 76–78].

Extending diversity surveys from caves to include mines and catacombs also reveal a range of cyanobacterial diversity in entrances and in artificially illuminated

Fig. 11.5. Phototrophic growth in St. Domitilla Catacombs, Rome, Italy [149, 150].

areas. From Greek settings, a total of 44 Cyanobacteria have been isolated [79], but some sites, like Perama Cave in Greece, have high diversity with 31 taxa. Mines and catacombs, in contrast, have low diversity, with only four taxa in Milos catacombs. Phototrophic biofilms associated with artificial lighting in the Catacombs of Saint Callixtus and Saint Domitilla in Italy (▸ Fig. 11.5) are dominated by *Gloeothece membranacea*, *Scytonema julianum*, and *Fischerella maior* [80], but phototrophic communities in different Roman catacombs are dominated by *Eucapsis*, *Leptolyngbya*, *Scytonema*, and *Fischerella* spp. [81]. Some of these taxa have never been identified outside of these habitats and are considered to be obligate cavernicole taxa unable to survive outside the caves or other low-light environments. *Scytonema julianum* are commonly found in caves subjected to artificial lighting.

11.2.2.5 Firmicutes

Firmicutes comprise the second major phylum after Actinobacteria in all studied caves, including show caves. Many studies confirm dominance in different geographic locations, although Lee et al. [1] find that Firmicutes account for ~3% of all groups. *Bacillus* is the most commonly detected genus by culture-dependent and -independent surveys in various Australian caves [82], and this genus represents over 49% of isolates from the Spanish Cave of Bats and over 44% in the Saint Agatha Catacombs in Malta [83]. The abundance of Firmicutes, as well as Actinobacteria, in culture-dependent studies should be attributed to the spore-forming capability of these two bacterial groups and their wide dissemination in caves through air currents. In fact, Fernandez-Cortes et al. [84] report that *Bacillus* spore cave dispersion patterns have a tendency to dominate airborne diversity after tourist visitation, but low dispersion occurs on days without visitation. From the comparison of microbial communities in three areas of Kartchner Caverns differentially impacted by tourism (i.e. high, medium, and low), Proteobacteria represent 77% of the total isolates in the high impacted area, but Firmicutes dominated the low and medium impacted areas (66% and 52% of the total isolates, respectively) [85]. Thermophilic Firmicutes species of the genera *Thermaerobacter*, *Thermolithobacter*, and *Thermoanaerobacter* are dominant in a cave from the Buda thermal karst system in Hungary [86]. The presence in this cave is likely attributed to prevailing environmental conditions and not direct human impact because this cave is not visited by tourists. Other show caves have low to unmeasured Firmicutes abundances, including the Frassasi caves, Acquasanta Terme, Nullarbor, and Tito Bustillo caves [8, 33, 87–90].

11.2.2.6 Nitrospirae

Members of the phylum Nitrospirae are characterized by their activity as nitrite oxidizers, a process mostly known from the proteobacterial classes Alpha-, Beta-, and Gammaproteobacteria [91]. The core of sprout-like aggregates found in Vjetrenica Cave in Bosnia and Herzegovina consists of members of a novel deep-branching lineage in the bacterial phylum Nitrospirae, and this lineage accounts for > 25% of the community [13]. The autotrophic, nitrite-oxidizing *Nitrospira moscoviensis* comprise 12% of the aquatic microbial mat community from the Nullarbor caves in Australia [87]. Nitrospirae also contribute 12.5% to the total microbial community in a tunnel excavated in granite in Porto, Portugal [21], and *Nitrospira* comprise the core of yellow colonies from different European caves [47]. Lower percentages of Nitrospirae-related sequences have been retrieved from microbial communities in some caves [35, 92] and lava caves [8].

11.2.2.7 Proteobacteria

Lee et al. [1] identify that Proteobacteria account for 50% of all bacterial groups from 60 caves surveyed using culture-independent methods. Some examples of the wide distribution and abundance of this phylum include the Wind Cave dominated by Gammaproteobacteria with Acidobacteria [19], various lava caves dominated by Acidobacteria, Alpha- and Gammaproteobacteria [8], and a ferromanganese deposit in a Portuguese tunnel being dominated by Gammaproteobacteria and Alphaproteobacteria [21]. Biofilms in sulfidic caves are primarily composed of Gammaproteobacteria and Epsilonproteobacteria [11, 93], and Beta-, Gamma-, Delta-, and Epsilonproteobacteria account for over 75% of clones from sulfidic stream biofilms in the Frasassi caves [90].

Biocidal treatment in the Lascaux Cave, France, resulted in a predominance of Betaproteobacteria and Gammaproteobacteria [94, 95]. The following section will describe the separate proteobacterial class diversity from show caves, tunnels, mines, and tombs.

Alphaproteobacteria are well represented in all subterranean environments, particularly the order Rhizobiales. A number of novel Rhizobiales species have been described from caves, such as *Martelella mediterranea* [96], *Aurantimonas altamirensis (=Aurimonas altamirensis)* [97], and some other from tombs, such as *Sphingopyxis italica* [98]. Alphaproteobacterial sequences have also been retrieved from caves and tombs in Spain, China, Slovenia, Czech Republic, Italy, France, Portugal, USA, etc. [8, 34, 35, 46, 47, 53, 94]. Recently, prosthescate bacteria (e.g. *Hyphomicrobium, Pedomicrobium, Prosthecomicrobium*, etc.) have been identified from subterranean environments [21, 53]. Five *Gluconacetobacter* species (e.g. *G. tumulicola, G. asukensis, G. tumulisoli, G. takamatsuzukensis* and *G. aggeris*) have been isolated from the Takamatsuzuka and Kitora stone chamber in Japan [99, 100]. *Acidiphilium* and *Acidocella* species are in extremely acidic caves and mines [14, 101]. *Wolbachia*, an endosymbiontic alphaproteobacterial genus that is associated with over 60% of insect species, have been found in the collembolan *Folsomia candida* collected in the Wind Cave [102].

Among the Betaproteobacteria in subterranean environments, the distribution of genera *Gallionella* and *Leptothrix* correspond to biogenic iron and manganese deposits [103, 104]. *Gallionella* has also been identified in the acidic water springs of an abandoned copper mine [28]. Reticulated filaments encrusted with a manganese sheath have been described from an ancient granite tunnel, but are yet affiliated with any known microorganism [21]. The formation of ochreous speleothems from lava tubes on Terceira Island (The Azores) was investigated by De los Rios et al. [105]. These formations are composed of ferrihydrite deposits intermixed with bacterial structures, but an evident population of living cells of *Gallionella* or *Leptothrix* has not been detected in the speleothems. Based on *Gallionella* and *Leptothrix* morphotypes observed by using backscattered electron mode of scanning electron microscopy, it is possible these bacteria may precipitate ferrihydrite in the ochreous speleothems. Other Betaproteobacteria can be involved in sulfur (e.g. *Thiobacillus*spp.) and ammonia oxidation (e.g. *Nitrosomonas* and *Nitrosospira* spp.) or methylotrophy (e.g. *Methylotenera, Methylophilus*, and *Methylovorus* spp.), as detected from Movile Cave [30], Niu Cave in China [35], lava caves from The Azores [106], Spanish caves [33, 34], etc. Betaproteobacterial sequences retrieved from Frasassi biofilms in sulfidic streams are related to *Thiomonas* spp. and *Thiobacillus denitrificans* [90].

Gammaproteobacteria likely constitute the dominant proteobacterial group in most caves. Their members are well represented in acidic (e.g. *Acidithiobacillus*) and halophilic (e.g. *Halothiobacillus, Halomonas*) environments, as well as in oligotrophic caves or caves with either natural or anthropogenic organic carbon inputs (e.g. *Pseudomonas, Ralstonia, Stenotrophomonas, Xanthomonas*, etc.) or with fecal contamination (e.g. *Enterobacter, Escherichia, Klebsiella, Proteus*, etc.). Representative species

Fig. 11.6. Bacterial colonization on a painting from the early fifth century BC in Tomba del Colle, Italy [53].

of Gammaproteobacteria have been detected in European, American, and Asian show caves [19, 23, 32–35, 47, 90, 94, 107], in Etruscan tombs (▶ Fig. 11.6) [46, 53] and mines [65]. Specific bacteria involved in iron and sulfur oxidation (e.g. *Acidithiobacillus ferrooxidans*), sulfur oxidation (e.g. *Beggiatoa, Thiovirga, Thiothrix, Thioploca, Thiocapsa, Allochromatium*), nitrification (e.g. *Nitrosococcus*), nitrogen fixation [89], and methanotrophy (e.g. *Methanobacterium, Methylococcus*) are commonly retrieved from clone libraries [30, 31, 33, 34, 90, 108, 109].

The class Deltaproteobacteria is composed of a variety of anaerobic sulfate-reducing bacteria usually present in caves and other subsurface environments. White biofilms coat surfaces in sulfidic streams in the Frasassi caves, being largely composed of Proteobacteria associated with sulfur cycling and dominated by *Beggiatoa*-like or *Thiothrix*-like cell morphologies [90]. Roughly half of the deltaproteobacterial clones relate to the sulfur-reducers *Desulfocapsa* and *Desulfonema* spp. The close association between sulfur-oxidizers and sulfur- and sulfate-reducers represents positive biological feedback to sulfuric acid speleogenesis and processes that create subsurface porosity in carbonate rocks. Other sulfate-reducing bacteria belonging to the genera *Desulfovibrio* and *Desulfomicrobium* can be found in Altamira Cave [110].

Epsilonproteobacteria are predominately described from active sulfidic caves and have not been identified in high abundances in any other type of subterranean habitat, except petroleum fields and groundwater [111]. This class represents a unique assemblage of microorganisms mainly known from phylogenetic studies of 16S rRNA genes as there have been no successful cultures from this group from caves, mines, or tunnels. Biofilms from an Italian cave near Acquasanta Terme, about 80 km from the

Frasassi caves, have Sulfurovumales-group members [88] and only share few species in common with the biofilms from the Frasassi system.

11.2.2.8 Other phyla

Chlorobi, Chloroflexi, Gemmatimonadetes, Planctomycetes, Verrucomicrobia, and other bacterial phyla are scarcely found in caves, and if they are, they only constitute minor fractions. Some notable exceptions include anoxygenic phototrophs (Chlorobium from microbial mats in anchialine caves in Bahamas [112], the anaerobic portion of the microbial mats from the Lower Kane Cave, with uncultured members of the Chloroflexi phylum [93], or the microbial community of Niu Cave, where Plantomycetes are the third major group of bacteria) [35]. Gemmatimonadetes and Verrucomicrobia are members of the bacterial communities inhabiting the cave walls of the Buda thermal karst system in Hungary, a nonshow cave [86].

11.2.3 Fungi

Fungi are by far the most fastidious cave microorganisms due to their high rate of spore production and air dispersion; this can lead to inherent difficulty to control an outbreak if one were to occur [113–115]. A review on fungi in caves and mines done by Vanderwolf et al. [116] reports on 1029 species from 518 genera. Several studies focus on fungal diversity in individual underground systems. For example, different niches from Domica Cave in Slovakia have 195 fungal taxa, with the main genera being *Penicillium, Aspergillus, Trichoderma, Cladosporium*, and *Chaetomium* [117]. A saline cave in Israel has abundant *Aspergillus, Penicillium*, and *Chaetomium* [118], and *Aspergillus* and *Penicillium* are abundant in an Indian cave [119]. Among 47 different fungal taxa isolated from air and guano of a Brazilian tropical cave, *Aspergillus, Penicillium, Cladosporium*, and *Fusarium* are abundant [120]. In all, the most frequently encountered genera, which are essentially cosmopolitan air-borne and soil-borne fungi, are *Aspergillus, Penicillium, Mucor, Fusarium, Trichoderma, Cladosporium, Alternaria, Paecilomyces, Acremonium, Chrysosporium*, etc. These genera are common in European and North American outdoor environments throughout the year [121–124] and influence spore content in cave and adit air.

Air circulation is important for the transport and dispersion of airborne microorganisms and nutrients from outside to inside a cave. In fact, most airborne fungi come from the external air that penetrates caves at entrances. For example, the genera *Penicillium, Aspergillus*, and *Mucor* are the most diverse taxa from ground sediments, rodent excrements (▸ Fig. 11.7), dripping waters, and fungal growth in Ardales Cave in Spain, a cave with Paleolithic paintings and that has been heavily impacted by human [125]. *Aspergillus, Penicillium*, and *Cladosporium* become the main fungal genera in the cave air after the entrance door is opened for tourists [84]. Similarly, the high

Fig. 11.7. *Beauveria felina* and *Phycomyces nitens* growing on rodent excrements in Ardales Cave, Spain [84, 125].

abundance of *Cladosporium* and *Epicoccum* spores in the air of Altamira Cave is also caused from outside air ventilation and wind currents. Identical 16S rRNA gene sequences for *Cladosporium* and *Epicoccum* from outside and inside the cave air are described [126]. Fungal spores have also been examined from the air of other caves, and from speleothems, sediments, and animal excrements (▸ Fig. 11.7) [9, 117, 127–129].

Despite recent efforts to understand fungal diversity and distribution in caves, however, more research needs to be done. Another area of research relates to the isolation of a few novel cave fungi, spurred by the new actinobacterial species that produce novel bioactive compounds [130]. New fungal isolates include *Chrysosporium speluncarum* [131], *Ochroconis lascauxensis* and *Ochroconis anomala* from Lascaux Cave [114], *Aspergillus thesauricus* and *Aspergillus baeticus* from Spanish caves [132], and *Pidoplitchkoviella terricola* from Slovakia [133]. In the future, the search for novel fungal taxa should focus on specific niches, including oligotrophic conditions and mineral formations, instead of aerobiological settings or colonies where organic matter is present (e.g. woods, guano, etc.) that will provide large lists of already well-known fungi.

Another source of fungi in caves is from arthropods and other animals [113, 134]. Since the winter of 2006–2007, *Pseudogymnoascus destructans* (formerly known as *Geomyces destructans*), a psycrophilic fungus commonly found in caves, has received considerable attention because it causes white-nose syndrome in bats from Europe and the USA [135]. The first occurrence of white-nose syndrome in bats in the USA was in a show cave, Howe Caverns in New York (USA).

11.3 Consequences of Microbial Growth and Biogeochemical Cycling

Archaea, Bacteria, and Fungi contribute to the biogeochemical cycling of elements, although knowledge and extension of the processes are still poorly understood. Microbes are essential for food web development in subsurface environments, particularly in the absence of light where photosynthesis is not possible. Microbes also play important roles in speleothem formation and rock deterioration.

In dark caves and mines, it has been important to investigate how microbes acquire energy and obtain organic carbon. The presence of chemolithoautotrophic bacteria from culture-dependent and -independent analyses provides a way to evaluate whether microbes are capable of producing their own energy and organic carbon to sustain an ecosystem, or if the ecosystem needs organic matter provided from an outside source. Sulfur- and sulfide-oxidizers, iron- and manganese-oxidizers, sulfate-reducers, nitrifiers, etc., appear abundant in some caves, although they are not evenly distributed. This means that their potential to provide sufficient energy to the rest of the ecosystem may not be large enough.

Most of the research on biogeochemical cycling in caves, particularly related to lithotrophy, has been done in caves that do not have tourism or artificial lighting. For example, in the Movile Cave [30, 136], primary productivity is driven by sulfur-oxidizing (e.g. *Thiobacillus*, *Thiothrix*, *Thioploca*, *Beggiatoa*), ammonia-, and nitrite-oxidizing bacteria (e.g. *Nitrosomonas*, *Nitrospira*). Similar sulfur-based chemolithoautotrophic primary productivity has been documented from Cueva de las Sardinas in Mexico [137], a cave used by local residents for religious ceremonies, and the Frasassi cave snottites dominated (>70% of the cells) by *Acidithiobacillus thiooxidans* [11]. Nitrite oxidation drives primary productivity in Weebubbie Cave, Nullarbor, Australia [87]. Thaumarchaeota may also play a significant role in carbon and nitrogen biogeochemical cycling [138, 139]. Evidence of a metabolically active Crenarchaeota community possibly involved in ammonia oxidation was reported by Gonzalez et al. [140] from the Altamira Cave.

Although not likely a chemolithoautotrophic process, because of the association with iron pigments (hematite) in rock art paintings, Acidobacteria in Spanish caves [32, 33] may be capable of Fe(III) reduction. One of the Acidobacteria cultivated so far, *Geothrix fermentans*, is a Fe(III)-reducing bacterium [141] and genomes from Acidobacteria strains contained genes enabling to take up iron from the environment [142]. The acidobacterial genomes (representatives of subgroups 1 and 3) also showed the ability to degrade a variety of sugars and complex polysaccharides, amino acids, alcohols, and to reduce nitrate and nitrite, to take iron from the environment, and to tolerate acidic conditions, all of which make them well adapted to diverse conditions in caves [142].

In many other caves, organic matter comes from plants from outside the cave [9, 143, 144]. Guano also contains considerable amounts of ammonia, which can support ammonia-oxidizing archaea [145]. Each cave has particular characteristics (e.g. host rock, environmental conditions, carbon input, etc.) that differentiate one cave system from other systems, such that even two different geologic formations within the same cave can harbor significantly different microbial communities [146]. This manifests itself not only in distinct ecosystem properties, such as food web structures, but also in the types of consequences of microbial growth in caves, particularly show caves.

Cyanobacteria are especially successful microorganisms in subsurface environments, like cave and mine adit entrances, due to the high humidity available and the very low light intensity. As primary producers, they can sustain ecosystems in caves, particularly near artificial lights. This primary productivity would not exist in deep caves without the addition of lights. Moreover, because of the geochemical conditions of a cave, and active photosynthesis, filamentous Cyanobacteria can calcify. *S. julianum* and *G. calcarea*, among others, are common in caves, tombs, and catacombs [147–150]. The distinct morphology of the calcite crystals in *S. julianum* and *G. calcarea* sheaths indicates that the calcification process is not merely an abiotic process but is biomediated [151]. Boring of limestone by Cyanobacteria has been observed in caves (Mariona Hernandez-Marine, personal communication). A complete review of Cyanobacteria in caves was recently published, with discussion of different aspects of environment, light, calcification, interaction with chemoorganotrophs, etc. [150].

Actinobacteria and Proteobacteria also play significant roles in biogeochemical cycles, including carbon, nitrogen, sulfur, iron, and manganese, as well as affect biomineralization processes. Specific to Actinobacteria, these microbes have also been linked to biodeterioration processes in show caves and tombs [43, 45, 46, 52, 53, 152]. Destructives processes include dissolution, and boring or pitting [153]. In contrast, biomineralization processes include the precipitation of calcite and other minerals (struvite, witherite, barite, etc.), whereby actinobacterial bioinduction of crystals has been reported from caves and catacombs (▸ Fig. 11.8) [45, 46, 52, 154–157]

Biodeterioration is also caused by fungal colonization, particularly in show caves, tombs, and catacombs. The classical example is from the Lascaux Cave in France, with several fungal outbreaks in the last 12 years, a topic widely discussed in the scientific literature [10, 113, 114, 148, 158–162] and also in this volume (Chapter 13). An interesting study of biodeterioration was carried out in Japan, in the Takamatsuzuka and Kitora earthen tombs, referred to as tumuli [163]. The tombs have magnificent mural paintings of the 7th century, and 262 strains from 79 moldy spots on the plaster walls were evaluated. Around 100 species have been identified, with the primary colonizers being *Fusarium*, *Trichoderma*, and *Penicillium*. From these species, 24 genetically diverse fusaria nest in clade 3 of the *Fusarium solani* species complex, which also includes the fusaria isolated from the Lascaux Cave [164]. *Penicillium* are also major dwellers in the tumuli. In fact, from 662 *Penicillium* isolates from 373 samples, 181 phenotypically cluster with *Penicillium paneum*, a species previously isolated from

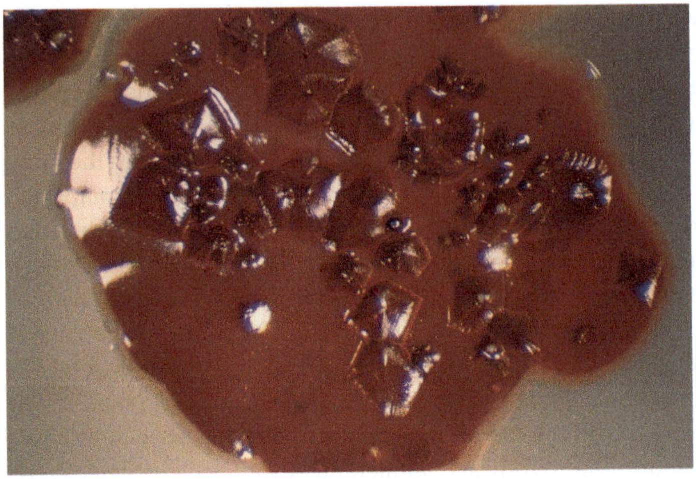

Fig. 11.8. *Rubrobacter* sp. bioinducing struvite crystals in a culture medium [52].

moldy rye breads, other foods, and baled grass silage but not from mural paintings on plaster walls [165]. In addition, novel species of *Candida*, *C. tumulicola*, and *C. takamatsuzukensis*, have been isolated from the tumuli [166].

RNA-based analysis makes it possible to identify species that are metabolically active in a subterranean environment. However, this approach has been sparsely implemented in the study of metabolically active cave fungi. A recent application of RNA-based methodology provides relevant information about the fungi colonizing Lascaux Cave and their evolution over time. In fact, a succession from *Ochroconis* spp. to *Acremonium* spp. and black yeasts has been observed over the past few years during which time black stains are affecting the walls and rock art of the cave [161]. Incidentally, although not explored as part of this chapter, yeasts are also abundant in caves [167, 168], as well as black yeasts [161]. Another succession that has been examined using RNA-based approaches is in the Japanese tumuli, where Sugiyama et al. [169] report that the predominant fungi initially are *Fusarium, Trichoderma*, and *Penicillium* spp. Then, in response to environmental changes over time, *Fusarium* and *Trichoderma* decrease and are replaced by *Acremonium*.

Lastly, fungi have also been implicated in the formation of calcitic nanofibers in caves, although this is a controversial topic. Went [170] associate *Verticillium lamellicola* hyphae with the active tip of stalactites in Lehman Caves, Nevada (USA), and suggests that hyphae act as the nucleation site for the crystallization of calcite. Calcitic nanofibers could also originate from the breakdown of fungal hyphae [171]. Other authors report an absence of fungi in certain cave speleothems, including moonmilk [5, 47, 157], but Gherman et al. [14] suggest that hair-like mineralization of ferric hydroxide on stalactite tips in a pyrite mine are deposited on fungal cell walls.

11.4 Cave Management

Humans are the main threat to the conservation of a cave. Cave managers are the people ultimately responsible for the conservation of caves, and poor management spawns well-known problems. Tourism pressure, often fostered by the local authorities for regional development, results in actions counterproductive to cave conservation. Increasing visitation leads to important environmental impacts, first caused by the adaptation of caves to visitation, such as installation of artificial lighting, and secondly by the visitors themselves. Severe impacts to caves are caused by the construction of concrete walls, boring to access galleries, and the formation of artificial halls at the time of discovery or later. Human visitation can introduce organic carbon and exotic microorganisms, which may result in a drastic shift from indigenous microbial communities to exogenous communities, in addition to the direct damage inflicted to the cave. Visitors of any kind (i.e. humans and wildlife) leave behind skin cells, hair, bacteria and fungi from hair and skin, earth from the neighboring agricultural soils from their feet, and occasionally vomit, feces, and urine. Modern examples of these are considered vandalism, but ancient examples provide us an opportunity to glimpse the diet of Paleolithic peoples. In a few cases, caves have had to be closed due to remarkable deterioration that occurred as a consequence of large number of visitors. Tourists were re-routed to a replica cave. Examples of these situations are provided in this volume, Chapters 13 and 14.

One of the major impacts on oligotrophic caves is that organic carbon brought into the cave from outside enriches the system in carbon. Fungi, which rarely are found living under oligotrophic conditions in caves, can then colonize and grow on any organically enriched material introduced in a cave. This was exemplified in the fungal outbreaks observed in the Lascaux Cave [10] and recently confirmed in Castañar de Ibor, an oligotrophic Spanish cave were the ecosystem was suddenly disrupted by a huge amount of detritus (i.e. vomit from a visitor) [115]. From no previous fungal growth, a sequence of growth and sporulation followed, initiated by *Mucor circinelloides* and *F. solani*, both species capable of occupying the habitat rapidly with explosive reproduction rates. Other secondary colonizers, *Mucor racemosus*, *Fusarium oxysporum*, *Chaetomium globosum*, *Mortierella alpina*, *Hypocrea lixii*, *Aspergillus ustus*, *Paecilomyces lilacinus*, *Cosmospora consors*, and a *Penicillium* sp. were isolated and identified from small white mycelia that appeared on sediments two months after the disturbance. Environmentally friendly methods were used to successfully control the outbreak [115].

Aerobiological sampling is now being introduced as a management tool to control microbial growth of fungi and cyanobacteria. Airborne fungi are common in caves and an index for estimating fungal contamination in show caves was recently proposed by Porca et al. [128]. But, in most show caves, artificial lighting represents a serious ecological disruption and also causes an aesthetic problem because of the promotion of cyanobacterial and algal growth (▸ Fig. 11.5), which increases populations of asso-

Fig. 11.9. *Scytonema julianum* on the ground sediments from the Tito Bustillo Cave, Spain [148].

ciated heterotrophic bacteria [172]. The combination of different analytical methods could allow managers to detect the extent of human-induced changes that, in certain cave areas, can be very hazardous [84, 126]. A number of different situations in which Cyanobacteria colonized caves and catacombs are described by Albertano [150].

In caves with rock art paintings, the Lascaux Cave is an extreme case because of being invaded by the alga species *Bracteacoccus minor* only 20 years after its discovery, as a result of the lighting used to show the paintings to visitors [173]. Tito Bustillo Cave in Spain, in the early 1990s, also suffered from green growth of Cyanobacteria and algae on both stalactites and stalagmites due to artificial lighting and *S. julianum*, a calcifying cyanobacterium colonies formed on ground sediments (▸ Fig. 11.9) [147].

Controlling the growth of phototrophic microbes in illuminated caves, especially show caves, has been well examined in the literature. Mulec and Kosi [174] review the pros and cons of physical, chemical, and biological methods to control phototrophic growth in caves. But, currently there is no ideal solution, although some cleaning procedures have been used successfully in Slovenian show caves [175]. A summary of applications is included in this volume, Chapter 12. Some suggestions may also be to rely on simple flashlights for visitors rather than on a permanent lighting system, as well as a change in cave management. The use of monochromatic lamps has been proposed as a strategy to control or inhibit phototrophic growth [176–178]. Green and blue monochromatic lights were used for *in situ* 3-month long exposure experimentsinside the Catacombs of Saint Callixtus and Domitilla [176]. There was a reduction in cyanobacterial growth. Roldán et al. [177] found that cultures of the cyanobacterium *Gloeothece membranacea* and the chlorophyta *Chlorella sorokiniana* are inhibited when subjected to green light. In 2005, yellow LEDs (λ 595 nm) were installed in the Frozen Niagara section of Mammoth Cave, Kentucky (USA), to control or eliminate phototrophic growth; 2 years later, no growth was observed on the LED-lighted areas [179]. Unfortunately, some cave restoration efforts in the last decade applied conventional biocides that are likely not suitable for the unknown and complex microbial communities growing on and beneath rock surfaces. Boston et al. [180] discuss the effect of different cleaning chemicals on microbial communities on mineral formations, but no biocides are mentioned. Restoration efforts can have an opposite effect, particularly if the restorers use substrates that support the growth of microorganisms and, consequently, accelerate the deterioration process [161]. A few treatments to clean stalactite surfaces in the Salpetre Caves at Collbató, Spain, were used over 1 year under different lighting conditions, including 70% alcohol/distilled water and then treated with benzalkonium chloride (1‰) or Preventol R50 (2%) [181]. Phototrophic microorganisms recolonized the stalactite illuminated with white light, but there was negligible growth on the stalactite exposed to green light. Replacement of conventional lighting by monochromatic lamps is a useful strategy to prevent photosynthetic growth in artificially illuminated caves.

Jurado et al. [5] suggest that the eradication of microorganisms in caves seems to be a dangerous process. Cleaning provokes unpredicted changes in cave's biodiversity, and a subsequent re-colonization would produce undesired bacterial and fungal outbreaks. In some caves, pristine bacterial colonization in cave rock surfaces may provide a natural advantage as biological control agents, and the presence of antifungal compounds produced by the bacteria would protect the surfaces from fungal colonization. Also, the use of biocides in caves should be carefully studied and controlled, depending on each case. Biocides to clean the fungal black stains in the Lascaux Cave resulted in a more diverse fungal community, composed of the frequent airborne genera *Aspergillus, Cladosporium, Trichoderma, Alternaria*, and *O. lascauxensis*, that was almost exclusively the only fungus present before treatment. These airborne fungi were secondary colonizers able to use dead organic matter from the microbes killed by the biocide. Consequently, some fundamental ecological aspects regarding cave ecosystems (e.g. microbial ecology, biogeochemistry, microbial outbreaks, etc.) still remain unanswered due to the wide variety of host rocks and microniches available for microorganisms. It appears that a general consensus has emerged in the last decade regarding the need to protect show caves by introducing appropriate management. This step forward in conservation should allow us to predict a better future for the show caves.

Acknowledgments

This work was funded by the "Research Programme in Technologies for the Assessment and Conservation of Cultural Heritage" (Consolider CSD2007-00058) and CSIC project 201230E125. Apologies to colleagues whose work has not been cited due to space limitations.

References

[1] Lee NM, Meisinger DB, Aubrecht R, et al. (2012) Caves and karst environments. In: Bell EM, ed. Life at Extremes: Environments, Organisms and Strategies for Survival. Wallingford, CAB International, 2012, 320–344.

[2] Barton HA, Northup DE. Geomicrobiology in cave environments: past, current and future perspectives. J Cave Karst Stud, 2007, 69, 163–178.

[3] Engel AS, Paoletti MG, Beggio M, et al. Comparative microbial community composition from secondary carbonate (moonmilk) deposits: implications for the *Cansiliella servadeii* cave hygropetric food web. Int J Speleol, 2013, 42, 181–192.

[4] Chelius MK, Beresford G, Horton H, et al. Impacts of alterations of organic inputs on the bacterial community within the sediments of Wind Cave, South Dakota, USA. Int J Speleol, 2009, 38, 1–10.

[5] Jurado V, Fernandez-Cortes A, Cuezva, S, et al. The fungal colonisation of rock art caves. Naturwissenschaften, 2009, 96, 1027–1034.

[6] Staudigel H, Anitori R, Davis R, Cornell L, Tebo BM. Dark oligotrophic volcanic ecosystems (DOVEs) in fumarolic ice caves of Mt. Erebus volcano. Am Geophys Union, Fall Meeting, 2011, B44B-08.

[7] Quintana ET, Badillo RF, Maldonado LA. Characterisation of the first actinobacterial group isolated from a Mexican extremophile environment. Anton Leew, 2013, 104, 63–70.

[8] Northup DE, Melim LA, Spilde MN, et al. Lava cave microbial communities within mats and secondary mineral deposits: implications for life detection on other planets. Astrobiology, 2011, 11, 601–618.

[9] Saiz-Jimenez C, Cuezva S, Jurado V, et al. Paleolithic art in peril: Policy and science collide at Altamira Cave. Science, 2011, 334, 42–43.

[10] Alabouvette C, Saiz-Jimenez C. Écologie Microbienne de la Grotte de Lascaux. Sevilla, Instituto de Recursos Naturales y Agrobiología, CSIC, 2011.

[11] Jones DS, Albrecht HL, Dawson KS, et al. Community genomic analysis of an extremely acidophilic sulfur-oxidizing biofilm. ISME J, 2012, 6, 158–170.

[12] Cuezva S, Sanchez-Moral S, Saiz-Jimenez C, Cañaveras JC. Microbial communities and associated mineral fabrics in Altamira Cave, Spain. Int J Speleol, 2009, 38, 83–92.

[13] Kostanjšek R, Pašić L, Daims H, Sket B. Structure and community composition of sprout-like bacterial aggregates in a Dinaric Karst subterranean stream. Microb Ecol, 2013, 66, 5–18.

[14] Gherman VD, Boboescu IZ, Pap B, Kondorosi E, Gherman G, Maróti G. An acidophilic bacterial-archaeal-fungal ecosystem linked to formation of ferruginous crusts and stalactites. Geomicrobiol J, 2013, 31, 407–418.

[15] Engel AS. Observations on the biodiversity of sulfidic karst habitats. J Cave Karst Stud, 2007, 69, 187–206.

[16] Jarrel KF, Walters AD, Bochiwal C, Borgia JM, Dickinson T, Chong JPJ. Major players on the microbial stage: why archaea are important. Microbiology, 2011, 157, 919–936.

[17] Offre P, Spang A, Schleper C. Archaea in biogeochemical cycles. Annu Rev Microbiol, 2013, 67, 437–457.

[18] Northup DE, Barns SM, Yu LE, et al. Diverse microbial communities inhabiting ferromanganese deposits in Lechuguilla and Spider Caves. Environ Microbiol, 2003, 5, 1071–1086.

[19] Chelius MK, Moore JC. Molecular phylogenetic analysis of *Archaea* and *Bacteria* in Wind Cave, South Dakota. Geomicrobiol J, 2004, 21, 123–134.

[20] Benson CA, Bizzoco RW, Lipson DA, Kelley ST. Microbial diversity in nonsulfur, sulfur and iron geothermal steam vents. FEMS Microbiol Ecol, 2011, 76, 74–88.

[21] Miller AZ, Hernández-Mariné M, Jurado V, et al. Enigmatic reticulated filaments in subsurface granite. Environ Microbiol Rep, 2012, 4, 596–603.

[22] Legatzki A, Ortiz M, Neilson J, et al. Bacterial and archaeal community structure of two adjacent calcite speleothems in Kartchner Caverns, Arizona, USA. Geomicrobiol J, 2011, 28, 99–117.

[23] Ortiz M, Neilson JW, Nelson WM, et al. Profiling bacterial diversity and taxonomic composition on speleothem surfaces in Kartchner Caverns, AZ. Microb Ecol, 2013, 65, 371–383.

[24] Metcalf WW, Griffin BM, Cicchillo RM, et al. Synthesis of methylphosphonic acid by marine microbes: a source for methane in the aerobic ocean. Science, 2012, 337, 1104–1107.

[25] Piñar G, Saiz-Jimenez C, Schabereiter-Gurtner C, Blanco-Varela MT, Lubitz W, Rölleke S. Archaeal communities in two disparate deteriorated ancient wall paintings: detection, identification and temporal monitoring by DGGE. FEMS Microbiol Ecol, 2001, 37, 45–54.

[26] Reitschuler C, Lins P, Wagner AO, Illmer P. Cultivation of moonmilk-born non-extremophilic Thaum and Euryarchaeota in mixed culture. Anaerobe, 2013, 29, 73–79.

[27] Macalady JL, Lyon EH, Koffman B, et al. Dominant microbial populations in limestone-corroding stream biofilms, Frasassi Cave system, Italy. Appl Environ Microbiol, 2006, 72, 5596–5609.

[28] Falteisek L, Cepicka I. Microbiology of diverse acidic and non-acidic microhabitats within a sulfide ore mine. Extremophiles, 2012, 16, 911–922.

[29] Nazaries L, Murrell JC, Millard P, Baggs L, Singh BK. Methane, microbes and models: fundamental understanding of the soil methane cycle for future predictions. Environ Microbiol, 2013, 15, 2395–2417.

[30] Chen Y, Wu L, Boden R, et al. Life without light: microbial diversity and evidence of sulfur- and ammonium-based chemolithotrophy in Movile Cave. ISME J, 2009, 3, 1093–1104.

[31] Schirmack J, Mangelsdorf K, Ganzert L, Sand W, Hillebrand-Voiculescu A, Wagner D. *Methanobacterium movilense* sp. nov., a hydrogenotrophic, secondary alcohol utilizing methanogen from the anoxic sediment of the subsurface lake in Movile Cave, Romania. Int J Syst Evol Microbiol, 2013, doi: 10.1099/ijs.0.057224-0.

[32] Schabereiter-Gurtner C, Saiz-Jimenez C, Piñar G, Lubitz W, Rölleke S. Altamira cave Paleolithic paintings harbor partly unknown bacterial communities. FEMS Microbiol Lett, 2002, 211, 7–11.

[33] Schabereiter-Gurtner C, Saiz-Jimenez C, Piñar G, Lubitz W, Rölleke S. Phylogenetic 16S rRNA analysis reveals the presence of complex and partly unknown bacterial communities in Tito Bustillo cave, Spain, and on its Palaeolithic paintings. Environ Microbiol, 2002, 4, 392–400.

[34] Schabereiter-Gurtner C, Saiz-Jimenez C, Piñar G, Lubitz W, Rölleke S. Phylogenetic diversity of bacteria associated with Paleolithic paintings and surrounding rock walls in two Spanish caves (Llonin and La Garma). FEMS Microbiol Ecol, 2004, 47, 235–247.

[35] Zhou JP, Gu YQ, Zou CS, Mo MH. Phylogenetic diversity of bacteria in an earth-cave in Guizhou province, Southwest of China. J Microbiol, 2007, 45, 105–112.

[36] Janssen PH. Identifying the dominant soil bacterial taxa in libraries of 16S rRNA and 16S rRNA genes. Appl Environ Microbiol, 2006, 72, 1719–1728.

[37] Barns SM, Cain EC, Sommerville, L, Kuske CR. *Acidobacteria* phylum sequences in uranium-contaminated subsurface sediments greatly expand the known diversity within the phylum. Appl Environ Microbiol, 2007, 73, 3113–3116.

[38] Zimmermann J, Gonzalez JM, Saiz-Jimenez C, Ludwig W. Detection and phylogenetic relationships of highly diverse uncultured acidobacterial community on paleolithic paintings in Altamira Cave using 23S rRNA sequence analyses. Geomicrobiol J, 2005, 22, 379–388.

[39] Meisinger DB, Zimmermann J, Ludwig W, et al. *In situ* detection of novel *Acidobacteria* in microbial mats from a chemolithoautotrophically based cave ecosystem (Lower Kane Cave, WY, USA). Environ Microbiol, 2007, 9, 523–534.

[40] Zimmermann J, Gonzalez JM, Saiz-Jimenez C. Epilithic biofilms in Saint Callixtus Catacombs (Rome) harbour a broad spectrum of Acidobacteria. Anton Leeuw, 2006, 89, 203–208.

[41] Groth I, Vetermann R, Schuetze B, Schumann P, Saiz-Jimenez C. Actinomycetes in karstic caves of Northern Spain (Altamira and Tito Bustillo). J Microbiol Methods, 1999, 36, 115–122.

[42] Groth I, Schumann P, Laiz L, Sanchez-Moral S, Cañaveras JC, Saiz-Jimenez C. Geomicrobiological study of the Grotta dei Cervi, Porto Badisco, Italy. Geomicrobiol J, 2001, 18, 241–258.

[43] Stomeo F, Portillo MC, Gonzalez JM, Laiz L, Saiz-Jimenez C. *Pseudonocardia* in white colonisations in two caves with Paleolithic paintings. Int Biodeterior Biodegrad, 2008, 62, 483–486.

[44] Jurado V, Laiz L, Rodriguez-Nava V, et al. Pathogenic and opportunistic microorganisms in caves. Int J Speleol, 2010, 39, 15–24.

[45] Cuezva S, Fernandez-Cortes A, Porca A, et al. The biogeochemical role of *Actinobacteria* in Altamira Cave, Spain. FEMS Microbiol Ecol, 2012, 81, 281–290.

[46] Diaz-Herraiz M, Jurado V, Cuezva S, et al. The actinobacterial colonisation of Etruscan paintings. Sci Rep, 2013, 3, 1440, DOI: 10.1038/srep01440.

[47] Porca E, Jurado V, Žgur-Bertok D, Saiz-Jimenez C, PašićL. Comparative analysis of yellow microbial communities growing on the walls of geographically distinct caves indicates a common core of microorganisms involved in their formation. FEMS Microbiol Ecol, 2012, 81, 255–266.

[48] Portillo MC, Saiz-Jimenez C, Gonzalez JM. Molecular characterization of total and metabolically active bacterial communities of "white colonisations" in the Altamira Cave, Spain. Res Microbiol, 2009, 160, 41–47.

[49] Imperi F, Caneva G, Cancellieri L, Ricci MA, Sodo A, Visca P. The bacterial aetiology of rosy discoloration of ancient wall paintings. Environ Microbiol, 2007, 9, 2894–2902.

[50] Schabereiter-Gurtner C, Piñar G, Vybiral, D, Lubitz W, Rölleke S. *Rubrobacter*-related bacteria associated with rosy discolouration of masonry and lime wall paintings. Arch Microbiol, 2001, 176, 347–354.

[51] Ripka K, Denner EBM, Michaelsen A, Lubitz W, Piñar G. Molecular characterisation of *Halobacillus* strains isolated from different medieval wall paintings and building materials in Austria. Int Biodeterior Biodegrad, 2006, 58, 124–132.

[52] Laiz L, Miller AZ, Jurado V, et al. Isolation of *Rubrobacter* strains from biodeteriorated monuments. Naturwissenschaften, 2009, 96, 71–79.

[53] Diaz-Herraiz M, Jurado V, Cuezva S, et al. Deterioration of an Etruscan tomb by bacteria from the order *Rhizobiales*. Sci Rep, 2014, 4, 3610, DOI: 10.1038/srep03610.

[54] Groth I, Saiz-Jimenez C. Actinomycetes in hypogean environments. Geomicrobiol J, 1999, 16, 1–8.

[55] Spear JR, Barton HA, Robertson CE, Francis CA, Pace NR. Microbial community biofabrics in a geothermal mine adit. Appl Environ Microbiol, 2007, 73, 6172–6180.

[56] Cheng J, Chen W, Huo-Zhang B, et al. *Microlunatus cavernae* sp. nov., a novel actinobacterium isolated from Alu ancient cave, Yunnan, south-west China. Anton Leeuw, 2013, 104, 95–101.

[57] Cheng J, Zhang Y-G, Chen W, et al. *Saccharopolyspora cavernae* sp. nov., a novel actinomycete isolated from the Swallow Cave in Yunnan, south-west China. Anton Leeuw, 2013, 104, 837–843.

[58] Herold K, Gollmick FA, Groth I, et al. Cervimycin A-D: a polyketide glycoside complex from a cave bacterium can defeat vancomycin resistance. Chem, 2005, 11, 5523–5530.

[59] Subramani R, Aalbersberg W. Marine actinomycetes: an ongoing source of novel bioactive metabolites. Microbiol Res, 2012, 167, 571–580.

[60] Tiwari K, Gupta RK. Rare actinomycetes: a potential storehouse for novel antibiotics. Crit Rev Biotechnol, 2012, 32, 108–132.

[61] Niyomvong N, Pathom-aree W, Thamchaipenet A, Duangmal K. Actinomycetes from tropical limestone caves. Chiang Mai J Sci, 2012, 39, 373–388.

[62] Roslev P, Bukh AS. State of the art molecular markers for fecal pollution source tracking in water. Appl Microbiol Biotechnol, 2011, 89, 1341–1355.

[63] Carmichael MJ, Carmichael SK, Santelli CM, Strom A, Bräuer SL. Mn(II)-oxidizing bacteria are abundant and environmentally relevant members of ferromanganese deposits in caves of the upper Tennessee River Basin. Geomicrobiol J, 2013, 30, 779–800.

[64] Carmichael SK, Carmichael MJ, Roble LA, Johnson K, Gao Y, Bräuer S. Evidence of sustained anthropogenic impact in Carter Salt Peter Cave, Carter County, Tennessee. J Cave Karst Stud, 2013, 73, 184–209.

[65] Rastogi G, Osman S, Vaishampayan PA, Andersen GL, Stetler LD, Sani RK. Microbial diversity in uranium mining-impacted soils as revealed by high-density 16S microarray and clone library. Microb Ecol, 2010, 59, 94–108.

[66] Tomova I, Lazarkevich I, Tomova A, Kambourova M, Vasileva-Tonkova E. Diversity and biosyntehtic potential of culturable aerobic heterotrophic bacteria isolated from Magura Cave, Bulgaria. Int J Speleol, 2013, 42, 65–76.

[67] Tomova I, Tomova A, Vasileva-Tonkova E, et al. *Myroides guanonis* sp. nov., isolated from prehistoric paintings. Int J Syst Evol Microbiol, 2013, 63, 4266–4270.

[68] Hufkens K, Scheunders P, Ceulemans R. Ecotones in vegetation ecology: methodologies and definitions revisited. Ecol Res, 2009, 24, 977–987.

[69] Martínez A, Asencio AD. Distribution of cyanobacteria at the Gelada Cave (Spain) by physical parameters. J Cave Karst Stud, 2010, 72, 11–20.

[70] Hernández-Mariné M, Canals T. Cianofíceas filamentosas cavernícolas. Stud Bot, 1994, 13, 227–229.

[71] Lamprinou V, Danielidis DB, Economou-Amilli A, Pantazidou, A. Distribution survey of cyanobacteria in three Greek caves of Peloponnese. Int J Speleol, 2012, 41, 267–272.

[72] Hernández-Mariné M, Canals T. *Herpyzonema pulverulentum* (Mastigocladaceae) a new cavernicolous atmophytic and lime-incrusted cyanophyte. Algol Stud, 1994, 75, 123–136.

[73] Hernández-Mariné M, Asencio AD, Canals A, Ariño X, Aboal M, Hoffmann L. Discovery of populations of the lime-incrusting genus *Loriella* (Stigonematales) in Spanish caves. Algol Stud, 1999, 94, 121–138.

[74] Beltran JA, Asencio AD. Cyanophytes from the L'Aigua cave (Alicante, SE Spain) and their environmental conditions. Algol Stud, 2009, 132, 21–34.

[75] Lamprinou V, Hernández-Mariné, M, Canals, T, Kormas, K, Economou-Amilli A, Pantazidou A. Two new stigonematalean cyanobacteria: *Iphinoe spelaeobios* gen. nov., sp. nov. and *Loriellopsis cavernicola* gen. nov., sp. nov. from Greek and Spanish caves. Morphology and molecular evaluation. Int J Syst Evol Microbiol, 2011, 61, 2907–2915.

[76] Lamprinou V, Skaraki K, Kotoulas G, Economou-Amilli A, Pantazidou A. *Toxopsis calypsus* gen. nov., sp. nov. (Cyanobacteria, Nostocales) from cave 'Francthi', Peloponnese, Greece – morphological and molecular evaluation. Int J Syst Evol Microbiol, 2012, 62, 2870–2877.

[77] Lamprinou V, Skaraki K, Kotoulas G, Anagnostidis, K, Economou-Amilli A, Pantazidou A. A new species of *Phormidium* (Cyanobacteria, Oscillatoriales) from three Greek Caves: morphological and molecular analysis. Fundam Appl Limnol, 2013, 182, 109–116.

[78] Roldán M, Ramírez M, del Campo J, Hernández-Mariné M, Komárek J. *Chalicogloea cavernicola* gen. nov. sp. nov. (Chroococcales, Cyanobacteria) from low light aerophytic environ-

ments: Combined molecular, phenotypic and ecological criteria. Int J Syst Evol Microbiol, 2013, 63, 2326–2333.
[79] Pantazidou A, Roussomoustakaki M. Biodiversity and ecology of cyanobacteria in a variety of hypogean ecosystems (Greece). In: 14th International Congress of Speleology. Athens, Hellenic Speleological Society, 2005, 624–628.
[80] Bellezza S, Albertano P, De Philippis R, Paradossi G. Exopolysaccharides in cyanobacterial biofilms from Roman catacombs. Algol Stud, 2005, 117, 117–132.
[81] Bruno L, Billi D, Belleza S, Albertano P. Cytomorphological and genetic characterization of troglobitic *Leptolyngbya* strains from Roman hypogea. Appl Environ Microbiol, 2009, 75, 608–617.
[82] Adetutu EM, Thorpe K, Shahsavari I, et al. Bacterial community survey of sediments at Naracoorte Caves, Australia. Int J Speleol, 2012, 41, 137–147.
[83] De Leo F, Iero A, Zammit G, Urzì C. Chemoorganotrophic bacteria isolated from biodeteriorated surfaces in cave and catacombs. Int J Speleol, 2012, 41, 125–136.
[84] Fernandez-Cortes A, Cuezva S, Sanchez-Moral S, Porca E, Jurado V, Saiz-Jimenez C. Detection of human-induced environmental disturbances in a show cave. Environ Sci Pollut Res, 2011, 18, 1037–1045.
[85] Ikner LA, Toomey RS, Nolan G, Neilson JW, Pryor BM, Maier RM. Culturable microbial diversity and the impact of tourism in Kartchner Caverns, Arizona. Microb Ecol, 2007, 53, 30–42.
[86] Borsodi AK, Knáb M, Krett G, et al. Biofilm bacterial communities inhabiting the cave walls of the Buda Thermal Karst System, Hungary. Geomicrobiol J, 2012, 29, 611–627.
[87] Holmes AJ, Tujula NA, Holley M, et al. Phylogenetic structure of unusual aquatic microbial formations in Nullarbor caves, Australia. Environ Microbiol, 2001, 3, 256–264.
[88] Jones DS, Tobler DJ, Schaperdoth I, Mainiero M, Macalady JL. Community structure of subsurface biofilms in the thermal sulfidic caves of Acquasanta Terme, Italy. Appl Environ Microbiol, 2010, 76, 5902–5910.
[89] Desai MS, Assig K, Dattagupta S. Nitrogen fixation in distinct microbial niches within a chemoautotrophy-driven cave ecosystem. ISME J, 2013, 7, 2411–2423.
[90] Macalady JL, Jones DS, Lyon EH. Extremely acidic pendulous cave wall biofilms from the Frasassi cave system, Italy. Environ Microbiol, 2007, 9, 1402–1414.
[91] Ehrich S, Behrens D, Lebedeva E, Ludwig W, Bock E. A new obligately chemolithoautotrophic, nitrite-oxidizing bacterium, *Nitrospira moscovensis* sp. nov. and its phylogenetic relationship. Arch Microbiol, 1995, 164, 16–23.
[92] Jones DS, Lyon EH, Macalady JL. Geomicrobiology of biovermiculations from the Frasassi cave system. J Cave Karst Stud, 2008, 70, 78–93.
[93] Engel AS, Meisinger DB, Porter ML, et al. Linking phylogenetic and functional diversity to nutrient spiraling in microbial mats from Lower Kane Cave (USA). ISME J, 2010, 4, 98–110.
[94] Bastian F, Alabouvette C, Saiz-Jimenez C. Bacteria and free-living amoeba in Lascaux Cave. Res Microbiol, 2009, 160, 38–40.
[95] Bastian F, Alabouvette C, Saiz-Jimenez C. Impact of biocide treatments on the bacterial communities of the Lascaux Cave. Naturwissenschaften, 2009, 96, 863–868.
[96] Rivas R, Sánchez-Márquez S, Mateos PF, Martínez-Molina E, Velázquez E. *Martelella mediterranea* gen. nov., sp. nov., a novel alpha-proteobacterium isolated from a subterranean saline lake. Int J Syst Evol Microbiol, 2005, 55, 955–959.
[97] Jurado V, Gonzalez JM, Laiz L, Saiz-Jimenez C. *Aurantimonas altamirensis* sp. nov., a member of the order Rhizobiales isolated from Altamira Cave. Int J Syst Evol Microbiol, 2006, 56, 2583–2585.
[98] Alias-Villegas C, Jurado V, Laiz L, Saiz-Jimenez C. *Sphingopyxis italica* sp. nov., isolated from Roman catacombs. Int J Syst Evol Microbiol, 2013, 63, 2565–2569.

[99] Tazato N, Nishijima, M, Handa Y, Kigawa R, Sano C, Sugiyama J. *Gluconacetobacter tumulicola* sp. nov. and *Gluconacetobacter asukensis* sp. nov., isolated from the stone chamber interior of the Kitora Tumulus. Int J Syst Evol Microbiol, 2012, 62, 2032–2038.

[100] Nishijima M, Tazato N, Handa Y, et al. *Gluconacetobacter tumulisoli* sp. nov., *Gluconacetobacter takamatsuzukensis* sp. nov. and *Gluconacetobacter aggeris* sp. nov., isolated from Takamatsuzuka Tumulus samples before and during the dismantling work in 2007. Int J Syst Evol Microbiol, 2013, 63, 3981–3988.

[101] Kimura S, Bryan CG, Hallberg KB, Johnson DB. Biodiversity and geochemistry of an extremely acidic, low-temperature subterranean environment sustained by chemolithotrophy. Environ Microbiol, 2011, 13, 2092–104.

[102] Moore JC, Saunders P, Selby G, et al. The distribution and life history of *Arrhopalites caecus* (Tullberg): Order: Collembola, in Wind Cave, South Dakota, USA. J Cave Karst Stud, 2005, 67, 119–119.

[103] Peck SB. Bacterial deposition of iron and manganese oxides inNorth American caves. Bull Natl Speleol Soc, 1986, 48, 26–30.

[104] Moore GW. 1981, Manganese deposition in limestone caves. In: Beck BF. ed. Proceedings of the 8th International Congress of Speleology, vol. II, 1981, 642–645.

[105] De los Rios A, Bustillo MA, Ascaso C, Carvalho MR. Bioconstructions in ochreous speleothems from lava tubes on Terceira Island (Azores). Sediment Geol, 2011, 236, 117–128.

[106] Hathaway JJM, Sinsabaugh RL, MLNE Dapkevicius, Northup DE. Diversity of ammonia oxidation (*amo*A) and nitrogen fixation (*nif*H) genes in lava caves of Terceira, Azores, Portugal. Geomicrobiol J, 2013, 31, 221–235.

[107] Campbell JW, Watson A, Watson C, Ball H, Pirkle R. *Escherichia coli*, other coliform, and environmental chemoheterotrophic bacteria in isolated water pools from six caves in Northern Alabama and Northwestern Georgia. J Cave Karst Stud, 2011, 73, 75–82.

[108] Herbert RA, Ranchou-Peyruse A, Duran R, Guyoneaud R, Schwabe S. Characterization of purple sulfur bacteria from the South Andros Black Hole cave system: highlights taxonomic problems for ecological studies among the genera *Allochromatium* and *Thiocapsa*. Environ Microbiol, 2005, 7, 1260–1268.

[109] Han B, Chen Y, Abell G, et al. Diversity and activity of methanotrophs in alkaline soil from a Chinese coal mine. FEMS Microbiol Ecol, 2009, 70, 196–207.

[110] Portillo MC, Gonzalez JM, Saiz-Jimenez C. 2006. Diversity of sulfate-reducing bacteria as an example of the presence of anaerobic microbial communities in Altamira Cave (Spain). In: Fort R, Alvarez de Buergo M, Gomez-Heras M, Vazquez-Calvo C. eds. Heritage, Weathering and Conservation. London, Taylor and Francis, 2006, 1, 367–371.

[111] Campbell BJ, Engel AS, Porter ML, Takai K. The versatile ε-proteobacteria: key players in sulphidic habitats. Nat Rev Microbiol, 2006, 4, 458–468.

[112] Gonzalez BC, Iliffe TM, Macalady JL, Schaperdoth I, Kakuk B. Microbial hotspots in anchialine blue holes: initial discoveries from the Bahamas. Hydrobiologia, 2011, 677, 149–156.

[113] Bastian F, Alabouvette C, Saiz-Jimenez C. The impact of arthropods on fungal community structure in Lascaux Cave. J Appl Microbiol, 2009, 106, 1456–1462.

[114] Martin-Sanchez PM, Nováková A, Bastian F, Alabouvette C, Saiz-Jimenez C. Two new species of the genus *Ochroconis, O. lascauxensis* and *O. anomala* isolated from black stains in Lascaux Cave, France. Fungal Biol, 2012, 116, 574–589.

[115] Jurado V, Porca E, Cuezva S, Fernandez-Cortes A, Sanchez-Moral S, Saiz-Jimenez C. Fungal outbreak in a show cave. Sci Total Environ, 2010, 408, 3632–3638.

[116] Vanderwolf KJ, Malloch D, McAlpine DF, Forbes GJ. A world review of fungi, yeast, and slime molds in caves. Int J Speleol, 2013, 42, 77–96.

[117] Nováková A. Microscopic fungi isolated from the Domica Cave system (Slovak Karst National Park, Slovakia). A review. Int J Speleol, 2009, 38, 71–82.
[118] Grishkam I, Nevo E, Wasser SP. Micromycetes from the saline Arubotaim Cave: Mount Sedom, the Dead Sea southwestern shore, Israel. J Arid Environ, 2004, 57, 431–443.
[119] Karkun A, Tiwari KL, Jadav SK. Fungal diversity of Mandeepkhol Cave in Chhattisgarh, India. Adv Biores, 2012, 3, 119–123.
[120] Taylor ELS, Stoianoff MAR, Ferreira RL. Mycological study for a management plan of a neotropical show cave (Brazil). Int J Speleol, 2013, 42, 267–277.
[121] Shelton BG, Kirkland KH, Flanders WD, Morris GK. Profiles of airborne fungi in buildings and outdoor environments in the United States. Appl Environ Microbiol, 2002, 68, 1743–1753.
[122] Cooley JD, Wong WC, Jumper CA, Straus DC. Correlation between the prevalence of certain fungi and sick building syndrome. Occup Environ Med, 1998, 55, 579–584.
[123] Pusz W, Ogórek R, Uklańska-Pusz, CM, Zagożdżon P. Speleomycological research in underground Osówka complex in Sowie Mountains (Lower Silesia, Poland). Int J Speleol, 2014, 43, 27–34.
[124] Pashley CH, Fairs A, Free RC, Wardlaw AJ. DNA analysis of outdoor air reveals a high degree of fungal diversity, temporal variability, and genera not seen by spore morphology. Fungal Biol, 2012, 116, 214–224.
[125] Hermosín, B, Nováková, A, Jurado, V, et al. 2010. Observatorio microbiológico de cuevas: evaluación y control de comunidades fúngicas en cuevas sometidas al impacto de actividades turísticas. In: Durán JJ, Carrasco F. eds. Cuevas: Patrimonio, Naturaleza y Turismo. Madrid, Asociación de Cuevas Turísticas, 2010, 513–520.
[126] Garcia-Anton E, Cuezva S, Jurado V, et al. Combining stable isotope ($\delta^{13}C$) of trace gases and aerobiological data to monitor the entry and dispersion of microorganisms in caves. Environ Sci Pollut Res, 2014, 21, 473–484.
[127] Docampo S, Trigo MM, Recio M, Melgar M, García-Sánchez J, Cabezudo B. Fungal spore content of the atmosphere of the Cave of Nerja (southern Spain): diversity and origin. Sci Total Environ, 2011, 409, 835–843.
[128] Porca E, Jurado V, Martin-Sanchez P. M, et al. Aerobiology: An ecological indicator for early detection and control of fungal outbreaks in caves. Ecol Indic, 2011, 11, 1594–1598.
[129] Wang W, Ma X, Ma Y, et al. Molecular characterization of airborne fungi in caves of the Mogao Grottoes, Dunhuang, China. Int Biodeterior Biodegrad, 2011, 65, 726–731.
[130] Cheeptham N. Cave Microbiomes: A Novel Resource for Drug Discovery. SpringerBriefs in Microbiology, New York, Springer, 2013.
[131] Nováková A, Kolarik M. *Chrysosporium speluncarum*, a new species resembling *Ajellomyces capsulatus*, obtained from bat guano in caves of temperate Europe. Mycol Progress, 2010, 9, 253–260.
[132] Nováková A, Hubka V, Saiz-Jimenez C, Kolarik M. *Aspergillus baeticus* sp. nov. and *Aspergillus thesauricus* sp. nov.: two new species in section *Usti* originating from Spanish caves. Int J Syst Evol Microbiol, 2012, 62, 2778–2785.
[133] Nováková A. *Pidoplitchkoviella terricola* – an interesting fungus from the Domica Cave (Slovakia). Int J Speleol, 2009, 38, 23–26.
[134] Jurado V, Sanchez-Moral S, Saiz-Jimenez C. Entomogenous fungi and the conservation of the cultural heritage: A review. Int Biodeterior Biodegrad, 2008, 62, 325–330.
[135] Wibbelt G, Puechmaille SJ, Ohlendorf B, et al. Skin lesions in European hibernating bats associated with *Geomyces destructans*, the etiologic agent of white-nose syndrome. PLoS ONE 2013, 8,e74105.
[136] Sarbu SM, Kane TC, Kinkle BK. A chemoautotrophically based cave ecosystem. Science 1996, 272, 1953–1955.

[137] Langecker TG, Wilkens H, Parzefall J. Studies on the trophic structure of an energy-rich Mexican cave (Cueva de las Sardinas) containing sulfurous water. Memor Biospeol, 1996, 23, 121–125.
[138] Brochier-Armanet C, Boussau B, Gribaldo S, Forterre P. Mesophilic crenarchaeota: proposal for a third archaeal phylum, the *Thaumarchaeota*. Nature Rev Microbiol, 2008, 6, 245–252.
[139] Walker CB, de la Torre JR, Klotz MG, et al. *Nitrosopumilus maritimus* genome reveals unique mechanisms for nitrification and autotrophy in globally distributed marine crenarchaea. Proc Natl Acad Sci USA, 2010, 117, 8818–8823.
[140] Gonzalez JM, Portillo MC, Saiz-Jimenez C. Metabolically active Crenarchaeota in Altamira Cave. Naturwissenschaften, 2006, 93, 42–45.
[141] Coates JD, Ellis DJ, Gaw CV, Lovley DR. 1999. *Geothrix fermentans* gen. nov., sp. nov., a novel Fe(III)-reducing bacterium from a hydrocarbon-contaminated aquifer. IJSEMJ, 1999, 49, 1615–1622.
[142] Ward NL, Challacombe JF, Janssen PH, et al. Three genomes from the phylum *Acidobacteria* provide insight into the lifestyles of these microorganisms in soils. Appl Environ Microbiol, 2009, 75, 2046–2056.
[143] Saiz-Jimenez C, Hermosin B. Thermally assisted hydrolysis and methylation of dissolved organic matter in dripping waters from the Altamira Cave. J Anal Appl Pyrol, 1999, 49, 337–347.
[144] Birdwell J, Engel AS. Characterization of dissolved organic matter in cave and spring waters using UV–Vis absorbance and fluorescence spectroscopy. Org Geochem, 2010, 41, 270–280.
[145] Chronáková A, Horák A, Elhottová D, Kristufek V. Diverse archaeal community of a bat guano pile in Domica Cave (Slovak Karst, Slovakia). Folia Microbiol, 2009, 54, 436–446.
[146] Barton HA, Taylor NM, Kreate MP, Springer AC, Oehrle SA, Bertog JL. The impact of host rock geochemistry on bacterial community structure in oligotrophic cave environments. Int J Speleol, 2007, 36, 93–104.
[147] Ariño X, Hernandez-Marine M, Saiz-Jimenez C. Colonization of Roman tombs by calcifying cyanobacteria. Phycologia, 1997, 36, 366–373.
[148] Saiz-Jimenez C. Biogeochemistry of weathering processes in monuments. Geomicrobiol J, 1999, 16, 27–37.
[149] Sanchez-Moral S, Luque L, Cuezva S, et al. Deterioration of building materials in Roman Catacombs: The influence of visitors. Sci Total Environ, 2005, 349, 260–276.
[150] Albertano P. Cyanobacterial biofilms in monuments and caves. In: Whitton BA, ed. Ecology of Cyanobacteria II: Their Diversity in Space and Time. Dordrecht, Springer, 2012, 317–343.
[151] Ortega-Calvo JJ, Ariño X, Hernandez-Marine M, Saiz-Jimenez C. Factors affecting the weathering and colonization of monuments by phototrophic microorganisms. Sci Total Environ, 1995, 167, 329–341.
[152] Saiz-Jimenez C. Microbiological and environmental issues in show caves. World J Microbiol Biotechnol, 2012, 28, 2453–2464.
[153] Cockell CS, Herrera A. Why are some microorganisms boring? Trends Microbiol, 2008, 16, 101–106.
[154] Sanchez-Moral S, Cañaveras JC, Laiz L, Saiz-Jimenez C, Bedoya J, Luque L. Biomediated precipitation of calcium carbonate metastable phases in hypogean environments: a short review. Geomicrobiol J, 2003, 20, 491–500.
[155] Sanchez-Moral S, Luque L, Cañaveras JC, Laiz L, Jurado V, Saiz-Jimenez C. Bioinduced barium precipitation in St. Callixtus and Domitilla catacombs. Ann Microbiol, 2004, 54, 1–12.
[156] Cañaveras JC, Hoyos M, Sanchez-Moral S, et al. Microbial communities associated to hydromagnesite and needle fiber aragonite deposits in a karstic cave (Altamira, Northern Spain). Geomicrobiol J, 1999, 16, 9–25.

[157] Cañaveras JC, Cuezva S, Sanchez-Moral S, et al. On the origin of fiber calcite crystals in moonmilk deposits. Naturwissenschaften, 2006, 3, 27–32.

[158] Bastian F, Jurado V, Nováková A, Alabouvette C, Saiz-Jimenez C. The microbiology of the Lascaux Cave. Microbiology, 2010, 156, 644–652.

[159] Saiz-Jimenez C. Painted Materials. In: Mitchell R, McNamara CJ, eds. Cultural Heritage Microbiology. Washington, DC, ASM Press, 2010, 3–13.

[160] Saiz-Jimenez C. Cave conservation: A microbiologist's perspective. In: Cheeptham, N. ed. Cave Microbiomes: A Novel Resource for Drug Discovery. New York, Springer, 2013, 69–84.

[161] Martin-Sanchez PM, Nováková A, Bastian F, Alabouvette C, Saiz-Jimenez C. Use of biocides for the control of fungal outbreaks in subterranean environments: The case of the Lascaux Cave in France. Environ Sci Technol, 2012, 46, 3762–3770.

[162] Saiz-Jimenez C, Miller AZ, Martin-Sanchez PM, Hernandez-Marine M. Uncovering the origin of the black stains in Lascaux Cave in France. Environ Microbiol, 2012, 14, 3220–3231.

[163] Kiyuna T, An K-D, Kigawa R, Sano C, Miura S, Sugiyama J. Mycobiota of the Takamatsuzuka and Kitora Tumuli in Japan, focusing on the molecular phylogenetic diversity of *Fusarium* and *Trichoderma*. Mycoscience, 2008, 49, 298–311.

[164] Dupont J, Jacquet C, Dennetiere B, et al. Invasion of the French Paleolithic painted cave of Lascaux by members of the *Fusarium solani* species complex. Mycologia, 2007, 99526–99533.

[165] An KD, Kiyuna T, Kigawa R, Sano C, Miura S, Sugiyama J. The identity of *Penicillium* sp. 1, a major contaminant of the stone chambers in the Takamatsuzuka and Kitora Tumuli in Japan, is *Penicillium paneum*. Anton Leeuw, 2009, 96, 579–592.

[166] Nagatsuka Y, Kiyuna T, Kigawa R, Sano C, Miura S, Sugiyama J. *Candida tumulicola* sp. nov. and *Candida takamatsuzukensis* sp. nov., novel yeast species assignable to the *Candida membranifaciens* clade, isolated from the stone chamber of the Takamatsu-zuka tumulus. Int J Syst Evol Microbiol, 2009, 59, 186–194.

[167] Vaughan-Martini A, Angelini P, Zacchi L. The influence of human and animal visitation on the yeast ecology of three Italian caverns. Ann Microbiol, 2000, 50, 133–140.

[168] Sugita T, Kikuchi T, Makimura K, et al. *Trichosporon* species isolated from guano samples obtained from bat-inhabited caves in Japan. Appl Environ Microbiol, 2005, 75, 7626–7629.

[169] Sugiyama J, Kiyuna T, An K-D, et al. Microbiological survey of the stone chambers of Takamatsuzuka and Kitora Tumuli, Nara Prefecture, Japan: a milestone in elucidating the cause of biodeterioration of mural paintings. In: Sano C, ed. International Symposium on the Conservation and Restoration of Cultural Property: study of environmental conditions surrounding cultural properties and their protective measures. Tokyo, National Research Institute for Cultural Properties, 2009, 51–73.

[170] Went FE. Fungi associated with stalactite growth. Science, 1969, 166, 385–386.

[171] Bindschedler S, Millière L, Cailleau G, Job D, Verrecchia EP. Calcitic nanofibres in soils and caves: a putative fungal contribution to carbonatogenesis. Geol Soc London Spec Publ, 2010, 336, 225–388.

[172] Albertano P, Urzì C. Structural interactions among epilithic cyanobacteria and heterotrophic microorganisms in Roman hypogea. Microb Ecol, 1999, 38, 244–252.

[173] Lefèvre M. La maladie verte de Lascaux. Stud Conserv, 1974, 19, 126–156.

[174] Mulec J, Kosi G. Lampenflora algae and methods of growth control. J Cave Karst Stud, 2009, 71, 109–115.

[175] Mulec J. Human impact on underground cultural and natural heritage sites, biological parameters of monitoring and remediation actions for insensitive surfaces; case of Slovenian show caves. J Nat Conserv, 2014, DOI: 10.1016/j.jnc.2013.10.001.

[176] Albertano P, Bruno L, Bellezza S. New strategies for the monitoring and control of cyanobacterial films on valuable lithic faces. Plant Biosystems, 2005, 139, 311–322.
[177] Roldán M, Oliva F, Gónzalez Del Valle MA, Saiz-Jimenez C, Hernández-Mariné M. Does green light influence the fluorescence properties and structure of phototrophic biofilms. Appl Environ Microbiol, 2006, 72, 3026–3031.
[178] Hsieh P, Pedersen JZ, Albertano P. Generation of reactive oxygen species upon red light exposure of cyanobacteria from Roman hypogea. Int Biodeterior Biodegrad, 2013, 84, 258–265.
[179] Smith T, Olson R. A taxonomic survey of lamp flora (Algae and Cyanobacteria) in electrically lit passages within Mammoth Cave National Park, Kentucky. Int J Speleol, 2007, 36, 105–114.
[180] Boston PJ, Northup DE, Lavoie KH. Protecting microbial habitats. Preserving the unseen. In: Hildreth-Werker V, Werker J. eds. Cave Conservation and Restoration. Huntsville, National Speleological Society, 2006, 61–82.
[181] Akatova E, Roldan M, Hernandez-Marine M, Gonzalez JM, Saiz-Jimenez C. On the efficiency of biocides and cleaning treatments in restoration works of subterranean monuments. In: Science and Cultural Heritage in the Mediterranean Area. Palermo, Regione Siciliana, 2009, 316–322.

Janez Mulec

12 The Diversity and Ecology of Microbes Associated with Lampenflora in Cave and Karst Settings

Abstract: Lampenflora is a community of organisms where phototrophs predominate and give a characteristic greenish appearance to the entire biofilm in artificially lighted subsurface environments. Its natural passive or active transport enhanced by human activities, and physiological characteristics in combination with available light and substrata enable surfaces colonization. Phototrophs are not very diverse, yet the total diversity of such community including the heterotrophs still needs to be evaluated. Although lampenflora can also be a source of new organisms, the lighted patches deep in caves and mines are not their primary habitat and can be considered only a transitory one. Phototrophs with their presence and biochemical activity do affect the integrity of the adhered substrata, but more important, with their primary production they support the heterotrophic organisms in a community, which can largely deteriorate the substrata. Different physical and chemical methods have already been applied by management in caves with increased visitation to restrict lampenflora, but its resilience, continual presence of inoculum, makes its control difficult. Different emission spectra of lamps affect lampenflora growth only partially, especially of those organisms which are able to synthesize different accessory photosynthetic pigments.

12.1 Introduction

Humans have been attracted to caves for a very long time. In early history, humans used natural cavities for rituals, shelters, storage, and as places of inspiration [1]. More recently, caves are recognized as important tourist sites because they contain attractive features like speleothems or preserved remnants of prehistoric man, such as burials, artifacts of daily life, drawings, and inscriptions [2]. But, all underground habitats like caves and mines are dark. As such, visitation of caves and other underground features are impossible without artificial lighting, either carried by the sport or recreational cave explorer or from installed lighting systems for casual tourists. Deep caves and mines, away from entrance areas, can now be subjected to intermittent or constant artificial lighting connected with tourism, mining, the maintenance of the underground infrastructure, such as underground hydropower plants, or long-term site exploration.

For caves, there are few exceptions to absolute darkness, including natural horizontal and vertical cave entrances that are subjected to natural sunlight or caves formed in ice where sunlight can penetrate deep through the (semi)transparent layers of ice. Weak light can also be associated with organic materials, which is caused by chemiluminescent biochemical reactions. Bioluminescence is generated by unicellu-

Fig. 12.1. A troglobiotic millipede grazing on biofilm of lampenflora covered with a water film under an active seep in Prometheus Cave, Georgia. Photograph taken by Andreea Oarga-Mulec (width of motif 4 cm).

lar and higher organisms and usually requires the presence of oxygen [3]. Light emission is also known in diatoms, such as *Melosira* from a tin mine in Cornwall, United Kingdom [4]. Some researchers have suggested that prehistoric man used bioluminescent torches covered with complex microbial communities that emitted light as a way to explain the absence of smoke on prehistoric paintings, which should otherwise be present if the artists used ordinary fire torches [5].

Artificial lighting causes light eutrophication, with its associated consequences like the development of special ecological communities called lampenflora, a German term. The French phrase is "la maladie verte." Lampenflora results in a phenomenon on natural and artificial substrata that is atypical of natural phototrophic conditions due to the proliferation of mainly phototrophic organisms near artificial light sources [6]. Originally considered an interesting phenomenon, lampenflora in some caves is still preserved today as a tourist attraction [6]. However, lampenflora can also develop on completely artificial urban subsurface environments, such as at Kungsträdgården metro station in Stockholm, Sweden [7]. In general, phototrophs (cyanobacteria and microalgae) are responsible for the greenish appearance, but bacteria, fungi, and protozoa are also abundant and all benefit from the primary production of the phototrophs. Mosses and ferns also frequently colonize illuminated patches around lamps. Vascular plants are also sometimes found around lamps, but almost exclusively as germinating shoots [6]. Cave invertebrates frequently graze on lampenflora (▸ Fig. 12.1).

At many sites in caves, abundant lampenflora is found at higher photosynthetic photon flux density (PPFD), longer illumination periods, higher humidity, active seeps, and the presence of sediment. The lampenflora phenomenon impacts the cave ecosystem and cave fauna directly and indirectly, modifies the aesthetic appearances, and also mechanically and biochemically leads to deterioration of the substrata. Therefore, lampenflora affects not only substrata integrity, particularly for speleothems, but also historical inscriptions and rock art or paintings. Many physical and chemical removal procedures have been applied to restrict lampenflora growth [8]. This chapter summarizes factors that enable lampenflora growth, its species composition and effect on adhered substrata, and remediation actions for the affected surfaces.

12.2 Photosynthesis and Artificial Lighting

The quantity of energy (PPFD, expressed in μmol photos/m^2/s) available for photosynthesis is photosynthetically active radiation between 400 and 700 nm. During photosynthesis, available light radiant energy is captured by organisms that have a chlorophyll pigment system. The main photosynthetic pigment in oxygenic phototrophs is chlorophyll *a* (Chl *a*), which has two absorption maxima at 440 nm (γ peak in the violet part of the spectrum) and 680 nm (α peak in the red part of the spectrum (▸ Fig. 12.2 (a)). To overcome the gap between these two peaks, eukaryotic phototrophs synthesize accessory pigments, such as chlorophyll b, c_1, c_2, c_3, xanthophylls, and carotenes [9]. Cyanobacteria capture light using Chl *a*, and between its absorption maxima, by phycobiliproteins [10]. In all cases, light irradiation energy is transferred to Chl *a*, because its peak in the red part of the spectrum is at a longer wavelength than any of the other pigments [11]. Oxygenic photosynthesis leads to the synthesis of ATP, NADH, and release of oxygen during the light phase [12]. The dark reaction involves carbon dioxide fixation to form glucose, which can be polymerized into starch [13]. The CO_2 fixed into biomass represents a carbon base for higher trophic levels.

The photon quantity available for photosynthesis does not correspond to the illumination level sensed by the human eye. The human eye has a sensitivity peak at 555 nm (yellowish) during photopic vision, that is, illumination >3 Lux and at 507 nm (greenish) during scotopic vision [14–16]. Toward the red part of the spectrum, the sensitivity of human vision rapidly decreases (▸ Fig. 12.2 (a)). Consequently, for the illumination of underground features, diverse lighting setups with different emission spectra are used, some of which overlap with photosynthetically active radiation. Halogen lamps emit a large proportion of irradiance in the infrared (IR) part of the spectrum, which is responsible for the heating and drying of the cave atmosphere. They were popular in the past, but have been replaced with systems that use light-emitting diodes (LEDs) because LEDs appear to have a smaller impact on the cave atmosphere, consume less electricity, do not irradiate IR light, and emissions can be finely tuned to create a desirable contrast with the illuminated objects. For example, warm white light can be created with an appropriate combination of red, green, and blue LEDs. Metal halide lamps have also been introduced into some caves because they have generally

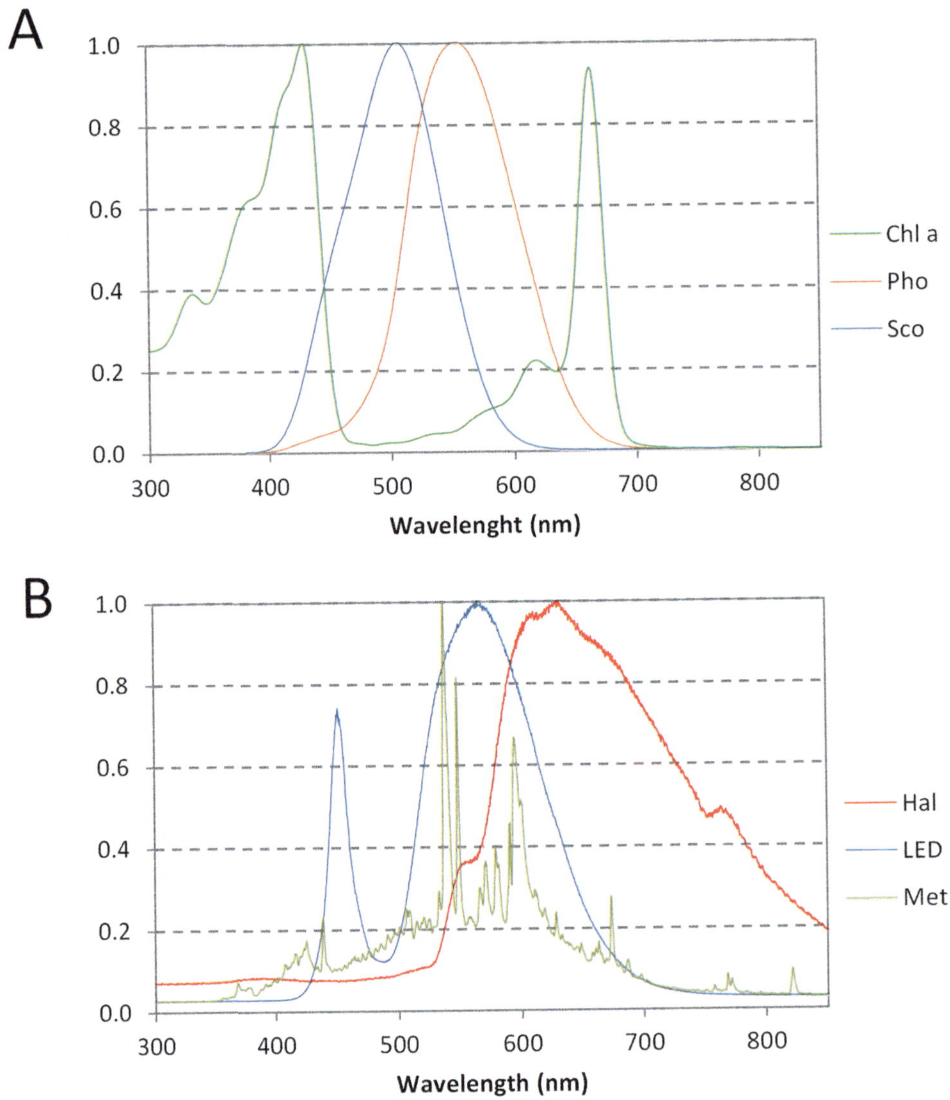

Fig. 12.2. Spectroscopic characteristics for (a) absorption spectrum of chlorophyll *a* (Chl *a*) [17], spectral luminous efficiency function for phototropic (Pho) and scotopic (Sco) vision [14–16], and (b) emission spectra of a halogen lamp (Hal), a cool LED) and a typical metal halide lamp (Met).

broader and different emission spectra than halogen and LED lamps (▸ Fig. 12.2 (b)). Other lamp types used in caves include classic incandescent and fluorescent lamps. In contrast to natural sunlight, artificial light shows no temporal oscillations in quality (i.e. wavelength) and availability of photons (as PPFDs). From this perspective, artificial light offers constant illumination conditions and primary productivity is continuous. But, in general, cave and mine managers do not pay attention to the light spectrum for artificial lighting that might enhance photosynthesis.

12.3 Species Composition

Phototrophs colonize aquatic habitats, but many can also live aerophytically under aeroterrestrial conditions and only require atmospheric humidity to sustain growth [18–20]. Many aerial environments exhibit suboptimum growth conditions, including poor PPFDs, which is crucial for phototrophic growth. Many other environmental parameters are also important for the successful development of phototrophic communities, including substrata characteristics, water and nutrient availability, CO_2 availability, and biological factors like availability of inoculum, competition, parasitism, predation, and organic excrements [21, 22]. Species composition should also be considered because succession processes play a role in the development of a community.

Early research determined that lampenflora is poorer in diversity, and cyanobacteria are less abundant than in aerophytic communities from cave entrances [23]. There is a clear distinction between aerophytic phototrophs illuminated by natural sunlight at cave entrances and lampenflora found in deep underground settings that are illuminated by artificial light. Older taxonomic reports of lampenflora were based only on classic microscopic identification, but taxonomic research transitioned to include cultures for species identification. Cultures became essential for microscopic identification because cell morphologies were frequently altered compared to non-cave isolates (i.e. smaller cells with an altered chloroplast structure) due to the poorly lighted environments. Recently, molecular studies have been shown to be invaluable and even required to further our knowledge of the total diversity of natural subaerial cyanobacteria and algae from caves [24], as well as of lampenflora [25].

Based on data from eight show caves [26–32], three major groups of microscopic phototrophs comprise lampenflora: cyanobacteria, Chlorophyta, and Chrysophyta (mostly diatoms) (▸ Fig. 12.3). Other groups, such as Cryptophyta, Euglenophyta, Eustigmatophyta, and Rhodophyta, are commonly represented by only a few specimens. In general, the phototrophs in lampenflora community are ubiquitous in a range of surface aquatic and terrestrial environments, are adaptable from soil sources, and can reproduce rapidly [27, 33]. Among the cyanobacteria, different species of *Aphanocapsa*, *Aphanothece*, *Chroococcus*, *Gloeocapsa*, *Lyngbya*, *Leptolyngbya*, and *Phormidium* are frequently encountered. For *Leptolyngbya* strains from poorly lighted underground habitats, morphological characteristics, such as the presence or absence of false branching, cell size, and constrictions at the cross wall, do not usually allow strains to be clearly distinguished. But, genetic sequencing of the non-functional internal transcribed spacer (ITS) region of rRNA and 16S rRNA has enabled isolates to be distinguished and intraspecies variation to be analyzed [34]. Chlorophytes often include *Chlorella*, *Scenedesmus*, and *Stichococcus*. The green alga *Trentepohlia aurea* is the most frequently encountered microorganism in lampenflora from show caves in Slovenia [26, 27]. Chrysophyta representatives frequently include the diatoms *Achnanthes*, *Cymbella*, *Fragilaria*, *Navicula*, and *Nitzschia*. Recently, the cyanobac-

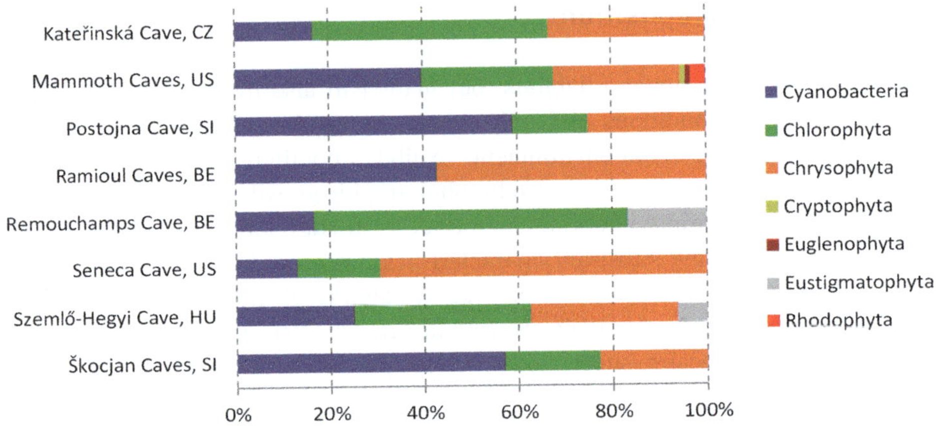

Fig. 12.3. Proportion of different groups of microscopic phototrophs in lampenflora in show caves: Kateřinska Cave [28] (Kateřinská jeskyně, Czech Republic), Mammoth Cave [29] and Seneca Cave [30] (USA), Postojna Cave [26, 27] and Škocjan Caves [26, 27] (Postojnska jama, Škocjanske jame, Slovenia), Ramioul Caves [31] and Remouchamps Cave [31] (La Grotte de Ramioul, La Grotte de Remouchamps, Belgium), and Szemlő-Henyi Cave [32] (Szemlő-hegyi-barlang Hungary). Data for the Mammoth Cave, Postojna Cave and Škocjan Caves were compiled from different sampling campaigns. Classification of eukaryotic algae was performed following Ettl and Gärtner [20].

teria *Iphinoe spelaeobios* and *Loriellopsis cavernicola* [35], and the diatom *Cholnokyella aerophila* [36] were described from caves using a combination of culturing and molecular methods. Regardless, and despite the long history of lampenflora studies in some caves, our knowledge of microbial diversity is still lacking in many respects. Part of the reason for this is because some authors focus only on certain groups of lampenflora algae, for example, diatoms. But, with an increased number of studies using molecular-based approaches, we are beginning to learn that cave subaerial habitats are populated by new, yet undescribed species.

12.4 Transport of Lampenflora Species and Their Relevance Underground

The existence of lampenflora in light-deprived parts of caves demonstrates a relatively constant and efficient transport of active phototrophic cells, spores, and thalli (vegetative tissue associated with shoots or twigs) into caves. Survival and successful colonization of the subsurface depend on the ecology and physiology of transported, incoming microorganisms, as well as nutrient and moisture conditions within the cave. To form lampenflora, the microorganisms must be able to survive transport, and adapt and multiply in the "new habitat".

There are several different modes of transport that bring external microbes into caves, including water flow, air currents, and the introduction by migratory animals and humans. Water seeping into caves might be one of the most important sources of external microbes [37], including phototrophs, organic carbon [38], and pollution. But, air currents can travel deep into the cave interior via natural and/or human-induced air movements. This mechanism not only provides a source of nutrients and particles into caves, but also serves as an inoculum source for many microbes, including phototrophs [39]. In active caves, it is also possible that aerosolization of microbial cells occurs, together with the formation of mist, when an underground river hits various obstacles, such as rocks or tree logs during its flow. Activities linked with mass tourism in caves, and the presence of existing lampenflora, also assist in the spread of microbes within caves. In some cases, intense lamps that produce heat (e.g. halogen) cause a weak air current, which is further enhanced by passing human tourists, and can notably contribute to the spread of lampenflora.

Several studies have been done that link phototroph occurrence in caves with transport processes. In Postojna Cave, Slovenia, the green aeroterrestrial alga *Klebsormidium flaccidum* was identified from cultures using Jaworski medium [40] inoculated with 5 mL of water from a permanent seep (at the time of sampling, the discharge was 14 mL/min). *K. flaccidum* has wide ecological tolerance and is frequently in the upper and lower soil layers, on rocks, stone walls, and tree bark [20], so its presence in cave seeps inside caves and near entrances is not surprising. Occasionally, unexpected phototrophs are identified in aerophytic cave habitats. At the cave entrance of Škocjan Caves, Slovenia, a green non-motile coenobial alga, *Pediastrum boryanum*, typical of freshwater aquatic habitats, was found among the phototrophic epilithic community. The aquatic organisms must live submerged under droplets of condensed water in a cave entrance. The cyanobacterium *Hapalosiphon intricatus* has been interpreted to survive in this way, too [41]. These phototrophs are infrequently observed in lampenflora communities.

Caves with lampenflora have been considered since the 1970s as an "ecosystem in formation" [42]. Lampenflora clearly constitutes an artificially induced underground ecosystem. Phototrophs form the base of a food chain and interact with inanimate components of the environment (e.g. air, water, and substrata). The accumulation of biomass as lampenflora overwhelms the more typical oligotrophic conditions frequently encountered in caves. Artificial light energy promotes the fixation of new biomass that can then become incorporated into biomass of true troglobiotic (i.e. obligate cave-adapted) organisms that are grazers (▸ Fig. 12.1). Similarly, the new ecological niche may also make troglophilic (i.e. not true cave fauna) animals more competitive than troglobiotic animals, which can have negative consequences to the cave ecosystem trophic structure.

12.5 Survival Strategies of Phototrophs

All underground spaces that are artificially and constantly illuminated are not ideal habitats for most phototrophs. Light is the main driver for lampenflora community and its dynamics, but many times the PPFD gradient is not reflected in a community structure with clear zonation of phototrophic organisms and quantity of fixed biomass [27]. Ideal candidates for the colonization of underground lighted places are typically aerophytic and soil phototrophs because many such phototrophs are adapted to low PPFDs in natural habitats. For example, many soil algae in lower soil layers survive with restricted levels of light. Desiccation tolerance, sometimes attributed to the biosynthesis of organic osmolytes, such as polyols [43], might also play an important role for the survival of phototrophs underground before they reach lighted and adequately humid places. Compatible solutes accumulated in cells are readily available for respiration and growth [44] after they reach a suitable niche. Diatoms, for example, frequently require energy and Ca^{2+} for successful colonization and growth [45]. In contrast to aquatic species that are more sensitive to drying, these soil-derived algae may rapidly reduce their photosynthetic activity under harsh desiccation conditions. But, after rewetting, they exhibit rapid metabolic recovery [43]. *Klebshormidium*, a genus of green algae that can also be found in lampenflora [27], acquires cross-walls in desiccated samples, which is probably responsible for high mechanical flexibility and may allow cells to maintain turgor pressure during dehydration.

The study of phototrophs from underground habitats has provided important information about their physiology, adaptations, and survival strategies. Lampenflora phototrophs have been identified in PPFDs as low as 0.33 μmol photons/m^2/s [8]. Cyanobacteria are observed in caves even at PPFDs ranging from 0.0008 to 0.06 μmol photons/m^2/s [46]. Ultrastructural analysis of phototrophs from natural, poorly lighted cave environments generally shows a dense thylakoid system in *Geitleria calcarea* [47], *Chroococcus* [48], *Gloeocapsa* [49], *Chroococcidiopsis*, *Cyanosarcina*, *Leptolyngbya*, *Phormidium*, and *Pseudocapsa* [50]. The resilience and ability of the photosynthetic apparatus of *Phormidium autumnale* to respond rapidly to different levels of PPFDs and to resume growth promptly after prolonged exposure to darkness might help this organism to colonize caves successfully [51].

Temperature is another important factor for growth of phototrophs. At low PPFDs, a slight increase in temperature can result in a considerable increase in biomass yield. For example, a temperature increase from 9 °C to 11 °C at 2.5 μmol photons/m^2/s for *Chlorella* sp. resulted in an approximately 30-fold increase in biomass. At typical cave temperatures, the light saturation point occurs at lower PPFDs than in other normally lighted environments. Photosynthetic light saturation in cyanobacteria from lampenflora was ~60 μmol photons/m^2/s [52]. The biosynthesis of accessory photosynthetic pigments below the light saturation point is noticeably elevated [27]. Chromatic adaptation of *Phormidium valderianum* in lampenflora was reported from an English cave [24].

Light absorption and its scattering on different cave surfaces affect the available light quantity and quality. In environments with extremely low PPFDs, few data exist concerning the extent to which some strains grow photoheterotrophically (mixotrophically) or possibly almost entirely heterotrophically [24]. Some phototrophs, such as *Chlorella*, are able to switch from photoautotrophic to heterotrophic metabolism and can even fix CO_2 heterotrophically [53–55]. The cyanobacterium *Phormidium* sp. in the Frasassi Caves, Italy, captured on average only between 6 and 10 µmol photons/m^2/s, whereas the photosynthetic compensation point was measured at 13 µmol photons/m^2/s [56].

Faster growth of competing phototrophs is reflected in the species composition of lampenflora, where eukaryotic algae represent a higher percentage of a community composition than in a community from a cave entrance. The study of microbial biofilms on buildings reveals that initial colonizers are eukaryotic algae, but cyanobacteria dominate in the later phases of species succession [57]. A similar trend is observed in the later stages of species succession in the lampenflora community [27]. In natural phototrophic communities, some organisms that are light-insensitive to high PPFDs, for example, large filaments of algae and cyanobacteria can physically protect underlying algae and cyanobacteria, particularly during succession.

In caves, protection against high light intensity can also provide non-cemented carbonate. Small cyanobacteria in cave entrances that are associated with speleothems can migrate into the non-cemented and soft part of a speleothem. This can be repeated until the entire speleothem structure is no longer completely cemented. The function of non-cemented $CaCO_3$ in speleothems is not just to protect sensitive microbes against high PPFDs, but also to retain moisture [58].

12.6 Colonization of Solid Surfaces

Different illuminated surfaces, such as flowstone, bedrock, sediments, artwork, inscriptions, and different artificial materials, are found underground and are subjected to colonization by heterotrophic and phototrophic organisms. In the karst caves, the most important sedimentary carbonate rocks are limestones and dolomites. These rocks differ in their mineral composition, mechanical and chemical properties, mechanism of formation, and age. The most important mineral in limestone is calcite ($CaCO_3$) and in dolomites, dolomite ($CaMg(CO_3)_2$) [59]. In the crystal structure, Ca^{2+} can be replaced with other ions, such as Ba^{2+}, Fe^{2+}, Mn^{2+}, Mg^{2+}, and Sr^{2+} [60]. The following trace elements can also be found in carbonate rocks: Al, Ba, Cd, Co, Cu, Fe, K, Mg, Mn, Na, Si, Sr, Ti, U, and Zn [61, 62]. In flowstone, various organic compounds are often present, such as fulvic and humic acids, amino acids, lipids, *n*-alkanes [63], and metal ions in different oxidation states, for example, Fe^{2+} and Fe^{3+}, which can further react [64]. Apart from C, H, O, at least Cu, Fe, Mg, Mn, Mo, N, P, and Zn are required for the growth of algae and cyanobacteria. For some species, the following elements are indispensable: Al, B, Br, Ca, Cd, Cl, Co, Cr, I, K, Na, Ni, S, Si, Ti, V, and W [65, 66]. These elements are provided in aeroterrestrial and subaerial habitats by dripping and seeping water, via rock weathering, or from the substratum to which they are attached. For diazotrophic species, the presence of nitrogen is not essential. Occasionally, a particular substratum composition or the *in situ* conditions are not optimum for the growth of phototrophs. In such cases, an increase in PPFD will not be directly reflected in an increase in biomass [27]. In some cases, nonphototrophic microbes might facilitate an easier interaction of phototrophs with the substratum [24]. At the initial stage of microbial colonization, surface texture, size and nature of mineral grains, microtopography [24], and small crevices, which retain higher humidity, are additional important factors.

12.7 Biodeterioration and Remediation

In the subsurface, all lighted surfaces with developed biofilms are subjected to biodeterioration, that is, the breakdown of materials by biochemical activities. Biodeterioration can become disturbing when damage to artwork and other substrata is clearly visible (▸ Fig. 12.4). Microorganisms have developed different strategies to uptake key elements from rock substrates for their growth, for example, acidolysis by biogenic organic acids and the synthesis of siderophores [67]. Negatively charged, highly hygroscopic exopolymers secreted by microbial biofilms strongly bind cations from the underlying minerals [68]. These high-molecular-weight polyelectrolytes comprised of polysaccharides (up to 90%), proteins, and nucleic acids also serve to retain water.

Fig. 12.4. Lampenflora colonizing flowstone with historical inscriptions in Postojna Cave. Photograph taken by Jurij Hajna.

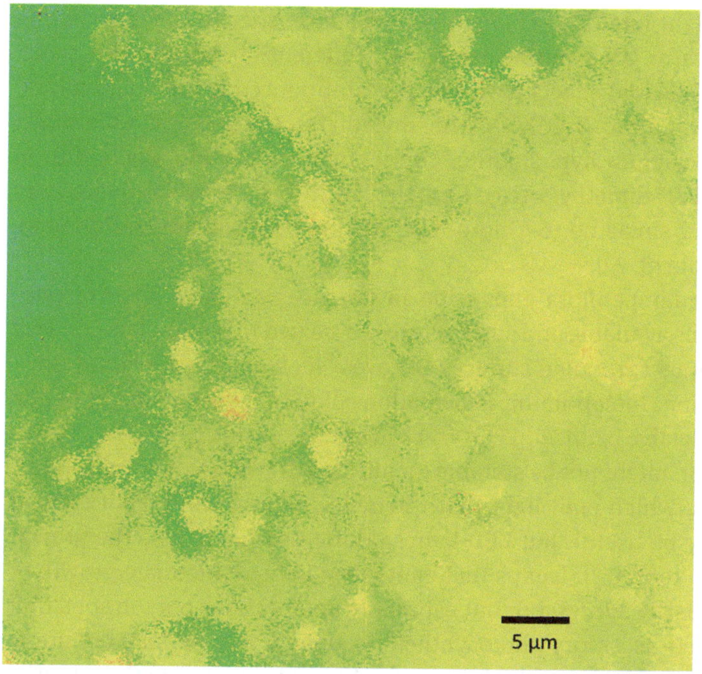

Fig. 12.5. Confocal microscopy of epilithic *Chlorella vulgaris* on limestone after 24 weeks of cultivation at 21.5 µmol photons/m^2/s with 8 h : 16 h light/dark period in a wet chamber under a metal halide lamp (see ▸ Fig. 12.2 (b), Met lamp). Note the dissolution visible as a ring extending around the cells into the limestone substratum composed of 97.7–98.7% $CaCO_3$ [71]. Photograph taken by Alenka Mauko Pranjić.

Contraction and expansion of the biofilm can weaken mineral structures [67]. Changes in rock integrity appear even with a single algal colonizer, such as *Chlorella vulgaris*, before the substratum acquires a greenish appearance (▸ Fig. 12.5). One result of biological activity is the oxidation of minerals and alteration in the mineral structures in the rock, which leads to degradation, weathering, increased porosity, and permeability [67]. Although heterotrophic and phototrophic biofilms can deteriorate stony substrata and artwork, biofilms can also promote the production of biogenic carbonates [69], which result from photosynthetic-mediated alkalization [52] and the encrustation of organisms [67].

In general, biodeterioration is generally a result of synergistic effects of phototrophs and heterotrophs [52], and the ancient rock art paintings in Lascaux Cave, France, and at other similar locations are not primarily threatened by algae, for example, *Bracteacoccus minor*, but by fungi and bacteria that have been introduced and spread by humans and their activities [70].

To control lampenflora growth in caves, several physical methods have been adopted: cleaning of phototrophic biofilms with brushes; shortening the lighting period; reducing the light intensity; changing the emission spectra of lamps; using ozone-producing lamps or UV emitting lamps [8]. Different chemicals with selective and non-selective biocidal effects have also been introduced to kill lampenflora, for example, atrazine [6-Chloro-*N*-ethyl-*N*'-(1methylethyl)-1,3,5-triazine-2,4-diamine], bromine compounds, calcium hypochlorite, cupric ammoniac compounds, diuron (*N*-3, 4-dichlorophenyl-*N*'-dimethyl urea), formalin, hydrogen peroxide, quaternary ammonium derivatives, simazine (6-Chloro-*N*,*N*'-diethyl-1,3,5-triazine-2,4-diamine), and sodium hypochlorite [8, 72].

The battle against lampenflora expansion in tourist caves is in favor of phototrophs, if light remains available and surfaces are not treated to remove lampenflora inoculum. For insensitive carbonate surfaces in caves, such as limestone and flowstone without cave fauna, the optimum remediation solution currently might be the application of an unselective oxidizer, 15% (v/v) hydrogen peroxide (pH 7.0 to 7.5). Application of this solution might not be suitable for affected drawings and inscriptions. The use of LED lamps, which emit light that resembles natural white light instead of halogen lamps, may be useful, but LED lamps do not notably reduce the growth of *C. vulgaris* on a long term [73]. Lamps that emit green light to attenuate growth of phototrophs are only partly successful, but especially not for those phototrophs that can rapidly modify their accessory photosynthetic pigments [74]. As lampenflora is composed of different phototrophs, the best method to reduce its growth in tourist caves and mines remains to use a very restricted lighting regime. Illuminated sites should be exposed to minimum PPFDs and the exposure to lighting should be as short as possible.

12.8 Conclusions

Lampenflora growing independently of sunlight deep in caves, which were early equipped with electric lighting, raised scientific interest. It is evident today that actively growing lampenflora phototrophs are aliens for such environments. With increased visitation of underground, cave management started to eliminate lampenflora prior knowing the exact community composition, what we still lack today. There is a chance to discover new species in this community, but for the phototrophic part of the community is obvious that artificially illuminated caves are not their primary habitat. Study of phototrophs from lampenflora clearly contributes to their ecology and physiology, but their main habitat is elsewhere than caves, that in karst just intercept their natural pathway of transport and dissemination. As primary producers they can sustain a heterotrophic part of the community, which further increase the biodeterioration of substrata, especially rock art paintings and historical inscriptions. Developed lampenflora is a result of light eutrophication and future studies should be focused on its effect on natural cave microbiota with the practical application for cave management to restrict its growth and minimize its effect.

Acknowledgments

The study was supported by the Slovenian Research Agency (J6-0152 and L1-5453).

References

[1] Bahn PG. Prehistoric Rock Art: Polemics and Progress: The 2006 Rhind Lectures for the Society of Antiquaries of Scotland. Cambridge, United Kingdom, Cambridge University Press, 2010.
[2] Ziegert H. Archaeology as History. Hamburg, Norderstedt: Books on Demand GmbH, 2002.
[3] Wilson T, Hastings J. Bioluminescence. Ann Rev Cell Dev Biol, 1998, 14, 197–230.
[4] Claus G. Bioluminiscence in *Melosira varians* Ag. IntJ Speleol, 1966, 2, 407–408.
[5] Gil AP, Ponte RA, Caminero DP, Beiras SL, de Arce Temes JE. Las bacterias luminosas como primitiva fuente de luz. Coruña, Spain, Actas del VI Congreso Español de Espelología., 1992, 61–64.
[6] Mulec J. Lampenflora. In: White WB, Culver DC. eds. Encyclopedia of Caves. 2nd edn. Amsterdam, Netherdlands, Elsevier, 2012, 451–456.
[7] Ivarsson LN, Ivarsson M, Lundberg J, Sallstedt T, Rydin C. Epilithic and aerophilic diatoms in the artificial environment of Kungsträdgården metro station, Stockholm, Sweden. Int J Speleol, 2013, 42, 289–297.
[8] Mulec J, Kosi G. Lampenflora algae and methods of growth control. J Cave Karst Stud, 2009, 71, 109–115.
[9] van den Hoek C, Mann D, Jahns H. Algae: an introduction to phycology. Cambridge, United Kingdom, Cambridge University Press, 2002.
[10] Oliver R, Ganf G. Freshwater blooms. In: Whitton BA, Potts M, eds. Ecology of cyanobacteria. Dordrecht, Netherlands, Kluwer Academic Publishers, 2000, 149–194.
[11] Kirk J. Light and photosynthesis in aquatic ecosystems. Cambridge, United Kingdom, Cambridge University Press, 1983.
[12] Kramer DM, Avenson TJ, Edwards GE. Dynamic flexibility in the light reactions of photosynthesis governed by both electron and proton transfer reactions. Trends Plant Sci, 2004, 9, 349–357.
[13] Sutton BG. The Path of Carbon in CAM Plants at Night. Funct Plant Biol, 1975, 2, 377–387.
[14] Judd DB, Wyszecki G. Color in Business, Science and Industry. 3rd ed, New York, Wiley, 1975.
[15] Sharpe L, Stockman A, Jagla W, Jagle H. A luminous efficiency function, V*(lambda), for daylight adaptation. J Vision, 2005, 5, 948–968.
[16] Vos JJ. Colorimetric and photometric properties of a 2° fundamental observer. Color Res Appl, 1978, 3, 125–128.
[17] Frigaard N, Larsen K, Cox R. Spectrochromatography of photosynthetic pigments as a fingerprinting technique for microbial phototrophs. FEMS Microbiol Ecol, 1996, 20, 69–77.
[18] Komárek J, Anagnostidis K. Cyanoprokaryota 1. Teil: Chroococcales. Heidelberg, Berlin, Germany, Spectrum Akademischer Verlag, 2000.
[19] Komárek J, Anagnostidis K. Cyanoprokaryota 2. Teil: Oscillatoriales. München, Germany, Elsevier Spektrum Akademischer Verlag, 2005.
[20] Ettl H, Gärtner G. Syllabus der Boden-, Luft- und Flechtenalgen. Stuttgart, Jena, New York, Gustav Fischer Verlag, 1995.
[21] Hynes HBN. Ecology of running waters. Liverpool, United Kingdom, Liverpool University Press, 1972.

[22] Werner D. The Biology of Diatoms. Oxford, United Kingdom, Blackwell Scientific Publications, 1977.

[23] Dobat K. Considérations sur la végétation cryptogamique des grottes du Jura Souabe (sud-ouest de l'Allemagne). Ann Spéléol, 1970, 25, 872–907.

[24] Pentecost A, Whitton B. Subaerial Cyanobacteria. In: Whitton BA, ed. Ecology of Cyanobacteria II. Dordrecht, Germany, Springer, 2012, 291–316.

[25] Cennamo P, Marzano C, Ciniglia C, Pinto G, Cappelletti P, Caputo P, Pollio A. A survey of the algal flora of anthropogenic caves of Campi Flegrei (Naples, Italy) archeological district. J Cave Karst Stud, 2012, 74, 243–250.

[26] Martinčič A, Vrhovšek D, Batič F. Flora v jamah z umetno osvetlitvijo. Biol Vestn, 1981, 29, 27–56.

[27] Mulec J, Kosi G, Vrhovšek D. Characterization of cave aerophytic algal communities and effects of irradiance levels on production of pigments. J Cave Karst Stud, 2008, 70, 3–12.

[28] Faimon J, Štelcl J, Kubešová S, Zimák J. Environmentally acceptable effect of hydrogen peroxide on cave "lamp-flora", calcite speleothems and limestones. Environ Pollut, 2003, 122, 417–422.

[29] Smith T, Olson R. A taxonomic survey of lamp flora (Algae and Cyanobacteria) in electrically lit passages within Mammoth Cave National Park, Kentucky. Int J Speleol, 2007, 36, 105–114.

[30] Dayner DM, Johansen JR. Observation on the algal flora of Seneca cavern, Seneca County, Ohio. Ohio J Sci, 1991, 91, 118–121.

[31] Garbacki N, Ector L, Kostikov I, Hoffmann L. Contribution à l'étude de la flore des grottes de Belgique. Belg J Bot, 1999, 132, 43–76.

[32] Rajczy M, Buczkó K. The development of the vegetation in lamplit areas of the cave Szemlő-Hegyi-Barlang. In: Proceedings of the 10th International Congres of Speleology. Budapest, Hungary, Hungarian Speleological Society, 1989, 514–516.

[33] Mulec J. Lampenflora in show caves in Slovenia. In: Prelovšek M, Zupan Hajna N, eds. Pressures and protection of the underground karst. Postojna, Slovenia, Karst Research Institute ZRC SAZU, 2011, 64–73.

[34] Bruno L, Billi D, Bellezza S, Albertano P. Cytomorphological and genetic characterization of troglobitic *Leptolyngbya* strains isolated from Roman hypogea. Appl Environ Microbiol, 2009, 75, 608–617.

[35] Lamprinou V, Hernández-Mariné M, Canals T, Kormas K, Economou-Amilli A, Pantazidou A. Morphology and molecular evaluation of *Iphinoe spelaeobios* gen. nov., sp. nov. and *Loriellopsis cavernicola* gen. nov., sp. nov., two stigonematalean cyanobacteria from Greek and Spanish caves. Int J Syst Evol Microbiol, 2011, 61, 2907–2915.

[36] Taylor JC, Lange-Bertalot H. *Cholnokyella aerophila* JC. Taylor & Lange-Bertalot gen. et spec. nov. A new diatom (Bacillariophyceae) from sandstone caves in South Africa. Nova Hedwigia, 2013, 97, 295–304.

[37] Gerič B, Pipan T, Mulec J. Diversity of culturable bacteria and meiofauna in the epikarst of Škocjanske jame Caves (Slovenia). Acta Carsol, 2004, 33, 301–309.

[38] Simon K, Pipan T, Culver D. A conceptual model of the flow and distribution of organic carbon in caves. J Cave Karst Stud, 2007, 69, 279–284.

[39] Mulec J, Vaupotič J, Walochnik J. Prokaryotic and eukaryotic airborne microorganisms as tracers of microclimatic changes in the underground (Postojna Cave, Slovenia). Microb Ecol, 2012, 64, 654–667.

[40] Warren A, Day JG, Brown S. Cultivation of algae and protozoa. In: Hurst CJ, Knudsen GR, McInerney MJ, Stetzenbach LD, Walter MV, eds. Manual of Environmental Microbiology. Washington, DC, USA, ASM Press, 1997, 61–71.

[41] Golubić S. Algenvegetation der Felsen: Eine ökologische Algenstudie im dinarischen Karstgebiet. In Elster HJ, Ohle W, eds., Stuttgart, Germany, E. Schweizerbart'sche Verlagsbuchhandlung, 1967.

[42] Dobat K. Ein Ökosystem in Aufbau: Die "Lampenflora Schauhöhlen." Umschau, 1972, 72, 493–494.

[43] Holzinger A, Karsten U. Desiccation stress and tolerance in green algae: consequences for ultrastructure, physiological and molecular mechanisms. Front Plant Sci, 2013, 4, 327.

[44] Karsten U, Schumann R, Mostaert A. Aeroterrestrial algae growing on man-made surfaces. In: Seckbach J, ed. Algae and Cyanobacteria in Extreme Environments. Amsterdam, Netherlands, Springer, 2007, 583–597.

[45] Cooksey KE, Cooksey B. Adhesion of fouling diatoms to surfaces: some biochemistry. In: Evans LV, Hoagland KD, eds. Studies in Environmental Sciences 28, Algal Biofouling. Amsterdam, Elsevier, 1986, 41–53.

[46] Martínez A, Asencio A. Distribution of cyanobacteria at the Gelada Cave (Spain) by physical parameters. J Cave Karst Stud, 2010, 72, 11–20.

[47] Couté A. Ultrastructure d'une cyanophycée aérienne calcifiée cavernicole: *Geitleria calcarea* Friedmann. Hydrobiologia, 1982, 97, 255–274.

[48] Cox G. Photosynthesis in the deep twilight zone: microorganisms with extreme structural adaptations to low light. In: Ford TD, ed. 7th International Congres of Speleology. Sheffield, United Kingdom, University of Leicester, 1977, 131–133.

[49] Cox G, Benson D, Dwarte DM. Ultrastructure of a cave wall cyanophyte-*Gloeocapsa* NS4. Arch Microbiol, 1981, 130, 165–174.

[50] Asencio A, Aboal M. Cell inclusions in the chasmoendolithic cyanophytes from cave-like environments in Murcia (SE Spain). Arch Hydrobiol Suppl Algol Stud, 2004, 113, 117–127.

[51] Montechiaro F, Giordano M. Effect of prolonged dark incubation on pigments and photosynthesis of the cave-dwelling cyanobacterium *Phormidium autumnale* (Oscillatoriales, Cyanobacteria). Phycologia, 2006, 45, 704–710.

[52] Albertano P. Cyanobacterial biofilms in monuments and caves. In: Whitton BA, ed. Ecology of Cyanobacteria II: Dordrecht, Germany, Springer, 2012, 317–343.

[53] Droop MR. Heterotrophy of carbon. In: Stewart WDP, ed. Algal Physiology and Biochemistry. Oxford, England, Blackwell Scientific Publications, 1974, 530–559.

[54] Lloyd D. Dark respiration. In: Stewart WDP, ed. Algal Physiology and Biochemistry. Oxford, England, Blackwell Scientific Publications, 1974, 505–529.

[55] Raven JA. Carbon dioxide fixation. In: Stewart WDP, ed. Algal Phyiology and Biochemistry. Oxford, England, Blackwell Scientific Publications, 1974, 434–455.

[56] Giordano M, Mobili F, Pezzoni V, Hein M, Davis J. Photosynthesis in the caves of Frasassi (Italy). Phycologia, 2000, 39, 384–389.

[57] Gaylarde P, Gaylarde C. Algae and cyanobacteria on painted buildings in Latin America. Int Biodeter Biodegr, 2000, 46, 93–97.

[58] Mulec J, Kosi G, Vrhovšek D. Algae promote growth of stalagmites and stalactites in karst caves (Škocjanske jame, Slovenia). Carbonate Evaporites, 2007, 22, 6–9.

[59] Ford D, Williams PW. Karst Hydrogeology and Geomorphology. West Sussex, England, Wiley, 2007.

[60] White W. Speleothems: general overview. In: White WB, Culver DC. eds. Encyclopedia of Caves. 2nd edn. Amsterdam, Netherlands, Elsevier, 2012, 777–786.

[61] Polyak V, Denniston R. Paleoclimate records from speleothems. In: White WB, Culver DC, eds. Encyclopedia of Caves. 2nd edn. Amsterdam, Elsevier, 2012, 577–585.

[62] Morse JW, Mackenzie FT. Geochemistry of Sedimentary Carbonates. Amsterdam, Elsevier, 1990.

[63] Hill CA, Forti P. Cave Mminerals of the World. 2nd edn. Huntsville, Alabama, USA, National Speleological Society, 1997.
[64] Hill CA, Forti P. Cave Minerals of the World. Huntsville, Alabama, USA, National Speleological Society, 1986.
[65] Stewart W. Algal Physiology and Biochemistry. Oxford, United Kingdom, Blackwell Scientific Publications, 1974.
[66] Rippka R. Isolation and purification of cyanobacteria. Method Enzymol, 1988, 167, 3–27.
[67] Warscheid T, Braams J. Biodeterioration of stone: a review. Int Biodeter Biodegr, 2000, 46, 343–368.
[68] Hoffmann L. Caves and other low-light environments: aerophitic photoautotrophic microorganisms. In: Bitton G, ed. Encyclopedia of Environmental Microbiology, New York, USA, Wiley, 2003, 835–843.
[69] Zammit G, Sanchez-Moral S, Albertano P. Bacterially mediated mineralisation processes lead to biodeterioration of artworks in Maltese catacombs. Sci Tot Environ, 2011, 409, 2773–2782.
[70] Saiz-Jimenez C. Painted material. In: Mitchell R, McNamarra CJ, eds. Cultural Heritage Microbiology: Fundamental Studies in Conservation Science. Washington DC, USA, ASM Press, 2010, 3–13.
[71] Gams I. International comparative measurements of surface solution by means of standard limestone tablets. Razpr Slov Akad Znan Umet Razr Naravosl vede, 1985, 26, 361–386.
[72] Bastian F, Jurado V, Novakova A, Alabouvette C, Saiz-Jimenez C. The microbiology of Lascaux Cave. Microbiology, 2010, 156, 644–652.
[73] Mulec J. Human impact on underground cultural and natural heritage sites, biological parameters of monitoring and remediation actions for insensitive surfaces: case of Slovenian show caves. J Nat Conserv, 2014, 22, 132–141.
[74] Roldán M, Oliva F, Gónzalez Del Valle MA, Saiz-Jimenez C, Hernández-Mariné M. Does green light influence the fluorescence properties and structure of phototrophic biofilms? Appl Environ Microbiol, 2006, 72, 3026–3031.

Pedro M. Martin-Sanchez, Ana Z. Miller, and Cesareo Saiz-Jimenez

13 Lascaux Cave: An Example of Fragile Ecological Balance in Subterranean Environments

Abstract: Lascaux Cave in France is famous for the exceptional quality of its Paleolithic paintings with 915 figures of animals and numerous geometric symbols. The cave has suffered three microbial crises, and since 1963 it is closed to the public due to the colonization of the alga *Bracteacoccus minor*. Further the fungi *Fusarium solani* and *Ochroconis lascauxensis* put the conservation of the paintings and engravings at risk. The microbial outbreaks were treated with biocides but this revealed to be a wrong decision, particularly the choice of benzalkonium chloride. Lascaux Cave is an example of how human activities have selected microbial communities that are particularly difficult to fight and are even more competitive in successional processes.

13.1 Introduction

Lascaux Cave is located in the department of Dordogne, France, near the town of Montignac, which is within the Vézère River basin in the northern Black Périgord region (▸ Fig. 13.1). Since its discovery in 1940, the cave has been highly regarded by the press and the general public because the artistic heritage is of exceptional quality. According to the inventory done by Aujoulat [1], there are 1963 representations with 915 figures of animals and numerous geometric symbols. Radiocarbon dating of fragments of reindeer antler, which were found during the excavations completed by Henri Breuil and Severin Blanc, established cave occupancy between 18,600 and 18,900 BP. The period is situated between the Upper Solutrean and the Early Magdalenian cultures [2]. The formal analysis of the figures found in the Lascaux Cave implies that they belong to the Solutrean tradition.

Lascaux Cave has a series of galleries that are accessible to humans, but the entire cave does not exceed 235 m in length (▸ Fig. 13.1). The total volume of the cave is 3300 ± 500 m^3 [3]. Traditionally, the cave has been divided into seven sectors: the Hall of the Bulls, the Axial Gallery, the Passageway, the Nave, the Chamber of the Felines, the Apse, and the Shaft of the Dead Man. There are three different passage areas. The first is associated with the entrance, made up of four vestibules 60 m long (SAS1-C1, C2 and C3, and SAS2), the Hall of the Bulls, and the Axial Gallery. The second passage area connects the Hall of the Bulls to the Passageway, the Nave, the Mondmilch Gallery, and the Chamber of the Felines, with a total length of about 110 m. Finally, the last area starts at the Apse and connects to the Shaft of the Dead Man and the Great Fissure, where a considerable amount of sediment marks the intersection with the Silted-up Chamber (▸ Fig. 13.1).

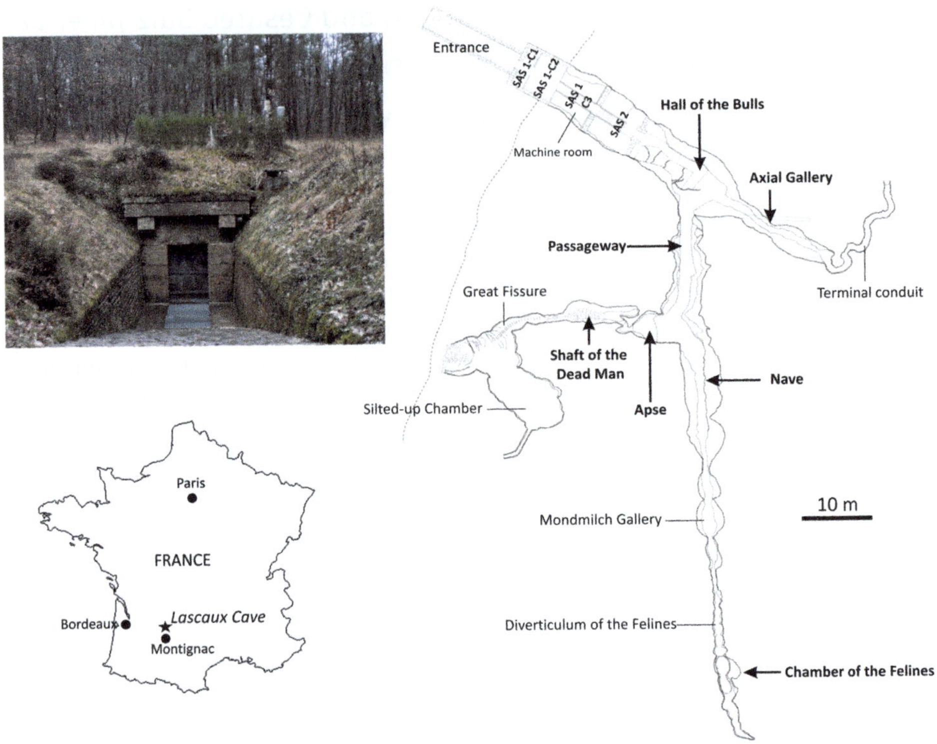

Fig. 13.1. Map of Lascaux Cave, including details of its entrance door, and the location of the cave in France. Map adapted from Aujoulat [1].

Lascaux Cave's notoriety is based on the paintings in the Hall of the Bulls, the Axial Gallery, and the Nave. The Hall of the Bulls is an extension of the entryway and is the largest room (320 m^3), with a length of 20 m, a width that varies between 5.5 and 7.5 m, and a maximum height of 6 m. The room's walls are almost entirely decorated for over approximately 30 m. The motifs in the Hall of the Bulls are the most famous and considered to be some of the most impressive examples of Paleolithic art. There are a total of 130 figures in the room, including 36 animal representations and 50 geometric symbols (▶ Fig. 13.2). The Hall of the Bulls and the Axial Gallery cave wall limestone is more compact and cohesive, compared to the vaulted ceiling in the Passageway with weaker calcareous sandstone. But, the rock shows evidence of multiple diagenetic episodes of dissolution and precipitation, as evidenced from the fill of the uneven cracks. The limestone walls in this area of the cave are also covered with a carbonate coating. The Axial Gallery is a conduit approximately 30 m long, located directly after the Hall of the Bulls. Paintings are distributed on both walls and the vaulted ceiling. There are 161 figures, with 58 figurative representations, mostly of animals (e.g. cows and bulls, horses, deer, ibexes, and a bison), 46 geometric figures,

Fig. 13.2. The Unicorn Panel in the Hall of the Bulls, Lascaux Cave, France. Photo N. Aujoulat © MCC-CNP.

and 57 undetermined figures that could be related to the other paintings or outlines of animals. The Nave is 20 m in length, has four panels situated on different levels of its left wall due to the downward slope in the gallery. There is only one panel on the right wall. This room is predominately decorated with animal figures, although there are also relevant geometric figures (i.e. the blazons under the Great Black Cow).

The climate of the area is mainly oceanic. The average year-round atmospheric temperature is 12.6 °C, with temperatures fluctuating between 3 °C in the winter to 20 °C in the summer. Rainfall averages 880 mm/year, although this is spread out over two main seasons, fall and spring. The cave interior temperature is very stable, being on average 12.5 °C at 99% relative humidity [4].

Karst development occurring from the Late Coniacian Stage left thousand-year-old clay and sand deposits on either side of the valley. These deposits, with abundant veins of iron and manganese oxides [5], are found above the cave and support a forest of oak, pine, and chestnut trees. The cave was discovered from a small crack in the sediment deposits that completely blocked the cave entrance. A large part of the cave is filled with these clayey deposits, which make up the lower sections of the walls and create lateral benches in most of the galleries. In general, these deposits are highly reactive due to their elevated water retention and cation exchange capacities. The clay found in the different areas of the cave has similar characteristics. X-ray diffraction analyses indicate the presence of clay minerals kaolinite, illite, and smectite [5]. Clays, and smectite specifically, are very sensitive to small climatic variations that cause the minerals to expand and contract, referred to as swelling clays.

Shortly after the cave's discovery, work done to adapt the cave for tourism completely eliminated the deposits above the cave. This caused an imbalance to the cave's climate [6] because the deposits buffered the internal atmosphere from the exterior climate, as well as served as a thermal and hydrological barrier. Because of this, and to maintain moisture balance, water is transported to a tank in the Machine Room, located at the entrance room (SAS1-C3), where it can then be pumped to the exterior (▶ Fig. 13.1). Limited natural water enters the cave, in the Chamber of the Felines and the upper area of the Mondmilch Gallery.

Overall, the mechanisms responsible for the climatic equilibrium in the interior of the cave are complex. Climatic research from 1963 shows that during 6 months of the year, the cave air water vapor condenses on the cave walls. During the rest of the year, evaporation occurs. These mechanisms are caused primarily by changes in the air and rock surface temperatures, and cause either carbonate corrosion or calcite deposition as due to changes in carbon dioxide (CO_2) levels [3].

13.2 Review of Historical Events, Conservation Efforts, and Scientific Research

13.2.1 Discovery and public exhibition

The cave was discovered on September 12, 1940, by four boys, Jacques Marsal, Georges Agnel, Simon Coëncas, and Marcel Ravidat. They noticed the small opening in the sediments and thought it could be a secret passage to the Lascaux Manor House, as told by local legend. The boys immediately notified their teacher, Léon Laval, who informed the prehistorian Henri Breuil of the discovery. One month later, Breuil led the first team to record the figures by using photography and making drawings. A few days after the discovery of the cave, it received a great influx of visitors. By December 1940, the cave was already classified a Historical Monument with the agreement of its owner, the Count de la Rochefoucauld-Montbel.

After World War II in 1947, the owner began to make the cave accessible to the public. The cave entrance was significantly leveled and rock and sediment deposits that had blocked access to the entrance were removed [6]. A monumental bronze door was built, as were stairs to descend directly to the Hall of the Bulls (▸ Fig. 13.1). The cave floor was lowered to make a walkway for tourists and lighting was installed throughout the route. On July 14, 1948, Lascaux Cave was officially opened to the public. From the moment the cave opened, the number of visitors quickly increased. By 1949, it was observed some black coatings on the cave walls. In 1950, two ante-rooms would be built before the Hall of the Bulls [6].

The first archaeological excavations by Abbot Breuil and Séverin Blanc began 1949 in the area of the Shaft of the Dead Man. By 1952, Abbot André Glory began to record the numerous engravings located mainly in the Apse, but also in the Passageway and the Chamber of the Felines. This work lasted until 1963, during which time he identified close to 1,500 figures and produced around 120 m^2 of tracings.

13.2.2 First microbial crisis (1955–1970)

By 1955, the cave was receiving about 30,000 visitors per year, which increased to 100,000 in 1960. Peaks of 1,800 visitors per day occurred in the summer, during which time CO_2 levels increased, as did the amount of water condensation and the cave air temperature. Preservation of the paintings was needed.

At the petition of the Commission on Historical Monuments, a powerful machine to regenerate the atmosphere was installed in the cave. The work was done between December 1957 and April 1958. The cave's vestibule was excavated to a depth of 5 m to make space for the large pieces of equipment put into the Machine Room. Burying the air extraction conduits meant digging out part of the floor on the two principal passage areas, from the Machine Room to the back of the Axial Gallery, and from the Hall of the Bulls to the Mondmilch Gallery (▸ Fig. 13.1). This work seriously affected the cave. About 440 m^3 of debris was cleared (around 1,200 tons), added to the 32 m^3 cleared in the initial construction between 1947 and 1948 [6]. This work also destroyed the most essential part of the Paleolithic archeological strata that existed in the cave, and that was never subject to any organized study. Construction was only supervised by Abbott Glory, who did not have the material or human resources appropriate for any such intervention.

In 1960, Max Sarradet observed the appearance of green stains along the walls and associated with algal colonies. This phenomenon came to be known as the green disease or "maladie verte," and was caused by increased CO_2 and the lighting necessary for visitation. Accelerated formation of a calcite film coating the walls was also observed, being attributed to water vapor condensation. This phenomenon came to be known as the white disease, or "maladie blanche" [6]. During this period, bacterial colonization started to be observed. By 1962, the number of visitors continued to increase as the green stains reached a critical level. The cave owner closed Lascaux Cave to the public in January 1963. In response to the owner's request for an expert ruling to try to remedy the confirmed disturbances, André Malraux, Minister of Culture, created a Scientific Committee to study the climatic and microbiological imbalance. Jacques Pochon, a microbiologist at the Pasteur Institute and member of the Scientific Committee, had already previously recognized a significant level of contamination from bacteria, actinomycetes, and many types of fungi on the cave walls and in the cave air.

Lefèvre [7] identified elevated levels of diverse organisms on the rocks, including different genera of algae, bacteria, fungi, protozoans, rotifers, and nematodes. He also noted that the unicellular green alga, *Bracteacoccus minor*, was the cause of the green disease and decided on specific treatments to fight it. A biocide treatment of an aqueous formaldehyde solution was applied on July 18–21, 1963, to target the algae throughout the cave, on the floor, bare rock, and even affected paintings (▸ Tab. 13.1) [7]. The compound was chosen because it was thought not to damage the metal oxides present in the natural pigments of the paintings [7]. An aqueous antibiotic solution, comprising of penicillin, streptomycin, and kanamycin, was also sprayed (▸ Tab. 13.1). All of these treatments were effective and eliminated the green stains in 4 months. However, in 1969, cave managers resumed treatments in some areas to fight new green colonies that appeared. Use of the previously applied biocide was recommended. As a preventive measure in the entryway to the cave and on one of the visitor walkways, the biocide was used until 1998 [4].

Table 13.1. Review of biocide treatments applied in Lascaux Cave, France.

Date	Biocide	Application mode	Target
July 1963	Penicillin, streptomycin and kanamycin; aqueous solutions	Spraying throughout the cave	Reduce bacteria, fungi and actinomycetes, previously to the treatment for algae
1963	Formaldehyde; substrate-dependent dilutions in water: 1:10 for sediments, 1:20 for rocks and 1:200 for paintings	Spraying throughout the cave	Green stains caused by algae (*Bracteacoccus minor*)
1969	Formaldehyde; aqueous solution	Spraying on affected areas	Green stains caused by algae
July 2001–December 2003	Vitalub QC50 (benzalkonium chloride 50%); solutions at 5%	Spraying on affected areas	White mycelial masses of *Fusarium solani*; Black stains
October 2001–December 2005	Quicklime; total amount 1,360 Kg	Covering all floors and affected lateral benches (total surface 250 m^2)	White mycelial masses of *Fusarium solani*
November 2001–July 2002	Vitalub QC50 (benzalkonium chloride 50%); solutions at 5%	Soaked compresses covering the affected areas	White mycelial masses of *Fusarium solani*
December 2001 (2 weeks)	Streptomycin sulfate 0.2% Polymyxin sulfate 0.15%	Spraying on affected areas	Bacterial colonizations (*Pseudomonas fluorescens*)
January 2008	Devor Mousse (benzalkonium chloride and myristalkonium chloride 10–25%; 2-octyl-2H-isothiazol-3-one 2.5%); solutions at 5%	Spraying on walls of Passageway, Apse and Nave	Black stains
August 2008–February 2009	Devor Mousse at 5% Parmetol DF12 (isothiazolinone) at 3%	Spraying and brush application on two black stains located in the Passageway (areas A and C)	Test to evaluate the efficacy of treatments against the black stains

In 1966, the air conditioning system was taken apart. A new climatic regulation system was installed in 1967 to create an artificial cold zone in the Machine Room. The objective was to stop the condensation and appearance of calcite films on the walls. The cold zone was meant to reproduce the effect of the natural deposits that had originally filled the cave entrance.

The cave has remained officially closed since April 18, 1963. In 1969, the Scientific Committee set a limit on the human presence in the cave: five people a day, for a maximum of 35 minutes, only five days a week. This measure continues in place today, with the exception of certain emergency situations [2].

13.2.3 Returning to the microbial balance (1970–2001)

In 1972, the Rochefoucauld Company donated the cave to the state. Slightly higher levels of microbes in the cave air were found in 1973, related to work done to build the cave replica. In 1979, Lascaux Cave and others located in the Vézère River valley were declared Heritage Sites by the United Nations Educational, Scientific and Cultural Organization (UNESCO). In 1983, Lascaux II was opened to the public. Lascaux II, a replica of the two main adorned halls, the Hall of the Bulls and the Axial Gallery, is located 350 m from the original cave and takes advantage of an old, partially buried quarry. Today, the Lascaux II tourist site receives more than 280,000 visits a year.

Between 1988 and 1999, Norbert Aujoulat resumed study of the parietal art present in Lascaux Cave, including the figures in the Hall of the Bulls and the Axial Gallery. The objective of this study was to identify the factors that influenced the construction sequence of these figures and iconographic ensembles [1].

From 1970, periodic checks were done on the microbiological presence in the cave air. This was initially done by the Pasteur Institute in Paris and later by the Historical Monument Research Laboratory, from the French Ministry of Culture and Communication (LRMH). Since 1994 the presence of microbes in the cave air became less stable, which coincided with the appearance of fungal colonies on the Hall of the Bulls floor. In 1996, some gray-green lichens (*sic*) were observed on the Hall of the Bulls walls [4]. This last observation seemed doubtful because of the ecology of the lichens, but unfortunately, no report has been found that identifies or gives further information about the observation. Then until 2001, the total count of the fungi and bacteria in the cave air has been kept at relatively low levels [4, 8].

13.2.4 Second microbial crisis (2001–2006)

The climatic regulation system installed in the cave in 1967 was replaced between 1999 and 2000. The new system was designed to carry out the same functions, but in practice the results became worse. Immediately after its installation in June 2001, white fungal mycelia began to appear in the entry hall SAS2. Initially, between July and September 2001, the affected areas were sprayed with a 5% solution of the biocide Vitalub QC50 (50% benzalkonium chloride), first dissolved in 90% ethanol and later in water. These treatments were unable to stop the outbreak, and the fungi quickly spread and reached the floor, vaulted ceilings, all of the lateral benches and walls of the Hall of the Bulls, the Axial Gallery, the Passageway, and the Nave. By September 2001, the LRMH identified the fungus *Fusarium solani*, very common in agricultural areas, to be responsible for this outbreak.

The LRMH developed a treatment plan with very aggressive biocides as an emergency measure [4, 8] (▸ Tab. 13.1). In November 2001, compresses soaked in 5% Vitalub QC50 were applied to the affected areas. Additionally, in October and November 2001, quicklime was applied to the floors and lateral benches of the cave to reduce the humidity, to slow fungal dispersion, and to produce an antifungal effect by drying the hyphae quickly. Approximately 1360 kg was spread out over 250 m^2 [9]. On August 22, 2002, the French Ministry of Culture and Communication created a Scientific Committee for the study and preservation of the cave, presided over by Marc Gauthier. The Vitalub QC50 spray was a normal practice through 2002 and 2003, in the fight against the outbreak of *F. solani*. The floors and lateral benches remained covered with quicklime until December 2005.

The cave habitat continued to be clearly disturbed throughout the treatment period and new microbiological problems appeared [10, 11]. Localized white, mucous colonies associated with the bacteria *Pseudomonas fluorescens* appeared and were treated during a few weeks with antibiotics: 0.2% streptomycin sulfate and 0.15% polymyxin sulfate [8, 12]. Since December 2001, 5 months after the beginning of the antibacterial treatments, the first localized black stains appeared on the walls and on the ceiling in the Machine Room and in the SAS2 room [13]. The black stains progressively expanded throughout the cave, and the new outbreak was treated with the same biocides (Vitalub QC50 sprays). Since January 2004, biocide applications were interrupted and treatment to fight white and black colonies involved cleaning with sponges, scalpels, and mechanized equipment (Gregomatic injector-extractor).

In March 2004, the Scientific Committee organized a complete conservation plan to manage the interventions and research necessary in the cave. In this way, a series of priority actions were established, among which three actions stand out because of their microbiological relevance:

(i) A Microbiology-Microclimate research program, with the objective being to characterize the microbial communities in Lascaux Cave and to look for possible correlations between microbial development and atmosphere and substratum physical parameters. This project began in 2006 with the participation of the microbiologists Claude Alabouvette and Fabiola Bastian of the "Institut National de la Recherche Agronomique" (INRA), Dijon, France, the climate specialist Adriana Bernardi, from the "Istituto di Scienze dell'Atmosfera e del Clima, CNR," Padua, Italy, and the LRMH.

(ii) Quicklime extraction from the floors and lateral benches. This work finished between December 2005 and October 2006.
(iii) Continued follow-up and mechanized colony and stain cleaning, depending on the intensity of the phenomenon and its location.

Dupont et al. [14] describe the isolation of fungal genera from white colonies, including *Chrysosporium*, *Gliocladium*, *Gliomastix*, *Paecilomyces*, *Trichoderma*, and *Verticillium*. They also conducted a molecular study of 36 strains of *Fusarium*, which were isolated mainly in the years 2004 and 2005, from the white colonies that covered the walls and lateral benches in the different rooms of the cave. All of the *F. solani* strains were identified as members of the species complex *F. solani*, which is comprised of more than 45 phylogenic and/or biological species [15]. Strain molecular characterization, based on the elongation factor 1α gene (EF-1α), ribosomal large subunit (LSU), and the ribosomal internal transcribed spacers (ITS), suggests that the *F. solani* complex belongs to three different phylogenic species, all of which are included within the clade 3 established by Zhang et al. [15]. This genetic variability found in Lascaux Cave could be indicative of there being different introduction events over time, including that there are diverse entry routes into the cave, such as from the ground above the cave through water or from small animals, or from the exterior and due to elevated human presence and activities [14].

The LRMH continued periodic testing of the cave air for microbiological contamination, while also increasing the control points significantly and analysis frequency. Noteworthy contamination fluctuations were observed, with significant air pollution from the end of 2004 through the beginning of 2005. Concentration levels of the genera *Penicillium* and *Aspergillus* increased due to higher human presence needed for control and maintenance [8]. Since 2005, the LRMH conducted numerous microbiological analyses to isolate and identify different colonies that appeared primarily in SAS1-C1, the Passageway, the Apse, and the Nave. The majority of the colonies analyzed were white and belonged to the genus *Fusarium*. However, colonies of other genera were isolated, including *Verticillium*, *Penicillium*, *Aspergillus*, *Gliomastix*, *Gliocladium*, *Cladosporium*, and *Alternaria* [4].

13.2.5 Third microbial crisis (2006–Present)

By 2006, white colony outbreaks were progressively reduced. But, the outbreak of black stains that first appeared at the end of 2001, with the white colonies and after emergency biocide treatments, advanced explosively and became the main problem threatening Lascaux Cave. There were two types of black stain phenomenon, differentiated by morphology and substratum type that affect stain diversity and evolution. The first type of black stain occurred on limestone walls and vaulted ceilings. The second type occurred on lateral benches and clayey sediments.

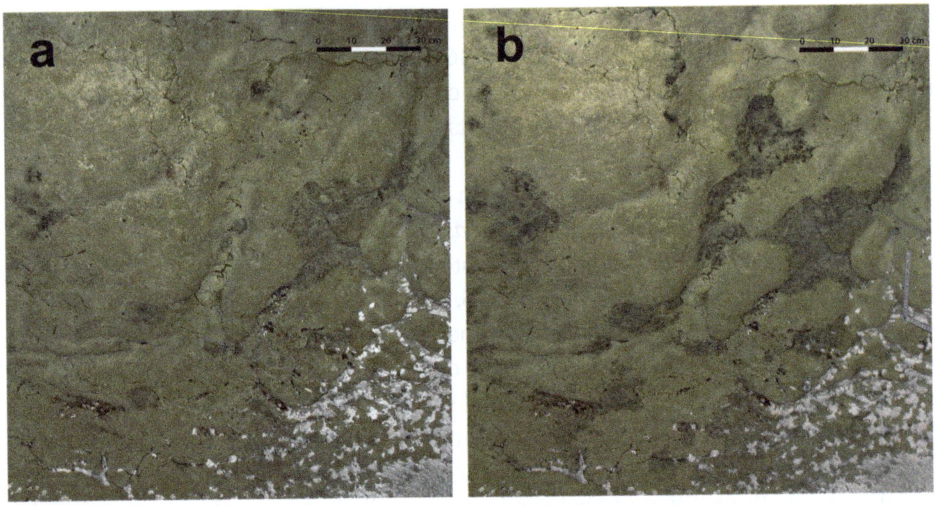

Fig. 13.3. Development of black stains in the vault of the Passageway. (a) December 2007. (b) June 2008 [16].

Starting December 2001, the first black stains appeared in the cave on bare limestone (▸ Fig. 13.3). They extended very quickly to colonize the majority of the vaulted ceilings in only 7 months. The dispersion pattern went from the Machine Room, to room SAS2, the Hall of the Bulls, the Axial Gallery, the Apse, the Nave, and the Mondmilch Gallery excluding the Passageway. Where the vaulted ceilings were covered by calcite, the black stains were limited to bare limestone. But, some of the colonies appeared on paintings in the Hall of the Bulls and on engravings in the Apse, as small black dots encrusted in the substratum or superficial gray-black and powdery features. The fungi *Gliomastix murorum* and *Cladosporium* sp. were initially identified by LRMH for some stains from the Hall of the Bulls [4].

Abundant, new black colonies were noted in March 2006 and continued through recently on the vaulted ceilings of the Passageway, the Nave, and the Apse. Resembling the black stains from the earlier outbreak, this new stain appeared very quickly, although its evolution was rather slow. By the summer of 2007, extensive staining developed and put the conservation of some paintings and engravings at risk. Some of the paintings with close contamination included the antlers of the Fallen Stag, the horns of the Great Black Cow, and the Frieze of the Swimming Stags. Two events could be considered the cause for stain appearance, specifically the increase in human activity in affected rooms starting in November 2005 to control, treat, and extract the quicklime, and the December 2006 dismantling of the "SAS Bauer" door situated between the Hall of the Bulls and the Passageway, originally installed in 1965. Door removal considerably increased the levels of water condensation in the cave, as well.

The other type of black staining appeared on lateral benches in all the rooms of the cave and clayey sediments between 2001 and 2004. Initially, these colonies were thought to be necrotic mycelium from the white colonies after the chemical treatment, as the two types of colonies were frequently associated. For that reason, treatments for both outbreaks were identical. Black stains were also observed in damp areas. From 2004 onward, black stains appeared separate from the white colonies on lateral benches throughout the cave. These stains continued growing and reached significant proportions by 2006 and 2007. The most obvious phenomenon was in the Passageway, where large surfaces of the benches suffered from progressive blackening. As with the black stains on bare limestone, the increase in blacking on benches and sediments was related to human activity and dismantling of the "SAS Bauer" door. It is important to point out that in rooms where the visitor numbers were lower, such as the Mondmilch Gallery and the Chamber of the Felines, the stains appeared on clayey substratum following a very different evolution process and the staining lessened over time.

Specifically, starting in January 2004, arthropods were detected in the Axial Gallery, on the Passageway lateral benches, and on the Mondmilch Gallery floor. A year later, in June 2005, the arthropod population increased considerably, being well distributed throughout the cave (Scientific Committee, June 2005 session). Most of the arthropods were found on black stains on clayey substratum and were identified as *Folsomia candida*, a collembolan [16]. The 2005 increase in black stains was attributed to *F. candida*. While efficiently feeding on the fungal mycelia, the collembola served as a dispersion vector because the fungal spores adhered to the bodies, or spores were dispersed in collembola excrement.

In April 2006 and January 2007, the INRA researchers took samples representative of different situations common in the cave: dry and damp areas, without visible colonies, with white and/or black colonies, and areas treated with biocide. Subsequent sample analysis, in collaboration with the "Instituto de Recursos Naturales y Agrobiologia" (IRNAS-CSIC, Seville, Spain), was intended as a way to create an extensive list of fungi and bacteria present in Lascaux Cave.

To do this, INRA and IRNAS-CSIC put together a DNA library based on different ribosomal genes, including 18S rRNA for fungi and 16S rRNA for bacteria and archaea. The study of the fungal communities [17], based on the analysis of 607 clones, revealed that the eight most common fungal phylotypes were entomophilous, *Geosmithia namyslowskii*, *Geosmithia putterillii*, *Isaria farinosa*, *Aspergillus versicolor*, *Tolypocladium cylindrosporum*, *Geomyces pannorun*, *Engyodontium album*, and *Clavicipitaceae* spp. Only seven of the clones were found to be corresponded to *F. solani*, likely due to the decrease after successive years of biocide treatment. The study of the bacterial communities [18], based on the analysis of 696 clones, revealed that the two most abundant genera were *Ralstonia* and *Pseudomonas*. The most abundant phylotypes were *Ralstonia mannitolilytica* and *Ralstonia pickettii*, both species having been described as human pathogens [19, 20]. Only five clones of *P. fluorescens*, a species considered responsible for the mucous colonies that appeared in 2001 after beginning the biocide treatments, were found [8, 12]. According to Bastian et al. [18], two possibilities could explain the scant representation of this species, including that the biocide treatments effectively eliminated the populations and that the original samples were incorrectly identified and really corresponded to a species from the genus *Ralstonia*. Biochemical identification of the species *Ralstonia* using commercial tests can be problematic and results are often confused with those of *P. fluorescens* [19, 21]. Other bacterial phylotypes corresponded to the genera *Afipia*, *Legionella*, and *Aquicella*, which can be described as free-living amoeba parasites. These protozoa are abundant in Lascaux Cave, as was demonstrated by Garcia-Sanchez et al. [22]. In that study, they identified four different species of amoebae, *Acanthamoeba astronyxis*, *Acanthamoeba castellanii*, *Acanthamoeba* sp. and *Hartmannella vermiformis*, in cave sediments. The abundant presence of bacteria with potentially pathogenic protozoa could be due to selection caused as a result of prolonged biocide use [23].

The black stain phenomenon in 2006, and especially in the summer of 2007, stirred up a great media debate due to the threat posed to the paintings and engravings, lack of understanding of the staining origins, and according to some authors [24–26], poor cave management since its discovery. This set off a debate between the International Committee for the Conservation of Lascaux (ICPL), primarily made up of American cave painting aficionados, and the organization in charge of cave management and conservation (i.e. the French Ministry of Culture and Communication). By August 2007, the LRMH and INRA research groups conducted parallel isolations on culture media to identify the most abundant strains and to find the fungi responsible for the black stain formation. Samples from the vaulted limestone ceilings of the Passageway and the Apse were analyzed, with many being collected from around the antlers of the Fallen Stag in the Apse. Using morphological criteria, the LRHM identified two majority genera, *Ulocladium* and *Gliocladium*. *Ulocladium* spp., which developed melanized colonies and, for that reason, were thought to be the main cause of the black stains. In contrast, *Gliocladium* spp. form white colonies [4]. The INRA obtained similar results, in that there were abundant olive green to black colonies of a dematiaceous fungus, similar to *Ulocladium* spp. described by the LRMH. However, the microscopic characteristics did not coincide with those described for the genus *Ulocladium*, and the molecular analysis of the 18S rRNA gene approximated it to the genus *Scolecobasidium* [27]. Later, analysis of the ITS regions determined that the closest species corresponded to *Scolecobasidium tshawytschae* [16]. This and other species in the genus *Scolecobasidium*, with ellipsoidal, clavate or fusiform conidia, were placed as the genus *Ochroconis* since 1973, but the genus *Scolecobasidium* remained restricted to the species with trilobate conidia [28]. Consequently, identification of primary fungi associated with black stains was still unclear in the years 2009 and 2010.

Taking into account the critical situation that the black stains had reached, a new biocide treatment was applied in January 2008 to the vaulted ceilings in the Passageway, the Apse, and the Nave. The biocide product selected by the LRMH, based on its efficacy in tests against *Ulocladium* and *Gliocladium* strains, was an aqueous solution of 5% Devor Mousse, which is a mixture of quaternary ammonium (10–25% benzalkonium chloride, 10–25% myristalkonium chloride) and 2.5% 2-octyl-2H-isothiazol-3-one (▸ Tab. 13.1). The biocide was sprayed for three consecutive days (8th–10th January) and the cave remained closed during the following 3 months. The LRMH analyzed diverse control parameters in different treated areas before and after the treatment. According to these analyses, the biocide treatments reduced air pollution, metabolic activity, and the viability of fungi present [4, 13]. However, the treatments were ineffective in controlling the black stains.

From reports sent to UNESCO by the ICPL in September 2007 and February 2008, UNESCO threatened the organizations in charge of cave management and conservation in July 2008 that they would include Lascaux Cave on the list of World Heritage Sites in Danger, unless the managers carried out the necessary measures to guarantee the cave's conservation [29]. Currently, the cave continues to be considered a World Heritage Site, and the French government regularly reports to UNESCO on its status, including work being done for conservation.

With the last biocide treatments, the debate about their efficacy and the risk they pose to the cave's ecosystems was reopened. Between August 2008 and February 2009, a test was done to evaluate the efficacy of different treatment measures, specifically biocide application and/or cleaning protocols. For this, four zones colonized by black stains on limestone rock in the Passageway were chosen. Each one of the treatment protocols was applied using a different treatment combination: (a) surface cleaning and biocide, (b) surface cleaning without biocide, (c) deep cleaning and biocide, and (d) deep cleaning without biocide. In zones A and C, three sprays of 5% Devor Mousse were applied with an additional application of 3% Parmetol DF12 (isothiazolinone) (▸ Tab. 13.1). According to the LRMH analysis [4], initially abundant colonies of the genera *Ulocladium* and *Curvularia* were isolated and with lesser frequency for *Verticillium* and *Penicillium*. The biocides used noticeably reduced the metabolic activity and the viability of the fungi, especially considering that a single fungus was isolated in areas treated with biocides.

13.3 Recent Research on the Black Stains Outbreak (2009–2013)

In May 2009, the Scientific Committee passed a research project entitled "Microbial Ecology in the Lascaux Cave," to be carried by researchers from the IRNAS-CSIC and the INRA. This 2-year project centered on the black stains and whose main results are summarized below.

13.3.1 *Ochroconis* associated with the black stains

Identification of previously described *Scolecobasidium* or *Ulocladium* fungi [4, 16] was carried out by analyzing 30 strains with similar morphologies that were isolated between 2007 and 2011 from different cave substrata. Their molecular characterization, analyzing the ITS regions and the RNA polymerase II subunit B (RPB2), and its morphological analysis, allowed for two new *Ochroconis* species descriptions [30] that were closely related to the genus *Scolecobasidium*. Twenty-eight of these strains belonged to a new species, described as *Ochroconis lascauxensis* (▸ Fig. 13.4). This species was found associated with the black stains on limestone, as well as on clayey sediments and in the air in different cave rooms. The two remaining strains show identical characteristics, described as another new species, *Ochroconis anomala*, whose distribution was quite restricted.

Due to the relevance of these identifications, *O. lascauxensis* has been recently included into the "2013 Top 10 New Species List" by the International Institute for Species Exploration at State University of New York [31]. These top 10 species were selected from more than 140 nominated species and out of 18,000 estimated species named in 2012 by an international committee of taxon experts.

With the objective of determining the distribution of *O. lascauxensis* in Lascaux Cave, a real-time PCR protocol was developed for the specific detection and quantification of this fungus in environmental cave samples [32]. A total of 51 samples collected between 2008 and 2010 were analyzed (e.g. black stains, uncolonized sediments, and air). The results showed that *O. lascauxensis* was extensively distributed throughout the cave and associated with black stains. Stain concentrations were generally higher than those detected in the sediments without apparent colonization. The greatest concentrations registered were of sediments from cave rooms primarily affected by black stains, the Hall of the Bulls, the Passageway, and the Apse. Likewise, the results obtained from the air samples coincided with the presence of black stains, with higher concentrations in the Passageway and at the beginning of the Axial Gallery, followed by adjacent area samples.

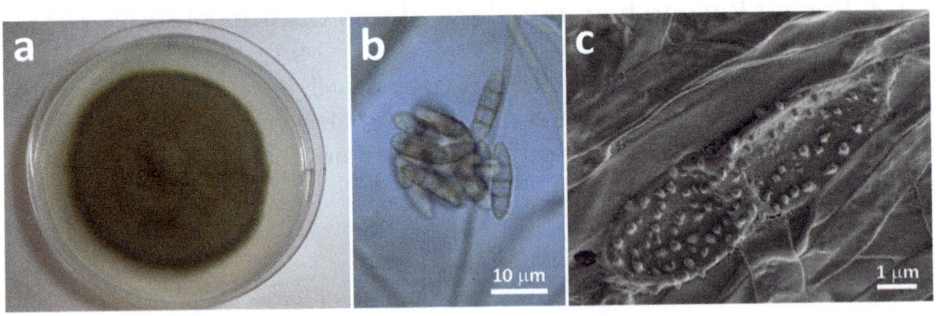

Fig. 13.4. *Ochroconis lascauxensis.* (a) Ten-day-old colonies on MEA media. (b) Conidia. (c) Scanning electron photomicrographs of conidia [30].

To determine the chemical nature of black stains in Lascaux Cave, four samples taken from different rooms and the melanin from *O. lascauxensis* were analyzed using surface-enhanced Raman spectroscopy (SERS) [33]. The *O. lascauxensis* melanin spectrum showed bands at 1610, 1305, and 1249 cm^{-1}, which were considered to be markers. In the SERS analysis of black stains taken directly from the cave, the biomarker bands were evident in the regions 1600–1610, 1300–1310, and 1240–1250 cm^{-1}, which suggested the *O. lascauxensis* melanin participated in black stain formation [33].

One of the stains analyzed was also abundantly colonized by collembolans and from which the biomarker bands had a much greater intensity than the rest, which confirmed that the collembolans enriched the fungal melanin stains. In effect, *F. candida* feeds on *O. lascauxensis* mycelia, but is not capable of digesting the recalcitrant compounds, like melanin, like the polysaccharides and proteins from the fungi. As such, melanin is excreted and persists on the cave walls as a residual compound that cannot be degraded by microorganisms. Analysis of excrement from collembolan fed with *O. lascauxensis* mycelia revealed SERS spectra with the characteristic melanin biomarker bands.

13.3.2 Evaluation of biocide treatment of black stains on limestone

Stain fungal communities in the Passageway, referred to as zones A and C, were found where biocide treatments were evaluated between August 2008 and February 2009. Stain C is shown in ▸ Fig. 13.3. Fungal community molecular characterization was done using denaturing gradient gel electrophoresis (DGGE) and DNA clone libraries [34]. The results showed the initial communities of the two stains before the treatments were clearly dominated by the fungus *O. lascauxensis*. Only two operational taxonomic units (OTUs) were detected in stain C, with the majority corresponding to *O. lascauxensis* (97% of the clones) and the other OTU being *Pochonia chlamydosporia* (3% of the clones). These results confirm that *O. lascauxensis* participated actively in the formation of the black stains that extended on the limestone walls and vaulted ceilings, between 2006 and 2008. This species has been isolated on repeated occasions in stain C and in other stains [30], and initially, *O. lascauxensis* acted as saprophytes that were specialized in organic matter decomposition. After treatment, the fungal community was much more diverse, with 12 OTUs detected. The proportion of *O. lascauxensis* was considerably reduced, although it continued to represent the majority (52.3% of the analyzed clones). The rest of the clone library was made up of the genera *Aspergillus*, *Cladosporium*, *Trichoderma*, *Alternaria*, *Rhodotorula*, and *Gymnascella*, all of which was frequently found in air cave samples [35, 36]. The DGGE profile results obtained in zone A were similar to zone C. In areas treated with biocides, there was rapid community succession and *O. lascauxensis* was partially replaced by other bacteria and fungi. These results cast doubt on the efficacy of applied treatments, according to what was established by Bastian et al. [18]. In all likelihood, the biomass of dead microorganisms and the biocide breakdown products provided the necessary organic matter to feed the secondary colonizers. The end result of the biocide treatments was an increase in the fungal diversity.

13.3.3 Black stain fungal communities on clayey sediment

Seventeen stain samples that appeared on the clayey walls and benches of the cave were collected between 2008 and 2010 and analyzed. The characterization of these fungal communities was done by using various techniques, including DGGE, DNA, and RNA clone libraries, and isolations on culture media [34]. The initial DGGE characterization of the communities showed clear differences in relation to the stains developed on limestone. In general, metabolically active fungal communities of this stain type were fairly diverse and included species of black yeasts from the family Herpotrichiellaceae. The fungus *Acremonium nepalense* was also common in the black stains. None of the samples had a DGGE pattern characteristic of *O. lascauxensis*, but quantifications done by real-time PCR and isolations on culture media revealed that the species was present in all of the samples [30, 32].

From clone libraries of three black stain samples taken in different areas of the cave in September 2010 (M1, M6, and M8), *O. lascauxensis* was only detected in M6 and represented 35.2% of the DNA clones. However, this species was not found to be metabolically active because there were no RNA clones detected. These results could be due to the presence of an elevated concentration of fungal spores, which would be present from previous activity.

The fungus *A. nepalense* was also isolated on repeated occasions from the different stains on the clayey substratum, but especially the black-colored stains (▸ Fig. 13.5). The abundance of this species was demonstrated through molecular analysis, DGGE, and clone libraries. Recently, different species of the genus *Acremonium* (section *Gliomastix*), associated with black stains that are deteriorating mural paintings located on funerary monuments in Japan, have also been detected [37].

The most abundant fungi in the clone libraries were species of black yeasts affiliated with the family Herpotrichiellaceae, with percentages of clones for the DNA analysis of 64.5% in M1, 41.8% in M6, and 50% in M8. For RNA, the black yeasts represented 56.5%, 71.6%, and 19.5% of the M1, M6, and M8 samples, respectively. Eight species of black yeasts were phylogenetically differentiated according to their ITS sequences. Two of the species were identified as *Exophiala castellanii* and *Exophiala moniliae*, described as human pathogens, but six of the species are still currently unknown. Black yeasts are especially relevant for their pathogenic potential, and especially the genus *Exophiala* that has been implicated in a wide number and variety of human infections [38]. *Exophiala* spp. can pose an important risk to the health of the staff that visit the cave. Therefore, prevention methods must be undertaken. Different black yeasts have also been associated with the formation of black stains on monu-

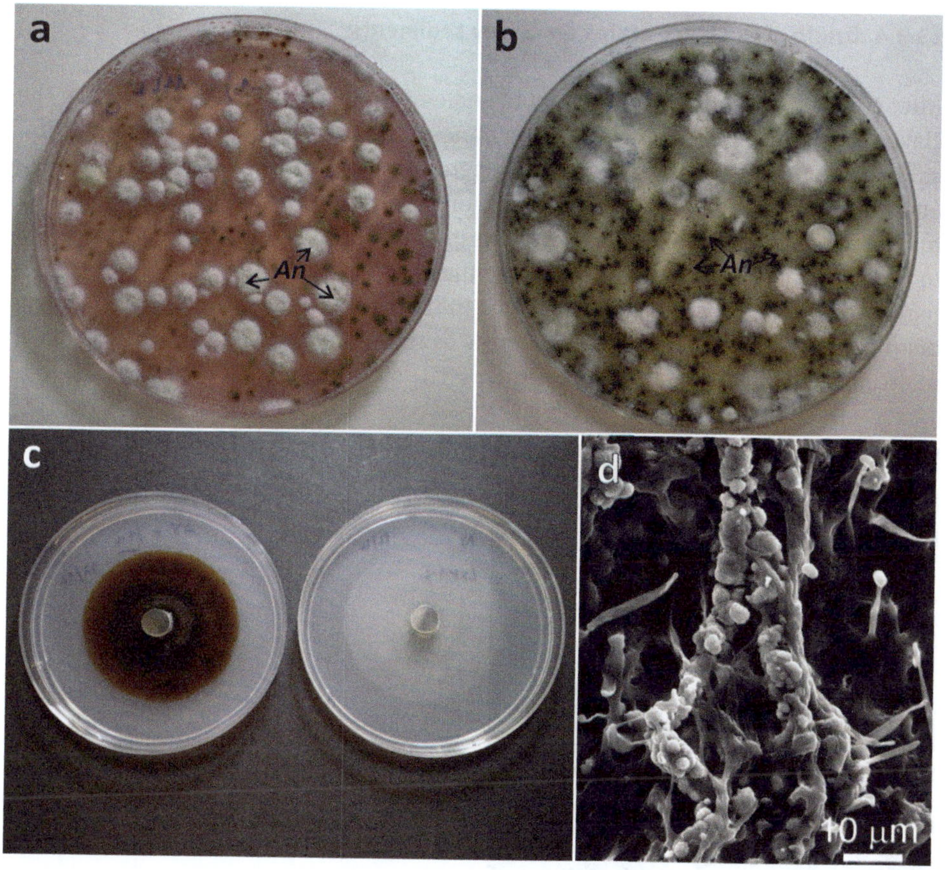

Fig. 13.5. Cultures of *Acremonium nepalense*. (a) Fungal colonies isolated from a black stain on clayey sediments of the Passageway and grown on DRBC medium at 15 °C for 25 days. *Acremonium nepalense* (*An*) developed abundant white-pink colonies in this medium. (b) Fungal colonies isolated from a black stain on clayey sediments of the Apse and grown on ECA medium at 15 °C for 25 days. *Acremonium nepalense* (*An*) developed abundant dark green colonies in this medium, which contains 0.4 mg/L $MnSO_4$. (c) Colonies of *A. nepalense* strain LXM1-1 grown for 10 days on AY solid medium supplemented (left) or not (right) with 1mM $MnSO_4$. (d) FESEM image of *A. nepalense* after grown for 20 days in AY-HEPES liquid medium with $MnSO_4$ and showing extensively mineralized hyphae with biogenic manganese oxides. Methodological details in [40].

ments, especially those belonging to the family Herpotrichiellaceae [39]. These black yeasts represent a greater preservation threat to the cave and its paintings.

In all, these studies confirm the need to use a combination of molecular techniques and culture-dependent techniques to make a complete description of the fungal communities on the black stains. The most practical and reliable approach turned out to be the comparative analysis of DNA and RNA by construction of clone libraries. However, these techniques also showed some limitations that made culturing necessary.

13.3.4 Origin of the black stains on clayey sediments

After completing a microscopic study of some of the black stains developed on the clayey sediment in the Passageway, the mineral substratum showed abundant fungal hyphae associated with the mineral aggregates of manganese oxide (▸ Fig. 13.6) and was the main cause of the black stains on clayey sediment [40]. The bacteria and fungi that oxidize manganese are very widespread in nature and have been described in different habitats. In recent years, identification of species with these properties has increased significantly and shows that the ability to oxidize manganese is widely distributed among bacteria [41]. Manganese oxidation caused by fungi could be as or more important [42].

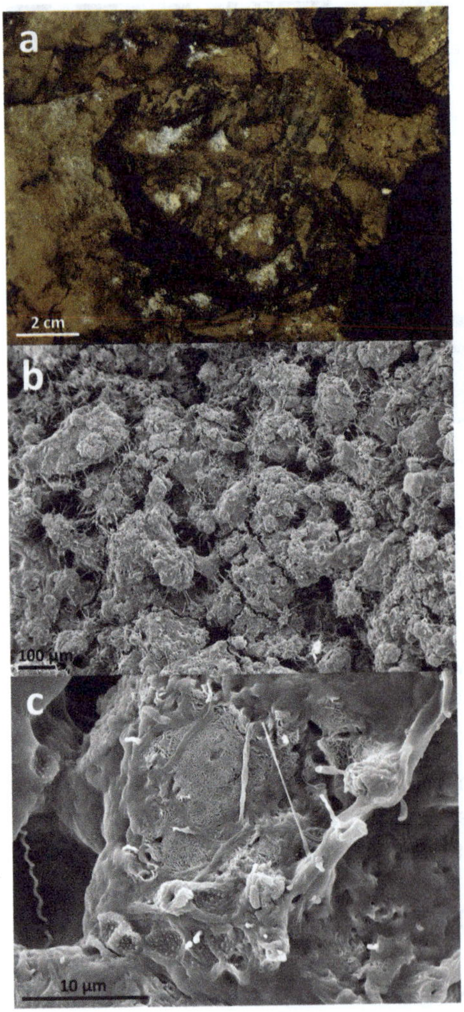

Fig. 13.6. Morphology of a black stain on clayey sediments the Passageway, Lascaux Cave, September 2010. (a) Macroscopic appearance. (b, c) FESEM images showing mineral aggregates associated with abundant fungal hyphae. At higher magnification, fungal structures closely related to the mineral substratum rich in manganese oxides and extracellular polymeric substances are evident.

From stains on clayey sediment (▸ Fig. 13.6 (a)), the morphology of manganese oxides coincided with structures previously described for biogenic manganese oxide, as observed by using field emission scanning electron microscopy (FESEM) (▸ Fig. 13.6 (b) and (c)) [42–44]. In the samples, abundant fungi associated with extracellular polymeric substances (EPS) surrounded manganese oxide mineral formations, which indicated that they could be the cause of manganese oxide precipitation. In ▸ Fig. 13.6 (c), the EPS matrix and associated manganese oxides are especially evident, which suggests that the EPS may serve as nucleation sites for manganese oxide precipitation.

Element distribution images generated by using energy-dispersive X-ray spectroscopy (EDS) indicated high Mn, Ca, O, and C values (▸ Fig. 13.7). These results corroborated that the black stains from clayey sediments are composed of manganese oxides. The presence of C is consistent with a microbial origin, and Ca is probably

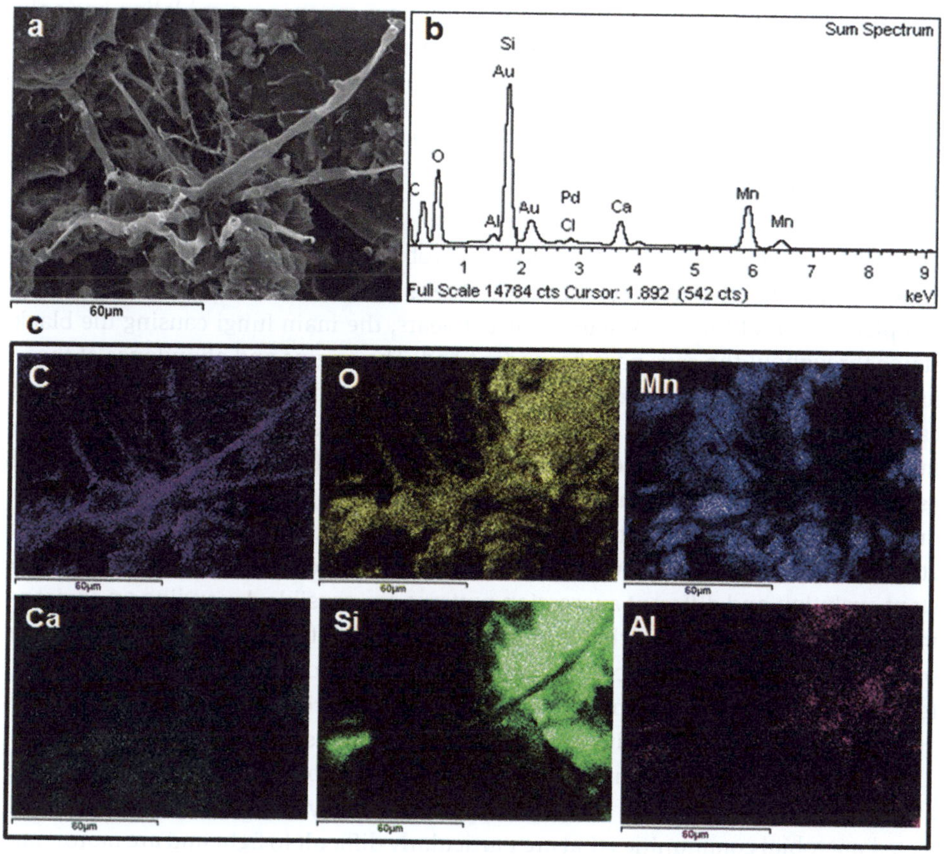

Fig. 13.7. (a) FESEM photomicrograph of a black stain from The Passageway, Lascaux Cave. (b) EDS spectrum from (a). (c) Element maps of the selected area that show the spatial distribution of carbon (c), oxygen (O) manganese (Mn), calcium (Ca), silicon (Si), and aluminum (Al).

associated with manganese oxides. White et al. [45] conclude that varying amounts of elements can be absorbed or incorporated into the manganese oxide structure based on the mineralogy and trace element chemistry of black manganese oxide deposits in caves. The high silicon content and the presence of aluminum in areas not enriched in carbon and manganese are consistent with the mineralogical composition of the clay substratum.

The capacity of the fungi isolated in the Lascaux Cave to oxidize manganese was later evaluated with culture media containing manganese, in the form of $MnSO_4$. Results indicated that *A. nepalense* participates actively in stain formation and induces Mn oxidation (II) to Mn oxides (III, IV) (▸ Fig. 13.5 (c) and (d)). In previous studies, this species was abundantly detected and found to be metabolically active in the stain types [34]. Miyata et al. [44] previously characterize manganese oxides produced by a species of *Acremonium*. Therefore, the precipitation of the biogenic manganese oxides in the black clayey stains is due to biomineralization caused by fungi, and possibly bacteria, present in the cave. Although up until now, manganese oxidation has only been associated with the fungus *A. nepalense*, a more in depth study should be done to detect other microorganisms that could be implicated in this process.

13.4 Conclusions

Despite the interest and years of study, the microbiology and microbial ecology of Lascaux Cave remained a mystery until 2009. These recent studies contribute to a more complete vision of the ecosystem. In recent years, the main fungi causing the black stains throughout the cave have been identified. The origins and abundance of these fungi are probably related to early biocide treatments and the introduction of organic carbon dissolved in dripping waters, especially during periods of intense rain. These findings should be considered by whoever is in charge of cave conservation that will establish adequate control guidelines to prevent the propagation of the earlier or new outbreaks. However, it is not possible to propose one specific solution to the black stain problem in Lascaux Cave. The microorganisms colonizing the cave are in constant evolution and different succession patterns are possible depending on environmental changes, among which it is important to highlight the influence and presence of carbon and nitrogen sources. Moreover, the cave's ecosystem has been totally altered by all of the different interventions performed since the cave's discovery, from adaptation work, massive visitation numbers, different systems of climatic regulation, and continuous biocide treatments. Currently, the microbial communities in the cave are not related to those that existed before its discovery. Human activities have selected microbial communities that are particularly difficult to fight and are more competitive in successional processes. Knowledge of what has happened to the cave, and of what and why it is happening now, could facilitate our understanding of patterns of microbial behavior to deal with conservation from a logical and scientific point of view.

Acknowledgments

Part of the research described in this chapter was supported by the DRAC Aquitaine, Ministry of Culture and Communication, France, project "Ecologie microbienne de la grotte de Lascaux." The Spanish Ministry of Science and Innovation, "Research Programme in Technologies for the Assessment and Conservation of Cultural Heritage," TCP CSD2007-00058, and CSIC project 201230E125 also contributed. AZM was supported by a Marie Curie grant. The facilities of Lascaux Cave staff, restoration team, DRAC Aquitaine, and the collaboration of Claude Alabouvette, Fabiola Bastian, and Alena Nováková are acknowledged.

References

[1] Aujoulat N. Lascaux: Le geste, l'espace et le temps. Paris, France, Editions du Seuil, 2004, 274 p.

[2] Geneste J-M. The major phases in the conservation of Lascaux Cave. In: Coye N, ed. Lascaux and Preservation Issues in Subterranean Environments. Proceedings of the International Symposium (Paris, February 26 and 27, 2009). Documents d'Archéologie Française 105. Paris, France, Éditions de la Maison des Sciences de l'Homme, 2011, 51–71.

[3] Malaurent P, Lacanette D, Brunet J, Riss J. Climatology of the subterranean environment at Lascaux: from a global study to the microclimatology of the cave walls. In: Coye N, ed. Lascaux and Preservation Issues in Subterranean Environments. Proceedings of the International Symposium (Paris, February 26 and 27, 2009). Documents d'Archéologie Française 105. Paris, France, Éditions de la Maison des Sciences de l'Homme, 2011, 121–142.

[4] Orial G, Bousta F, Francois A, Pallot-Frossard I, Warscheid T. Managing biological activities in Lascaux: identification of microorganisms, monitoring and treatments. In: Coye N, ed. Lascaux and Preservation Issues in Subterranean Environments. Proceedings of the International Symposium (Paris, February 26 and 27, 2009). Documents d'Archéologie Française 105. Paris, France, Éditions de la Maison des Sciences de l'Homme, 2011, 219–251.

[5] Lastennet D, Lopez B, Denis A, Mertz J-D. Hydrogeology, infiltration water characteristics and support properties at Lascaux. In: Coye N, ed. Lascaux and Preservation Issues in Subterranean Environments. Proceedings of the International Symposium (Paris, February 26 and 27, 2009). Documents d'Archéologie Française 105. Paris, France, Éditions de la Maison des Sciences de l'Homme, 2011, 95–120.

[6] Geneste J-M. De la découverte à la première crise bioclimatique de 1963. Monumental 2006, 2, 62–67.

[7] Lefèvre M. La maladie verte de Lascaux. Stud Conserv 1974, 19, 126–156.

[8] Orial G, Mertz J-D. Étude et suivi des phénomènes microbiologiques. Monumental 2006, 2, 76–86.

[9] Sire M-A. De l'élimination des champignons au constat d'état. Monumental 2006, 2, 68–75.

[10] Allemand L. Qui sauvera Lascaux? La Recherche 2003, 363, 26–33.

[11] Allemand L, Bahn PG. Best way to protect rock art is to leave it alone. Nature 2005, 433, 800.
[12] Bastian F, Orial G, François A, Alabouvette C. La grotte de Lascaux un écosystème complexe où bactéries et champignons interagissent. Biofutur 2007, 283, 28–31.
[13] Sire M-A. Lascaux. À la recherche d'une nouvelle stratégie de conservation préventive. Les Dossiers d'Archéologie 2008, 15, 54–63.
[14] Dupont J, Jacquet C, Dennetiere B, et al. Invasion of the French Paleolithic painted cave of Lascaux by members of the *Fusarium solani* species complex. Mycologia 2007, 99, 526–533.
[15] Zhang N, O'Donnell K, Sutton DA, et al. Members of the *Fusarium solani* species complex that cause infections in both humans and plants are common in the environment. J Clin Microbiol 2006, 44, 2186–2190.
[16] Bastian F, Jurado V, Novakova A, Alabouvette C, Saiz-Jimenez C. The microbiology of Lascaux Cave. Microbiology 2010, 156, 644–652.
[17] Bastian F, Alabouvette C, Saiz-Jimenez C. The impact of arthropods on fungal community structure in Lascaux Cave. J Appl Microbiol 2009, 106, 1456–1462.
[18] Bastian F, Alabouvette C, Jurado V, Saiz-Jimenez C. Impact of biocide treatments on the bacterial communities of the Lascaux Cave. Naturwissenschaften 2009, 96, 863–868.
[19] Daxboeck F, Stadler M, Assadian O, Marko E, Hirschl AM, Koller W. Characterization of clinically isolated *Ralstonia mannitolilytica* strains using random amplification of polymorphic DNA (RAPD) typing and antimicrobial sensitivity, and comparison of the classification efficacy of phenotypic and genotypic assays. J Med Microbiol 2005, 54, 55–61.
[20] Stelzmueller I, Biebl M, Wiesmayr S, et al. *Ralstonia pickettii* – innocent bystander or a potential threat? Clin Microbiol Infec 2006, 12, 99–101.
[21] Vaneechoutte M, De Baere T, Wauters G, et al. One case each of recurrent meningitis and hemoperitoneum infection with *Ralstonia mannitolilytica*. J Clin Microbiol 2001, 39, 4588–4590.
[22] Garcia-Sanchez AM, Ariza C, Ubeda JM, et al. Free-living amoebae in sediments from the Lascaux Cave in France. Int J Speleol 2013, 42, 9–13.
[23] Bastian F, Alabouvette C, Saiz-Jimenez C. Bacteria and free-living amoeba in Lascaux Cave. Res Microbiol 2009, 160, 38–40.
[24] Di Piazza M. The crisis in Lascaux: update March 2007. Rock Art Res 2007, 24, 136–137.
[25] Fox JL. Some say Lascaux Cave paintings are in microbial "crisis" mode. Microbe 2008, 3, 110–112.
[26] Bahn PG. Insider: Killing Lascaux. Archaeology 2008, 61, 18.
[27] Bastian F, Alabouvette C. Lights and shadows on the conservation of a rock art cave: the case of Lascaux Cave. Int J Speleol 2009, 38, 55–60.
[28] de Hoog GS, von Arx, JA. Revision of *Scolecobasidium* and *Pleurophragmium*. Kavaka 1973, 1, 55–60.
[29] Butler D. French bid to save rock art. Nature 2010, 467, 375.
[30] Martin-Sanchez PM, Novákóva A, Bastian F, Alabouvette C, Saiz-Jimenez, C. Two new species of the genus *Ochroconis, O. lascauxensis* and *O. anomala* isolated from black stains in Lascaux Cave, France. Fungal Biol 2012, 116, 574–589.
[31] 2013 Top 10 New Species List, International Institute for Species Exploration, State University of New York. Accessed 5 September 2014 at www.esf.edu/top10/2013.
[32] Martin-Sanchez PM, Bastian F, Alabouvette C, Saiz-Jimenez C. Real-Time PCR detection of *Ochroconis lascauxensis* involved in the formation of black stains in the Lascaux Cave, France. Sci Total Environ 2013, 443, 478–484.
[33] Martin-Sanchez PM, Sanchez-Cortes S, Lopez-Tobar E, et al. The nature of black stains in Lascaux Cave, France, as revealed by surface enhanced Raman spectroscopy. J Raman Spectrosc 2012, 43, 464–467.

[34] Martin-Sanchez PM, Nováková A, Bastian F, Alabouvette C, Saiz-Jimenez C. Use of biocides for the control of fungal outbreaks in subterranean environments: The case of the Lascaux Cave in France. Environ Sci Technol 2012, 46, 3762–3770.
[35] Porca E, Jurado V, Martin-Sanchez PM, et al. Aerobiology: an ecological indicator for early detection and control of fungal outbreaks in caves. Ecol Indic 2011, 11, 1594–1598.
[36] Martin-Sanchez PM, Jurado V, Porca E, Bastian F, Lacanette D, Alabouvette C, Saiz-Jimenez C. Airborne microorganisms in Lascaux Cave (France). Int J Speleol 2014, 43, 295–303.
[37] Kiyuna T, An KD, Kigawa R, Sano C, Miura S, Sugiyama J. Molecular assessment of fungi in "black spots" that deface murals in the Takamatsuzuka and Kitora tumuli in Japan: *Acremonium* sect. *Gliomastix* including *Acremonium tumulicola* sp. nov. and *Acremonium felinum* comb. nov. Mycoscience 2010, 52, 1–17.
[38] Revankar SG, Sutton DA. Melanized fungi in human disease. Clin Microbiol Rev 2010, 23, 884–928.
[39] Urzì C. Microbial deterioration of rocks and marble monuments in Mediterranean Basin: A review. Corros Rev 2004, 22, 441–457.
[40] Saiz-Jimenez C, Miller AZ, Martin-Sanchez PM, Hernandez-Marine M. Uncovering the origin of the black stains in Lascaux Cave in France. Environ Microbiol 2012, 14, 3220–3231.
[41] Tebo BM, Bargar JR, Clement BG, et al. Biogenic manganese oxides: Properties and mechanisms of formation. Annu Rev Earth Planet Sci 2004, 32, 287–328.
[42] Santelli CM, Webb SM, Dohnalkova AC, Hansel CM. Diversity of Mn oxides produced by Mn(II)-oxidizing fungi. Geochim Cosmochim Acta 2011, 75, 2762–2776.
[43] Spilde MN, Northup DE, Boston PJ, et al. Geomicrobiology of cave ferromanganese deposits: a field and laboratory investigation. Geomicrobiol J 2005, 22, 99–116.
[44] Miyata N, Tani Y, Maruo K, Tsuno H, Sakata M, Iwahori K. Manganese (IV) oxide production by *Acremonium* sp. strain KR21-2 and extracellular Mn (II) oxidase activity. Appl Environ Microb 2006, 72, 6467–6473.
[45] White WB, Vito C, Scheetz BE. The mineralogy and trace element chemistry of black manganese oxide deposits from caves. J Cave Karst Stud 2009, 71, 136–143.

Soledad Cuezva, Valme Jurado, Angel Fernandez-Cortes, Elena Garcia-Anton, Miguel Angel Rogerio-Candelera, Xavier Ariño, David Benavente, Juan Carlos Cañaveras, Cesareo Saiz-Jimenez, and Sergio Sanchez-Moral

14 Scientific Data Suggest Altamira Cave Should Remain Closed

Abstract: Altamira Cave in Cantabria, Northern Spain, is famous by the collection of Paleolithic paintings and engravings. In 1977, the cave was closed due to the progressive deterioration of the paintings after decades of continuous and massive visitation but reopened to the public with a limited number of daily visitors in 1982. In 2002, the cave had to be closed again due to increasing growth of phototrophic microorganisms on the paintings of the Polychrome Hall. After a decade of studies, the data indicated the need for the cave to remain closed to visitors. But, public and political pressures forced the Spanish Ministry of Culture to reopen the cave. The extensive dataset collected over the past 15 years do not support a reopening and the cave should remain closed to protect the Paleolithic paintings from the ravages of tourism.

14.1 Introduction

Altamira Cave in Cantabria, Northern Spain, is listed as a UNESCO World Heritage site because of its world famous collection of Paleolithic paintings and engravings (▸ Fig. 14.1). The paintings were made with two main pigments: red pigments, composed of Fe-rich clays and Fe-oxides with trace amounts of terrigenous grains, and black pigments composed of clays, carbonate, and abundant burnt wood-charcoal fragments. Altamira Cave (270 m in length) is one of the many caves in the upper vadose area of the tabular polygenic karstic system that developed on Cretaceous, beige-colored calcarenitic, marine limestones. The cave is situated on a topographical high-point (152 m above sea level, m.a.s.l.), and has a depth of 3–22 m (averaging 8 m) below the surface. The Polychrome Hall, where most paintings are located, is situated 60 m from the cave's entrance and is on a lower topographic level (146.5 m.a.s.l.) than the surrounding halls. The rock layer over this hall has an average thickness of 7.5 to 8 m. The fossiliferous packstone-grainstone limestones, which have less than 2% porosity, generally moldic and interparticle, are partly dolomitized. The proportion of terrigenous clasts (quartz and K-feldspar), Fe-oxides and hydroxide (mainly goethite), and clays (illite, smectite, and kaolinite) ranges from 5% to 10% in weight [1].

At present, the cave is open to visitors. The cave has a single entrance and a main passage that ranges in height from 2 to 12 m and width from 6 to 20 m. The main entrance is closed by a metal gate with a highly insulating heat insulation core (slotted

Fig. 14.1. Polychrome Hall ceiling in Altamira Cave, Spain.

surface <4%), which acts as the initial barrier to stop the exchange of energy and matter with the outside. In addition, a second door isolates the Entrance Hall from the rest of the cave, primarily the Polychrome Hall and Walls Hall. The solid doors prevent the free circulation of air, and most of the galleries have an artificial concrete cover over the bedrock outcrop on the surface, with varying thickness from 4 to 70 cm. This was done to avoid water infiltration through the cracks into the cave. The concrete was also covered with an organic layer to allow the growth of grasses. Unfortunately, exact records of what was done were not kept, but this activity resulted in the cave being sealed by artificial material.

In the past, and for millennia, the Paleolithic rock art in the cave was preserved due to the special environmental conditions, characterized by low rates of water infiltration and precipitation of mineral deposits, and the maintenance of very stable microclimate due to the limited air exchange with the external atmosphere. Throughout the last century, the cave suffered many environmental changes, as well as changes caused by construction that diminished conditions needed to conserve the paintings.

14.2 History

Marcelino Sanz de Sautuola visited Altamira Cave for the first time in 1875, after being notified by Modesto Cubilla who discovered the cave around 1868. A few years later, Sanz de Sautuola returned to Altamira Cave, accompanied by his daughter Maria, and she was the first to see the famous paintings in the Polychrome Hall. In 1880, Sanz de Sautuola published the discovery in a book entitled, *Brief Notes on Some Prehistoric Artifacts in the Province of Santander*, attributing the paintings to specifically the Paleolithic period (▸ Fig. 14.2). This publication was widely criticized by the scholars of the time because they believed that the paintings were more recent and instead should be attributed to shepherds of the region. Others, members of scientific institutions, argued that they could be attributed to Roman soldiers who had taken refuge in the cave during the Cantabrian Wars (29–19 B. C.). In a meeting held in 1886 at the Spanish Natural History Society, it was alleged that the paintings did not have the characteristics of the Stone Age and that their "expression was that of an average disciple of the modern school." After these debates, the discoveries at Altamira Cave plunged into obscurity.

In 1902, the French prehistorian E. de Cartailhac published *Les cavernes ornées de dessins, La grotte d'Altamira, Espagne, Mea Culpa d'un sceptique*, which recognized the paintings' antiquity. In 1906, and after 4 years of excavations in the cave, Cartailhac and Breuil published *La Caverne d'Altamira* a definitive validation of the paintings' origin. During the excavations, they had to remove tons of rock that had fallen from the cave ceiling.

Fig. 14.2. Polychrome Hall ceiling, as drawn by Sanz de Sautuola in 1880.

Between 1924 and 1935, more work was done to repair and reinforce the cave to facilitate both the access to, and route within, the cave, and to stabilize the rock because more ceiling rocks had collapsed, and to reduce water condensation and infiltration. Among other things, electric lights were installed. All of this resulted in mass visits to the cave, which reached 50,000 people in 1955. At the same time, articles began to be published in journals that alerted people to the discoloration of the paintings, a circumstance that was reported by the Province of Santander's Committee of Historical and Archaeological Monuments back in 1903.

Visitation to the interior of the cave increased to 177,000 people in 1973. By October 1977, Altamira Cave was closed to the public due to the progressive deterioration of the paintings after decades of continuous visitation. The Spanish Government took over the ownership of the cave in 1978, and Altamira Cave reopened to the public with a limited number of daily visitors in 1982.

In 1999, Sanchez-Moral et al. [1] reported the common occurrence of micropits and irregular (i.e. etched) surfaces that affected both the host rock with and without paintings on the Polychrome Hall ceiling. Micropits are less than 1 cm in diameter and consist of circular to ovoid depressions that formed by preferential etching of weaker grains, fissures, cleavage planes, etc. Another survey carried out in the late 1990s observed scarce colonization of the paintings by phototrophic communities due to the absence of permanent lighting, except in the Polychrome Hall and its surroundings where abundant colonies of mosses, algae, and cyanobacteria occurred near remnants of the old lighting, particularly near or around the ground spotlight. These colonies were inactive, but upon culture in the laboratory, the microorganisms grew rapidly.

Phototrophic colonization was likely greater in the past because former colonies were noted around old lights and some light green spots were found on the Polychrome Hall ceiling. Specifically, on a wall near a spotlight, colonies of moss protonemata were evident, but did not develop into the plant. A chlorophyta, *Bracteacoccus* cf. *minor*, was identified in the cave [2]. Additionally, the cyanobacterium *Scytonema julianum* was detected near a female deer painting being illuminated by a spotlight. There was a well-developed community of cyanobacteria in the outer entrance to the cave, which could make dispersal to the cave's interior by air currents possible. *S. julianum* grows inside natural caves and tombs, in the photic zone where enough light arrives, as described by Ariño et al. [3]. Abundant, slimy growth of cyanobacteria *Gloeothece* spp. was also noted from ground sediments near another old spotlight. *Gloeothece* excretes large amounts of exopolysaccharides, possibly as a response to stress conditions. Diatoms were identified from this sample, as well, likely associated with water availability due to the exopolysaccharides.

The phototrophic microorganisms identified inside Altamira Cave were typical of growth noted from soils and rock surfaces. Their presence is, without a doubt, due to anthropogenic modifications to the cave, including mainly lighting and enhanced dispersal of propagules by visitors and air currents. Most of the species identified exist at the cave's exterior, where a microbial film of mosses and cyanobacteria coats the

Fig. 14.3. Colonization by phototrophic microorganisms on the ceiling of the Polychrome Hall, Altamira Cave.

rock surface (e.g. *Gloeothece* sp., *Pseudocapsa* spp., *Plectonema* spp., and *S. julianum*). Inside the cave, and from a survey carried out in 2000, most of the phototrophic microorganisms observed in the Polychrome Hall were found to be of resting forms or to have reduced metabolic activity. However, some of these microorganisms are able to use organic carbon in periods of darkness. This is confirmed by the identification of metabolically active cyanobacteria from the cave, as reported by Portillo et al. [4]. Cañaveras et al. [5] predict that, in the event of a change in the lighting conditions – such as permanent or extensive lighting, as subsequently occurred – growth of these communities would take place.

In 2002, the cave had to be closed due to increasing growth of green colonies on the paintings of the Polychrome Hall (▸ Fig. 14.3). The phototrophic colonization was a consequence of extended periods of time when artificial lighting was used in the Polychrome Hall for photographic documentation of the ceiling. This was necessary for the construction of a replica. The green colonies, similar to those reported from Lascaux Cave in 1963 [6], were accompanied by extensive development of white microbial colonies on the walls and ceilings of the corridors and that also reached the red paintings in the Polychrome Hall. After closing the cave, green patches on or around the paintings weakened or disappeared with time and the corrosion rate of the paintings' host rock decreased. Closing the cave clearly benefited the conservation of paintings.

In 2010, controversy arose over Altamira Cave. After a decade of studies [1, 7–9], the data indicated the need for the cave to remain closed to visitors. But, public and political pressures forced the Spanish Ministry of Culture to reopen the cave [10]. Consequently, reopening of Altamira Cave ignores sound scientific reports and historical evidence that visitors affect Altamira Cave and others.

14.3 Altamira Cave Environmental Conditions

An exhaustive microclimatic study was carried out over a period of 15 years (1997–2012) to understand environmental conditions in the cave. The main conclusion was that change in cave physicochemistry was caused by visitation. The results matched data from 1997 to 1998 and published by Sanchez-Moral et al. [1]. However, the earlier studies considerably underestimated the real impact generated by visitor entry to the paintings on the ceiling, based on experimental visits conducted in 2007. Conclusions obtained in this 2007 study include:

- Before the new doors were installed, warm, wet air entered the cave during the summer season (June–September) and caused the formation of water microparticle clouds (hydroaerosols) that moved toward the cave's interior. Each hydroaerosol acts as a potential adherence nucleus for dust particles (solid aerosols) and spores in the air. Hydroaerosols also increase liquid water and organic matter availability by favoring the development and dissemination of microbial colonies.
- Visitation causes a substantial increase in both thermal, hygrometric oscillations, and CO_2 concentrations in cave interior's air. Visitation also changes temperature evolution between different areas of the cave, such as the Crossing and Polychrome Hall, and occasionally changes the local air circulation (► Fig. 14.4).
- Mobilization of interior air from the Polychrome Hall and mixing with air from nearby corridors were caused by the warming of Polychrome Hall air tempera-

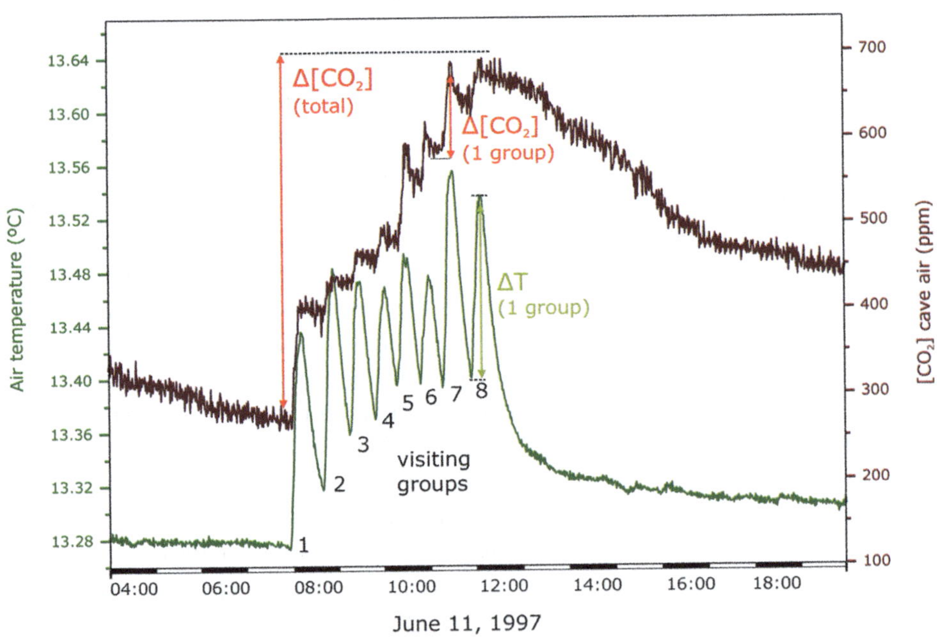

Fig. 14.4. The effect of visits on cave air temperature and carbon dioxide concentration in the Polychrome Hall, Altamira Cave. Bar 10 cm.

tures, and especially the sharp thermal increase caused near the ceiling in the entryway, when visitors stayed in the Hall. The thermal increase on the ceiling doubled what was measured by previous monitoring systems.

- The continual and almost daily opening of the access door, as well as the entry of consecutive groups into the Polychrome Hall, caused cumulative microenvironment disturbances and increased the average interior cave temperature (▸ Fig. 14.4).
- Experimental visitation in March 2007 demonstrated how visitors caused air mass mobilization and circulation on the ceiling of the Polychrome Hall. The air displacement happens throughout the visitor route and contributes to the progression of an air mass with hydroaerosols toward the interior of the cave. The hydroaerosols increase condensation and dispersion of microbes into the interior.
- Visitation prompted micro-corrosion processes that directly dissolved the host rock underlying the paintings. Visitors generate water vapor and increase cave air humidity, which increases water condensation, especially on the ceilings. Condensation starts almost immediately when visitors exit the Polychrome Hall. Condensation microniches became points for dissolution, as shown previously [1]. Dissolution is visible to the naked eye on the ceiling of the Polychrome Hall. These microniches also became places for colonies of white bacteria to develop at an alarming rate on the ceiling in the most interior parts of the Hall and to affect pigmented areas (▸ Fig. 14.5).

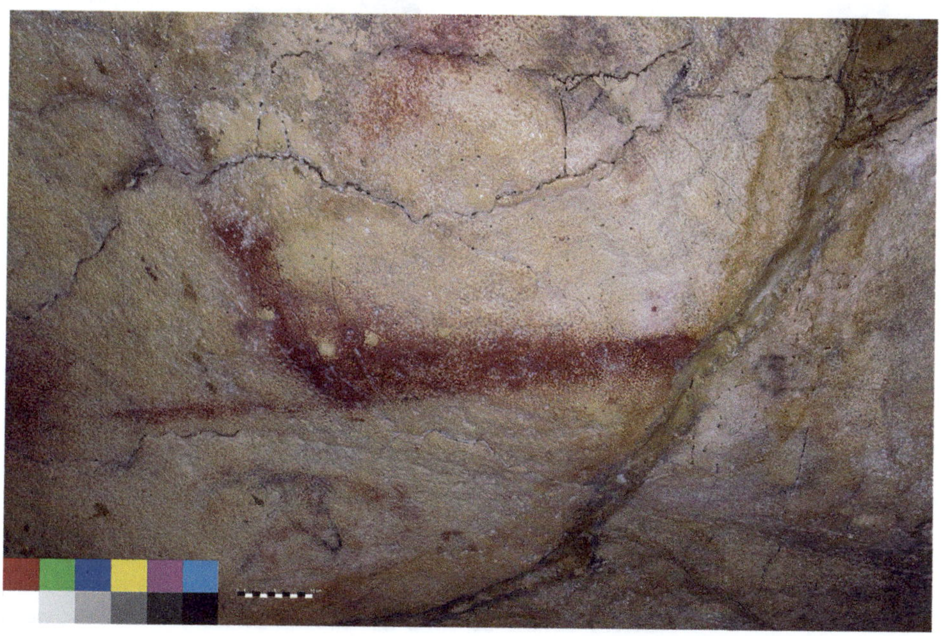

Fig. 14.5. White colonies on the ceiling of the Polychrome Hall, Altamira Cave. Bar 10 cm.

14.4 Altamira Cave Since 2009

One of the most important problems in Altamira Cave, as well as other caves open to the public, is biological colonization of solid surfaces and paintings. In Altamira Cave, microbial growth, both phototrophic and heterotrophic, was observed from the very moment of entry. Visitors and air currents from outside the cave strongly impacted the cave's ecosystem and microbial colonization. Starting at the external frame of the entrance door, it is colonized by calcifying cyanobacteria. Toward the galleries, colonization of the ceilings and walls is very evident but not related to phototrophic microorganisms, but rather to heterotrophic bacteria. In general, the density of microbial colonization decreases from very abundant colonies at the entrance, the ceiling of the

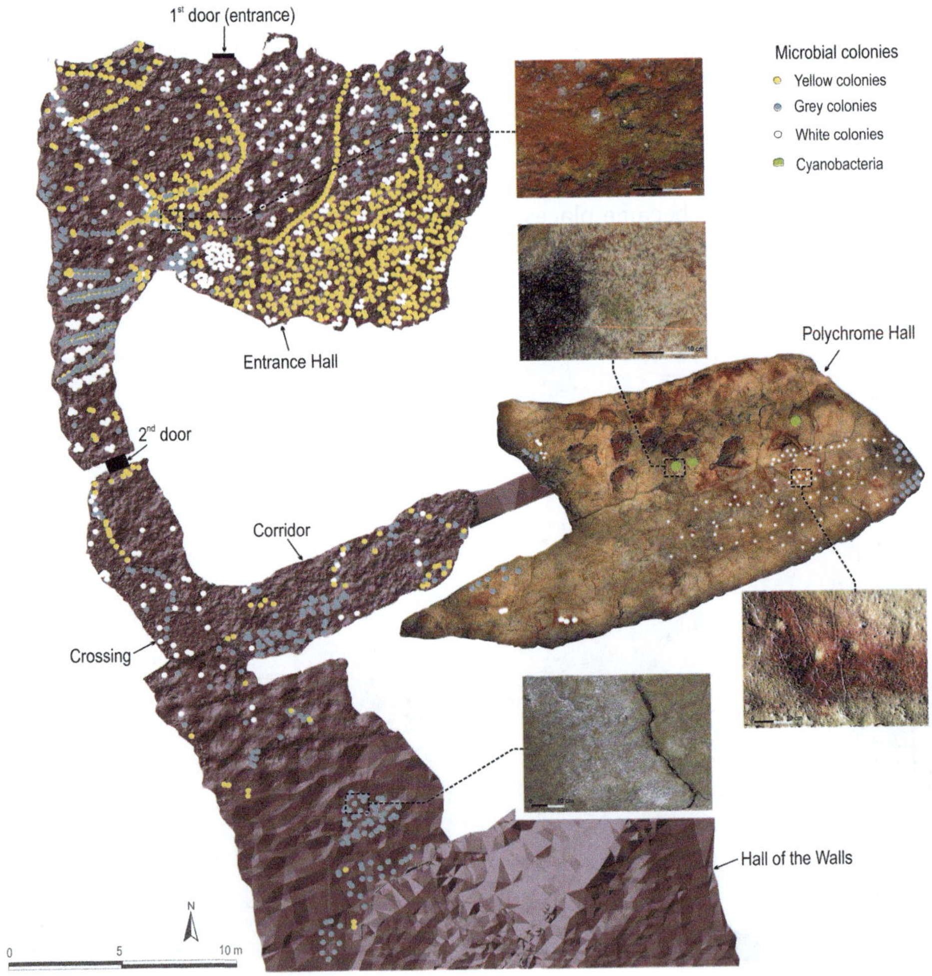

Fig. 14.6. Distribution of bacterial colonies on the ceiling of Altamira Cave (modified after Saiz-Jimenez et al. [10]).

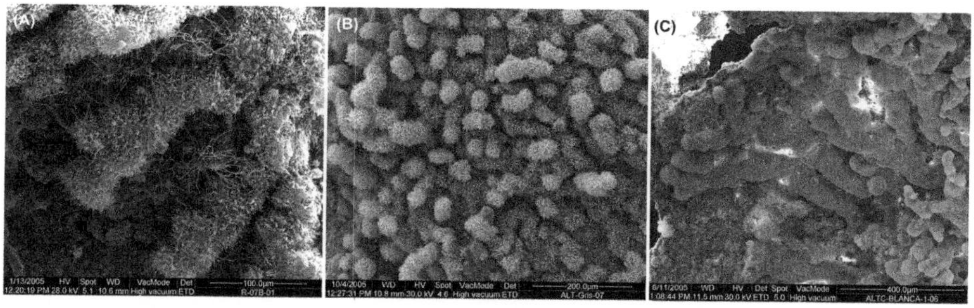

Fig. 14.7. Scanning electron microscopy photomicrographs from (a) yellow, (b) gray, and (c) white colonies from Altamira Cave.

so-called Kitchen, the entry hall of the cave, and in the galleries connecting to the Polychrome and Walls halls to less abundant colonies toward the end of the cave's galleries. An aerobiological study on June 2009 shows that the concentration of airborne bacteria inside the cave (up to 1800 colony forming units/m^3) is higher than outside the cave (980 colony forming units/m^3), and the concentration might corresponds to the presence of bacteria on the walls and ceilings of the halls and galleries close to the entrance (▸ Fig. 14.6). The higher concentration of airborne bacteria was found in the gallery leading to the Polychrome Hall. Samples from the transection Entrance-Polychrome Hall show the most abundant bacterium in the air of Altamira Cave to be *Micrococcus luteus* (85.3% of all colony forming units/m^3), followed by different species of *Pseudomonas* (6.8%) and *Bacillus* (1.8%) [11]. Wind currents promote the entry and transport of bacteria into the cave, as well as aerosol detachment from rock surfaces with condensed water that can remove and suspend bacteria from the cave walls and ceilings.

This is because the walls and ceilings are covered with different colored stains resulting from microbial colonies or small microbial mats, observed by naked eye as white, yellow, and gray colonies (▸ Fig. 14.6). When examined using environmental scanning electron microscopy (ESEM), Altamira Cave microbial colonies exhibit specific habits (▸ Fig. 14.7). Each one shows morphological and microstructural variability of different constituent elements (i.e. microorganisms, extracellular polymeric substances, mineral biofabrics, etc.). Based on all of these macro- and microscopic observations, the three colony types were examined to take into account the surface appearance and variations in color, morphology, texture, etc., as well as to map their distribution in the cave [8, 10].

14.4.1 Yellow colonies

The yellow colonies are round spots, 2 to 4–5 mm in diameter, that have circular to irregular contours and an apparent radial organization. Occasionally, overlapping neighboring colonies produce large, irregular, centimeter-size patches. Water drops condense on colony surfaces. SEM photomicrographs reveal that these colonies form by bundles of microorganisms having variable morphologies, including filamentous, spheroidal, and globular. No associated mineral deposits are found on the microbial structures. Porca et al. [12] report that these colonies are mainly dominated by Acidobacteria, Deltaproteobacteria, Chloroflexi, Alphaproteobacteria, Gammaproteobacteria, and Actinobacteria which constitute over 75% of the community, determined by OTU abundance (from an average of 3,000 clones).

14.4.2 Gray colonies

The intensity of gray colony color varies, as does colony size and texture. Generally, the gray colonies have circular contours and little relief. They are smaller, at 0.5 to 2–3 mm in diameter, than the yellow colonies. Water drops do not usually condense on the surfaces. Unlike the yellow colonies, the gray colonies are comprised of microbial bundles clearly radially arranged, starting from a central area that generates circular to irregular contours. The microorganisms have a variety of morphologies commonly embedded by EPS, including filamentous, spheroidal, or globular-coccoids. Associated mineral deposits are common. Two main groups, with significant size and hue variability are distinguished:

- *Light-gray colonies*: They are small (1 mm in diameter) and distributed mainly on the ceiling and around cracks in all the walls of the entry area, with further proliferation in the access to Polychrome and Walls Halls.
- *Dark gray colonies*: They show a more intense gray hue, are larger (1.5 to 2.5 mm in diameter), and are disseminated on ceilings and cave walls. These colonies show circular and irregular contours, sometimes with some radial arrangement and have little relief on the rock surface. These colonies have variable color intensity, even within the same colony. Intensity increases generally from light gray in the innermost area of the colony to a darker gray in its central part of the colony. The size and/or hue variability seems to correspond to the level of development and microbial metabolic activity, as well as to respond to variations in environmental conditions.

Cuezva et al. [13] describe the distribution of major phylogenetic groups in the gray colonies, which are dominated by Actinobacteria (77.5%). However, predominant microorganism involved in gray colony formation may be an unknown species belonging to the order Nitriliruptoridae. The gray colony community leads to rock dissolution and the generation of bioinduced calcium carbonate crystals as a by-product using the same methodology as described.

14.4.3 White colonies

The white colonies have variable size and microstructure, with a well-defined radial arrangement, and high relief on the rock surface. Drops of water commonly condense on the colony surface. At the microscopic scale, the white colonies have defined branches with radial and divergent arrangement toward the colony exterior. The branches constitute a dense network of thin, interwoven bacterial filaments. Two types of white colonies are distinguished:

- In the Polychrome Hall, the colonies are tiny (<1 mm in diameter) and are located preferentially inside small corrosion holes on bare rock, particularly in an area with red pigments.
- In the entrance area, round white colonies were larger, being 1.5 to 3 mm in diameter. When the colony diameter is around 1.5 mm, the colonies show less relief and the radial internal organization cannot be observed by the naked eye. This colony type was the most abundant and most widely disseminated on ceilings and walls of the cave, from the entrance area to the access gallery to the Polychrome Hall.

Portillo et al. [4] show that white colonies on rocks are mainly Actinobacteria, Alphaproteobacteria, Bacteroidetes, and Gammaproteobacteria, all of which accounted for over 68% of the community. When the white colonies are on red-pigmented (i.e. hematite) areas, the composition varies, likely due to the influence of iron on the microbial community. In this case, Actinobacteria represent 62.5% of the community alone, which contrasts where it is only 21.3% of the community if the colony is on a bare rock. The influence of iron on actinobacterial growth is stressed by Diaz-Herraiz et al. [14].

14.5 Fungi in Altamira Cave

Microbial growth and concomitant colony formation is likely a consequence of local enrichment in organic carbon, such that the formation (or non-formation) of colonies depends on the availability of nutrients and the micro-environmental conditions. SEM studies show that all rock surfaces and speleothems are covered by a dense network of bacterial filaments, but fungal hyphae were not observed.

In the past, limited rat excrements supported fungal growth, not only in Altamira Cave, but also in the nearby Stalactites Cave. Fungi are silent enemies waiting for a weakness in the Altamira Cave management. Fungi, which colonize different types of materials, have various origins: plant saprophytes introduced from the exterior of the cave through airborne spores, entomopathogenic fungi introduced by arthropods, and coprophilous fungi growing on rodent feces. In 2007 and later, periodic surveys show the growth of fungi on different materials introduced into the cave [10, 15] (► Tab. 14.1). In 2007, when microclimatic equipment was installed, despite being carefully cleaned

Table 14.1. Fungi in Altamira Cave, Spain.

Fungi in the air	
Location	**Taxonomic identification**
Kitchen Hall	*Epicoccum nigrum*
	Cladosporium cladosporioides
	Acremonium sp.
	Cladosporium sp.
	Penicillium sp.
Crossing	*Cladosporium macrocarpum*
	Epicoccum nigrum
	Cladosporium sp.
	Metacordyceps chlamydosporia
Polychrome Hall	*Epicoccum nigrum*
	Cladosporium sp.
	Penicillium chrysogenum
	Aspergillus sp.
Walls Hall	*Epicoccum nigrum*
	Cladosporium sp.
Grave Hall	*Cladosporium* sp.
Great Hall	*Cladosporium* sp.
Well Hall	*Cladosporium cladosporioides*
	Cladosporium sp.
	Trichoderma gamsii
Horse Tail Gallery	*Cladosporium* sp.

Fungi colonizing rock tablets exposed to the cave air		
Location	**Sample collected from**	**Taxonomic identifications**
Kitchen Hall	Rock tablets	*Lecanicillium psalliotae*
		Cladosporium cladosporioides
		Emericellopsis sp.
		Penicillium sp.
		Paecilomyces marquandii
Polychrome Hall	Rock tablets	*Penicillium polonicum*
		Penicillium sp.
		Trichurus sp.
	Metal surface	*Zasmidium cellare*
Walls Hall	Rock tablets	*Acremonium strictum*
		Penicillium polonicum
		Mortierella elongata
		Trichurus sp.
		Penicillium sp.
	Methacrylate frame	*Aspergillus versicolor*
		Fusarium oxysporum
	Metal surface	*Bionectria ochroleuca*

Fig. 14.8. Foot of a tripod used to hold microclimatic sensors that were colonized by fungi; the fungi are feed upon by collembola.

before installation, it was rapidly colonized by fungi (▸ Fig. 14.8). In 1996, however, different microclimate monitoring equipment installed in the cave was not colonized by fungi. The 2007 outbreak may have been caused by enhanced human activity inside and outside the cave, predominately due to archaeological excavations underway within the cave in October 2006. The removal of sediments likely increased the number of airborne spores.

Jurado et al. [15] installed a methacrylate frame holding tablets made of different rocks, including Altamira Cave host rock, to evaluate rock colonization inside the cave. The tablets were first sterilized in an autoclave for 30 min and the frame was cleaned with ethanol. The tablets were inserted in the methacrylate frame and were exposed to the air cave in three different locations, the entry hall, Polychrome Hall, and Walls Hall, on June 21, 2007. Two months later, fungal mycelia hung from the tablets. In September 2007, the upper rock surface of the tablets was covered by fungal mycelia and the lower surface showed hanging mycelia. The fungi identified on the rock tablets are shown in ▸ Tab. 14.1. During the test, tablets were naturally wetted by condensation and dripping water. Dripping water, on average, contains 5.5 mg of total organic carbon per liter at the sampling time. A culture medium, prepared with dripping water as a sole carbon source and purified agar, supported fungal isolate growth. Because no specific fungal DNA was amplified from a set of nine samples collected from different walls and halls, with and without bacterial colonies, we assume that fungal colonies were absent on the wall surfaces. It was hypothesized that an important number of actinobacteria, as well as other bacterial groups, synthesize bioactive compounds with antifungal properties in the cave. Therefore, it is possible that natural bacterial colonization on the cave walls plays an important role in protecting the walls from fungal development [15]. Consequently, fungal colonization of the exposed tablets, and any other inorganic or organic material, is due to airborne fungal spores being able to colonize the rock in the absence of bacteria with antifungal properties on the substratum.

To stop fungal colonization of equipment in 2007, periodic cleanings were needed, such as mechanical removal of the ground sediments adjacent to the instruments colonized by fungi, as well as hydrogen peroxide washings. The use of biocides, already employed in Lascaux Cave, was deliberately avoided prior to this point because the benzalkonium chloride applied in Lascaux Cave between 2001 and 2004 was used by microorganisms as a carbon and nitrogen source and caused further fungal outbreaks [16]. Instead, hydrogen peroxide oxidizes and mineralizes fungal biomass to yield carbon dioxide, oxygen, and water, all elements already present in the cave.

At present, the air of Altamira Cave has abundant bacteria and fungal spores that will grow as soon as nutrient sources are available. The concentration of fungal spores is generally higher in the halls and galleries close to the entry door (e.g. Kitchen Hall, the Crossing, and Polychrome Hall) and decreases toward the interior of the cave (Walls Hall) [17]. The aerobiological study done to evaluate the concentration of bacteria in the cave air also examined fungal spores. A Duo SAS (Surface Air System) model 360 sampler (International PBI, Milan, Italy) containing Petri dishes with Dichloran Rose Bengal Chloramphenicol agar was used to concentrated fungal spores. At the time of the study, the maximum concentration of fungal spores in the air was in the Polychrome Hall, at 390 colony forming units/m^3 [11]. Upon identification and considering all the cave halls and galleries sampled together, the spores of the genus *Cladosporium* represented 80.3% of all colony-forming units/m^3 in the cave air, followed by the genus *Epicoccum*, with 14.7% abundance. Minor percentages, around or below 2%, were obtained for *Acremonium*, *Penicillium*, and *Aspergillus*, among other genera. In the halls near the entrance (e.g. Kitchen Hall, the Crossing, Polychrome Hall, and Walls Hall), *Cladosporium* spores attained a concentration of 79.3%, followed by *Epicoccum* (15.9%). *Cladosporium* and *Epicoccum* are cosmopolitan, air-borne, and soil-borne fungi. *Cladosporium* spores are present in European and North American outdoor environments throughout the year and constitute between 40% and 80% of the total spores in the air [18].

The comparatively high abundance of *Cladosporium* in Altamira Cave likely denotes an external influence. In previous studies on cave air contamination, an index was created to classify caves into five categories, ranging from no problems to completely ecologically disturbed [17]. Altamira Cave falls into the third category, labeled as a cave threatened by fungi. The abundance of *Cladosporium* and *Epicoccum* all over Altamira Cave may also be a consequence of air exchange and ventilation from the outdoors. In fact, *Cladosporium* spp. and *Epicoccum nigrum* sequences from isolates collected outdoors were identical to those collected inside the cave [11]. This confirms that the presence of these two genera in the Altamira Cave air was due to transport from the exterior because of ventilation and wind currents [11].

14.6 Why Should Altamira Cave Remain Closed?

The extensive dataset collected over the past 15 years should be enough to recognize a critical situation for Altamira Cave, particularly if there is a management change or if the cave is reopened. The data do not support a cave reopening [10], but unfortunately scientific reports are not always considered by politicians. In 2007, some changes were made in the cave, as deduced from microclimatic modeling. The old metallic door was replaced with a new door equipped with a thermal insulation system. A second door was placed 20 m from the main entrance to reduce the entry of airborne particles, including plant aerosols and microbial spores, and to reduce water condensation rate in the Entrance Hall that would affect microbial metabolic activity. But, experimental visits in 2007 demonstrate that visitors increase cave air humidity and cause condensation immediately after exiting the Polychrome Hall. Visitation also induces microcorrosion processes, primarily on the ceilings, that directly affect the rock underlying the paintings. In the event of the cave re-opening, condensation and limestone dissolution will create preferential habitats for bacterial colonies to develop at an alarming rate on the ceiling of the inner-most parts of the cave and in areas with paintings. Moreover, human activity inside the cave or in its surroundings (e.g. due to archaeological excavations) also greatly impact the production and dispersion of bacterial and fungal spores in the cave air by enhancing colonization and growth of fungi. This was evident after excavations in 2006 that triggered new fungal growth on the sensors and the rest of the microclimatic control equipment in 2007. Although there has been a progressive decrease in the dissolved organic carbon and nitrogen content in dripping waters due to elimination of agricultural and cattle activities above the cave, animal excrement, organic matter, and humus produced over decades above the cave still represents a way for increased organic carbon to enter the cave in the dripping waters and sustains bacteria and fungi [19]. In conclusion, Polychrome Hall is the most fragile part of the cave due to its special morphogeometric characteristics (e.g. small gallery dimensions and low level of ventilation) and hydrokarstic dynamic (e.g. very slow infiltration levels) [1, 10]. Reopening the cave to visitors significantly impacts the preservation of paintings in the Polychrome Hall.

14.7 Final Remarks

Today's leisure tourism, which includes massive visitation of caves, should be looked upon as a potential risk for the conservation of cultural heritage sites. Archaeologists, environmentalists, and microbiologists agree on the beneficial effect of closing subterranean sites for their conservation [10]. This opinion is widespread and has been extended in recent times to some Egyptian tombs, including Tutankhamen's, which will be closed to visits with the tourism rerouted to a replica. At present, a reproduction of the Altamira Cave paintings can be visited at the new Altamira Museum [20].

Fig. 14.9. Replica of the Polychrome Hall ceiling in the Altamira Museum.

The purpose of the replica (▸ Fig. 14.9) was to divert pressure from the natural cave to the Museum. Use of the replica should continue, with the cave remaining closed, to protect the Paleolithic paintings from the ravages of tourism. However, in spite of the recommendations provided by scientists based on results from 15 years of cave monitoring [10, 21], the Ministry of Culture, responsible for cave management, decided to reopen the cave for visits. First, throughout 2014–2015, an experimental visiting regime was implemented that consisted of the weekly entrance of six people during 37 minutes. This was done despite the same environmental control having been conducted previously with conclusive findings that warned against the maintenance of such visits. The touristic use of the cave has been permanent since April 10, 2015. The reason for such a reopening was the appointment of a new scientific committee that prepared a report and concluded that cave deterioration was due to natural processes, not to the presence of humans. Unfortunately during the time of experimental visits, some red pigments from the bison paintings detached from the ceiling and fell to the ground. A red to pink stain of 3 cm^2 [22] formed, a phenomenon never observed before, and which confirmed the critical situation of the cave, irrespective of whether the deterioration was due to natural causes or to the human visitation.

Acknowledgments

This research was supported by the Spanish Ministry of Sciences and Innovation, project CGL2010-17108/BTE and CSIC project 201230E125. EG-A was supported by a CSIC JAE-Predoctoral grant. SC benefited from a postdoctoral Juan de la Cierva fellowship. AF-C was funded by a JAE-Doc postdoctoral fellowship. The Altamira Cave Research Centre and Museum staff are acknowledged for their collaboration throughout the research period. This is a TCP-CSD 2007–00058 paper.

References

[1] Sanchez-Moral S, Soler V, Cañaveras JC, Sanz-Rubio E, Van Grieken R, Gysels K. Inorganic deterioration affecting the Altamira Cave, N Spain: quantitative approach to wall-corrosion (solutional etching) processes induced by visitors. Sci Total Environ 1999, 243–244, 67–84.
[2] Lefèvre M. La maladie verte de Lascaux. Stud Conserv 1974, 19, 126–156.
[3] Ariño X, Hernandez-Marine M, Saiz-Jimenez C. Colonization of Roman tombs by calcifying cyanobacteria. Phycologia 1997, 36, 366–373.
[4] Portillo MC, Saiz-Jimenez C, Gonzalez JM. Molecular characterization of total and metabolically active bacterial communities of "white colonizations" in the Altamira Cave, Spain. Res Microbiol 2009, 160, 41–47.
[5] Cañaveras JC, Sanchez-Moral S, Soler V, Saiz-Jimenez C. Microorganisms and microbially induced fabrics in cave walls. Geomicrobiol J 2001, 18, 223–240.
[6] Lefèvre M, Pochon J, Laporte G, Chalvignac MA. Sur la decontamination bacterienne et algale de la grotte de Lascaux. C R Acad Sci Paris, 1964, 258, 6576–6578.
[7] Heras Martín C, Lasheras JA, Sanchez-Moral S, Bedoya J, Cañaveras JC, Soler V. The preservation of the cave of Altamira (1880–2002). In: Actes du XIVème Congrés UISPP, Section 18.4: Conservation, Restoration, Protection de l'Art Paléolithique. British Archaeological Reports, International Series, 2004, 1313, 21–31.
[8] Cuezva S, Sanchez-Moral S, Saiz-Jimenez C, Cañaveras JC. Microbial communities and associated mineral fabrics in Altamira Cave, Spain. Int J Speleol 2009, 38, 83–92.
[9] Lasheras JA, Sánchez-Moral S, Sáiz-Jiménez C, Cañaveras JC, de las Heras C. The conservation of Altamira Cave: a comparative perspective. In: Coye N, ed. Lascaux and Preservation Issues in Subterranean Environments. Proceedings of the International Symposium (Paris, February 26 and 27, 2009). Documents d'Archéologie Française 105. Paris, France, Éditions de la Maison des Sciences de l'Homme, 2011, 169–182.
[10] Saiz-Jimenez C, Cuezva S, Jurado V, et al. Paleolithic art in peril: Policy and science collide at Altamira Cave. Science 2011, 334, 42–43.

[11] Garcia-Anton E, Cuezva S, Jurado V, et al. Combining stable isotope ($\delta^{13}C$) of trace gases and aerobiological data to monitor the entry and dispersion of microorganisms in caves. Environ Sci Pollut Res 2014, 21, 473–484.

[12] Porca E, Jurado V, Zgur-Bertok D, Saiz-Jimenez C, Pasic L. Comparative analysis of yellow microbial communities growing on the walls of geographically distinct caves indicates a common core of microorganisms involved in their formation. FEMS Microbiol Ecol 2012, 81, 255–266.

[13] Cuezva S, Fernandez-Cortes A, Porca E, et al. The biogeochemical role of Actinobacteria in Altamira Cave, Spain. FEMS Microbiol Ecol 2012, 81, 281–290.

[14] Diaz-Herraiz M, Jurado V, Cuezva S, et al. The actinobacterial colonization of Etruscan paintings. Sci Rep 2013, 3, 1440, DOI: 10.1038/srep01440.

[15] Jurado V, Fernandez-Cortes A, Cuezva S, et al. The fungal colonization of rock art caves. Naturwissenschaften 2009, 96, 1027–1034.

[16] Martin-Sanchez PM, Nováková A, Bastian F, Alabouvette C, Saiz-Jimenez C. Use of biocides for the control of fungal outbreaks in subterranean environments: The case of the Lascaux Cave in France. Environ Sci Technol 2012, 46, 3762–3770.

[17] Porca E, Jurado V, Martin-Sanchez PM, et al. Aerobiology: An ecological indicator for early detection and control of fungal outbreaks in caves. Ecol Indic 2011, 11, 1594–1598.

[18] Spieksma FTHM. Outdoor atmospheric mould spores in Europe. In: XVIth European Congress of Allergology and Clinical Immunology. Bologna, Monduzzi 1995, 625–630.

[19] Saiz-Jimenez C, Hermosin B. The nature of the organic matter present in dripping waters from Altamira Cave. J Anal Appl Pyrol 1999, 49, 337–347.

[20] Lasheras JA, Heras Martin C. Cueva de Altamira and the preservation of its Palaeolithic art. Coalition 2006, 12, 7–13.

[21] Sanchez-Moral S, Cuezva S, Garcia-Anton E, et al. Microclimatic monitoring in Altamira Cave: Two decades of scientific projects for its conservation. In: Saiz-Jimenez C, ed. The Conservation of Subterranean Cultural Heritage. Leiden, CRC Press/Balkema, 2014, 139–144.

[22] Lasheras JA, Heras C de las, Prada A. Altamira and its future. In: Saiz-Jimenez C, ed. The Conservation of Subterranean Cultural Heritage. Leiden, CRC Press/Balkema, 2014, 145–164.

Index

D

N

O

P

Q

R

S

Printed and bound by CPI Group (UK) Ltd, Croydon, CR0 4YY
13/09/2026
14960851-0002